ALLEZEIT WACH 1842

Günter Kirschling

Quality Assurance and Tolerance

With 278 Figures

Springer-Verlag
Berlin Heidelberg NewYork
London Paris Tokyo
Hong Kong Barcelona Budapest

Professor Dr.-Ing. Günter Kirschling

Universität Kassel, FRG

Translated from the first German Edition "Toleranzen und Qualitätssicherung" (Springer-Verlag) by Sue Greener, Springfield, Five Ash Down, Uckfield, East Sussex TN22 3AH, England.

ISBN 978-3-642-48691-3 ISBN 978-3-642-48689-0 (eBook)
DOI 10.1007/978-3-642-48689-0

Library of Congress Cataloging-in-Publication Data

Kirschling, Günter. [Qualitätssicherung und Toleranzen. English] Quality assurance and tolerances / Günter Kirschling. p. cm. Translation of: Qualitätssicherung und Toleranzen. Includes bibliographical references and index.
ISBN-13: 978-3-642-48691-3
1. Quality control— Statistical methods.
I. Title. TS156.K49213 1991 658.5'62'015195—dc20.

Softcover reprint of the hardcover 1st edition 1991

Typesetting: Thomson Press (India) Ltd., New Delhi; Offsetprinting: Kutschbach, Berlin;

61/3020-543210—Printed on acid free paper.

Foreword of the First German Edition

It has been well known for decades that it is possible to calculate the tolerances of single dimensions within chains using statistical criteria. However, in practice, this is rarely done.

Measurement tolerances which are part of a series are almost exclusively calculated using conventional arithmetic. This is generally preferred, because the statistical method for calculating tolerances is complicated and difficult to comprehend.

This book therefore offers a comprehensive presentation of tolerance problems and their solution by statistical methods, successfully presenting all calculated solutions in clear figures or graphical form. These figures consist of with additional material on tolerance zones and information on the distribution assumptions of position, shape and width with further information on respective proportional non-conformities.

Trials with students and engineers in industry have shown that the use of these figures, as an aid to exact calculated solutions, make the approach to this complicated matter considerably easier.

This book is particularly appropriate for those working on the one hand in the field of development and contruction and on the other hand in production and quality control, especially in mechanical engineering and related fields. Readers in the quality assurance field who are familiar with the underlying principles of statistics with quantitative characteristics will have no difficulty in understanding the contents of this book. For readers who have either no knowledge or an insufficient knowledge of statistics we have provided Chaps. 2 to 5. These set out the theoretical principles of statistics necessary for an understanding of the tolerance analyses and process analyses in following chapter and are illustrated with practical examples.

Tolerance problems involve not only the accurate calculation of tolerances with respect to safeguarding the function of the dimensional chains, but also their realisation in production. This theme is dealt with extensively in Chaps. 13 to 15 and should be of particular interest to those who work in Production areas and in Quality Assurance.

Modern machining processes are distinguished by their high efficiency levels with the result that these processes show significant trends with machine tool wear. Because of that the familiar technical expertise on process control has to be extended in many ways using quality control charts see Chap. 15.

It is strongly recommended that the reader keeps constantly to hand a scientific calculator with $\bar{x}/s$-automatic facility, so as to be able to calculate mathematically the solutions given in examples and exercises. For this reason all calculated parameters, characteristics and estimate rates are given to six or more decimal places, a "precision" which is not necessary or practicable either in theory or in practice but which allows readers to check the accuracy of their calculations.

Since the literature offers little on the subjects of "process capability process control and process correction" and what is offered is partly contradictory, in many respects odd definitions and solutions may be found. I am conscious therefore that some figures and examples may contain obscurities or deficiencies; I am always grateful for advice on these.

I would like to express sincere thanks to Prof. Dr.-Ing. P. Th. Wilrich, Freie Universität Berlin, for his critical examination of the manuscript and for his suggestions for improvement.

I would also like to thank Springer-Verlag for their interest in the production of this book.

Melsungen, 1988 Günter Kirschling

Contents

Symbols

d_n	anticipated value of $\bar{R}/\sigma$
EG	cut off point
N	range of basic totals or of clearances or of a probability distribution as a result of a simulated discrete distribution using integral x-values Nominal size
N_j	extent of the j-th class in a probability distribution
n	random sample test range
n_j	extent of the j-th class in a random sample test
p	margin of error (exceeding limit value) in basic totals or in a probability distribution
$\hat{p}$	margin of error in the random sample test
$G(u)$	area value (probability total) from $-\infty$ to u
P_a	probability of acceptability
$Q(u)$	area value from u to $+\infty$
u	standardised, normally distributed random sizes ($\mu = 0$; $\sigma = 1$)
x	characteristic
x_i	i-th observed value of x
x_j	middle value of the j-th class
$g(x)$	probability of value x
$G(x)$	probability total up to and including x.

Other symbols in Table 2.1 and in the text.

1 Introduction

Employees working in areas of development and design as well as those in production and quality control are constantly confronted by the fact that it is impossible to manufacture technical products in accordance with exact theoretical dimensions. In fact—depending on the variability of technical resources—some deviations from the theoretical dimensions are unavoidable, and these deviation may be either assignable (systematic) or unassignable (random).

This is especially true for technical products which are manufactured in large batches or mass produced. At the same time, these products must be interchangeable, i.e. they must fulfill their given function despite unspecified mating in subsequent assembly or fabrication processes.

The designer must take account of unavoidable deviations by specifying direct or indirect limit values for all dimensions, between which the actual dimensions of the component must lie. In other words:

Each dimension has tolerances.

The selection of tolerances presents no problem to the designer if close tolerances i.e. binomial dimensional chains are involved. For decades there has been the ISO tolerance system for tolerances on fit, and this has been updated and expanded by many national and international standards. In this system, many concepts and their interrelationships are specified. Also specified are the values for tolerances (tolerance groups) in relation to nominal value ranges (tolerance progressions). Tolerances are also listed by zone.

It is also straightforward for the designer to specify tolerances when dimensions do not interact with other dimensions (indirect working dimensions more exact definitions later). These dimensions are governed by general tolerances which are set as standard.

There is a problem, however, in selecting tolerances according to the value and position of the tolerance range when there are three or more components in the dimensional chain. In these cases, there are no standards and the designer has to rely on his own judgement, often without knowing the decision criteria. Consequently many designers apply the principle:

"Tolerances should be as wide as possible and as narrow as necessary"

with the rider:

"if in doubt, better too narrow"

and there usually is some doubt.

Production and quality control personnel take account of unavoidable deviations by maintaining a quality control system. In each manufacturing process they take care to ensure that the actual value lies within the allowed tolerances. In manufacturing jargon: they ensure that tolerances are maintained as far as this is economically viable.

Daily there will be substandard components (components with actual dimensions outside the tolerance range) appearing in production batches. Then a decision must be taken to either:

—allow this batch to be released regardless for assembly or
—check through every component in this batch for non-conformity.

Production and quality control personnel are often not sufficiently well versed in or may be totally ignorant of the relevant decision criteria.

The aim of this book is to provide information on:

—effects of deviation progression law on deviations in the accumulative values of linear dimensional chains
—deviations of random distributions in contrast to normal distribution and their effects
—statistical tolerance calculation method
—special points which should be taken into account in receiving and production testing of dimensions and which belong to a statistically calculated linear dimension chain
—methods of process analysis and assessment of results
—process control with special references to the existence of trends.

Successful tolerance enlargements of individual dimensions by statistical tolerance calculation can only be carried out with mutual trust and cooperation between design, production and quality assurance personnel.

The following text deals exclusively with dimensions of length. Mathematical relationships applying to these dimensions can also be applied to other quality characteristics. Also only linear dimension chains are discussed. The same principles can be applied to non-linear dimension chains such as surface area and volume.

A direct reference to current standards is avoided wherever possible as they are so often subject to change. Reference standards are given in the bibliography (see Chap. 16).

2 Concepts

Table 2.1 lists the most important concepts of quality control and statistics in addition to the most important concepts relevant to dimensions and tolerances.

Table 2.1. Concepts of quality control, statistics, dimensions and tolerances.

Term	Definitions, comment
Quality	Totality of features and characteristics of a product or an activity that bear on its ability to satisfy stated on implied needs. Comment: readers are advised against the use of the term "good" as a synonym for "quality" in view of the international standard.
Quality structure	Model for the interlocking of all quality effective measures and results in the development and application phases of a product or activity (quality loop).

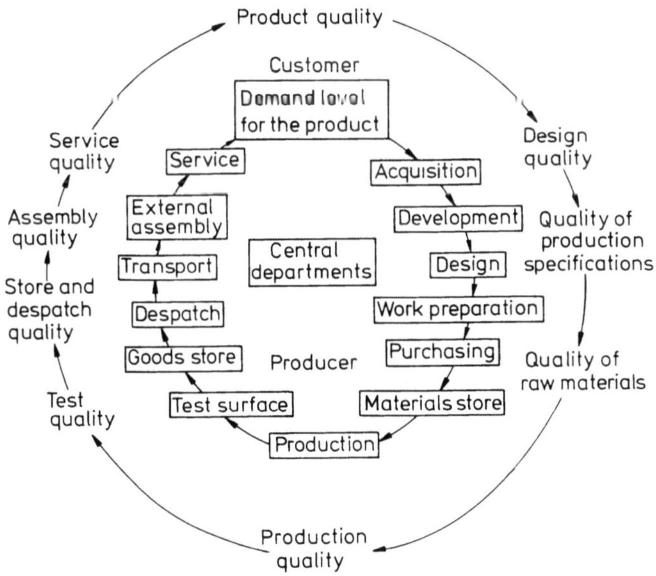

Example of a quality structure (outer structure) extended by the operational structure (inner structure).

Table 2.1. (Continued)

Term	Definitions, comment
Design quality	The excellence of the design in relation to ease of manufacture and to the customer's requirements.
Manufactured quality	The fidelity with which the product conforme to the design.
Nonconformity	The nonfulfilment of specified requirements. Comment: In technology for example a given requirement for a quantitative characteristic may be a fixed tolerance range between the limit sizes (G_u, G_o). If the characteristic size lies outside the tolerance range then this is a nonconformity. The total of the limit size intervals ($x - G_o$ or $G_u - x$) can be crucial in the decision as to what should happen to a nonconforming unit. Efficiency is not necessarily compromised by one nonconformity. The basic difference between "nonconformity" and "defect" is that specified requirements may differ from the requirements for the intended use.
Test characteristic	Characteristic on which a quality test may be performed.
Quality control system	The organisation of operational techniques and the activities which sustain a quality of product that will satisfy given needs, also the use of such techniques and activities.
Process quality audit	An analysis of elements of a precess and appraisal of completeness, correctness of conditions and probable effectiveness.
Product quality audit	A quantitative assessment of conformance to required product characteristics.
Quality control	The operational techniques and the activities which sustain a quality of product or service that will satisfy given needs also the use of such techniques and activities.
Quality design	Selection of quality characteristics and establishment of their required and permissible sizes for a product or activity with regard to given requirements for use and their implementation.
Quality assurance	All those planned or systematic actions necessary to provide adequate confidence that a product or service will satisfy given needs.
Statistical quality control	That part of quality control in which statistical techniques are used.
Production accuracy	Degree of consistency between the sizes of a production characteristic, achieved by the repeated application of established production experience. Comment: the size used for production accuracy is usually the standard deviation.
Quality verification [Testing]	The provision of evidence or proof that requirements for quality have been met. Comment: verification may be done by test; a critical trial on examination of one or more properties or character.
Statistical quality verification	That part of quality verification in which statistical techniques are used.
Test design	Design of quality testing.
Test plan	Result of test design.

Table 2.1. (Continued)

Term	Definitions, comment
Test specification	The document that describes in detail the methods of conducting tests including if necessary the criteria of assessing the results.
Test instruction	Instruction on how to perform a quality test.
Test procedure	The mode of conducting a quality test process.
Receiving test	Quality test of a delivered product.
Process test	Quality test of a process using either the characteristics of the process itself or its results.
Production test	Quality test during production processes.
Final test	Last of the quality tests before handover to the customer.
Quality technology	Application of scientific and technical knowledge as well as development technology for quality control.
Quality costs	The expenditure incurred by prevention and appraisal activities and by the losses due to internal and external failures.
Dimension	Measurable extent of "length", "breadth", "thickness", "area", "volume" etc.
Nominal size N	Size (magnitude) according to specification and applicable condition. Comment: The nominal size will often be expressed as a rounded number. Tolerance limits will be related to specified nominal size.
Desired size S	Size from which the actual size should deviate as little as possible. Author's comment: In the absence of a stated "desired size" the desired size is the median size.
Reference size	With regard to tool wear and tear, the predetermined size in the tolerance range which is different from the desired size. Comment: The reference size lies either above or below the desired size according to the direction of the actual size trend from tool wear.
Limit size G	Minimum or maximum permissible size.
Minimum size G_u	Minimum size permissible.
Maximum size G_o	Maximum size permissible.
Mean size C	Arithmetical mean size between the maximum and minimum sizes. Comment: The position of the tolerance field can be established by the mean size and the tolerance, e.g. by tolerance calculation according to statistical criteria. In this case, the intervals between the mean size and the limit sizes are designated as 'limit deviations".
Tolerated size M	Nominal size with assigned limit deviations, where the limit deviations can either be determined purely with reference to the nominal size or with the aid of general tolerances.
Tolerance on fit M_P	Tolerated size for a fit surface or related fit surfaces.
Internal tolerance on fit	Size for internal fit surface.
External tolerance on fit	Size for external fit surface.
Actual size I	The size established by measurement.
Test size	The tolerated size, which is applied in test planning to determine the range of the test.

Table 2.1. (Continued)

Term	Definitions, comment
	Comment: test sizes may be shown within a rounded frame e.g. (25) (Zeppelin shape size).
Deviation A	Size minus nominal size.
Limit deviation	Lower limit deviation or upper limit deviation.
Lower limit deviation A_u	Minimum size minus nominal size.
Upper limit deviation As_o	Maximum size minus nominal size.
Actual deviation A_i	Actual size minus nominal size.
Size tolerance T	The total variability allowed in a characteristic of a unit of product, e.g. maximum size minus minimum size or upper limit deviation minus lower limit deviation. Comment 1: Also referred to in brief as "tolerance". Comment 2: Examples of size tolerances would be either the tolerance of a tolerated size or the tolerance of a tolerance on fit.
Range of tolerance	The zone of values in which a measurable characteristic is in conformity within specified limit. Comment: The range of tolerance is specified by the tolerance itself and its relation to a reference size for the limiting values. Reference sizes can be the nominal size, desired size or mean size.
Tolerance zone	Interval between the minimum and the maximum sizes. Comment: The name "tolerance zone" is frequently used in relation to graphical representation.
General tolerance	Tolerance according to a tolerance system, whose application to the relevant size will be established by a generally accepted notation in a drawing, if necessary with reference to the chosen tolerance class. Comment: Formerly "free size tolerance".
Fit part	Part with one or more fit surfaces.
Mating	Joining related fit parts.
Fit surface	Any surface with surfaces toleranced to fit, when mated with another fit part.
Internal fit surface	Fit surfaces on inner mould parts. Comment: e.g. borehole.
External fit surface	Fit surfaces on outer mould parts. Comment: e.g. axle diameter.
Fit P	Size of internal fit surface minus size of external fit surface before mating.
Clearance P_S	Positive fit.
Oversize P_o	Negative fit.
Limiting fit	Minimum or maximum fit.
Minimum fit P_u	Fit with the minimum size of internal fit surface and the maximum size of external fit surface.
Maximum fit P_o	Fit with the maximum size of internal fit surface and the minimum size of extrnal fit surface.
Actual fit P_i	The fit established by measurement.

Table 2.1. (Continued)

Term	Definitions, comment
Fit tolerance P_T	Maximum fit minus minimum fit or the sum of the size tolerances for the sizes of internal fit surface and external fit surface. Important note: This definition is based on an arithmetical calculation.
Individual tolerance T_i	Tolerance of the *i*-th size of an arithmetically calculated, linear dimensional chain, consisting of *k* elements.
Arithmetical closing tolerance T_a	Sum of all *k* individual tolerances T_i.
Statistical individual tolerance T_{si}	Tolerance of the *i*-th size of a statistically calculated, linear dimensional chain, consisting of *k* elements.
Statistical closing tolerance T_s	Statistically calculated closing tolerance, calculated on the assumption of particular distributions in the individual actual sizes.
Quadratic closing tolerance T_q	Special case of T_s, calculated on the assumption that all individual actual sizes are normally distributed. $T_q = \sqrt{\sum T_{si}^2}$, all T_{si} equal size: $T_q = \sqrt{k}\, T_{si}$.

3 Probability Calculation

3.1 Definition of Probability

Assuming that the whole population is known, the probability of drawing a non-conforming or a conforming item can be calculated according to the following definition:

$$P(x) = \frac{\text{Number of favourable cases}}{\text{Number of possible cases}}.$$

This is called "probability" and is applicable to all the usual games of chance such as coin-tossing, dice, cards or roulette. In this case

non-conforming item = characteristic

conforming item = non-characteristic

This definition of probability is also applicable, if a probability distribution (Rectangular Distribution, Trapezoid Distribution or Normal Distribution) is derived from this so that a discrete distribution is simulated (Chap. 4).

Exercise 3.1

Given: In a lot of $N = 50$ items, $d = 3$ items are non-conforming.
Required: Probability of drawing one item and finding it non-conforming.
Solution: $P(x) = 3/50 = 0.06 = 6\%$.

Assuming that the whole population is not known—and this is the more common case in operating conditions—probability must be estimated by means of samples of determined relative frequency. Here the estimate will be more exact, the greater the sample size.

The large number rule (Bernoulli) states:

$$\lim_{n \to \infty} \left(\left| \frac{x}{n} - P_A \right| < \varepsilon \right) = 1,$$

with

x Number of non-conforming items in the sample,
n sample size,
P_A probability d/N,
ε optionally small number, other than zero.

Explanation. The probability that the difference between the observed relative frequency x/n and the probability $P_A = d/N$ according to the total will become smaller than ε, tends with an increasing n towards 1 or 100%.

In theory n must tend to infinity, which is only possible with a finite N if the sample is drawn "with putting back". In practice, however,

$$\left.\begin{array}{ll}\text{—occasionally} & n = 100 \\ \text{—usually} & n = 1000 \\ \text{—rarely} & n \geqq 10000\end{array}\right\} \text{classified in statistics as } \infty.$$

3.2 Addition Formula (OR Formula)

On the assumption that the results A_i, for which the individual probability are $P_i = P_i(A_i)$, are mutually exclusive, the total probability is that the results will occur as alternatives (OR?).

$$P_{ges} = \Sigma P_i = P_1 + P_2 + \cdots + P_k$$

Exercise 3.2

Given: In one lot of $N = 50$ items, $d = 3$ items are non-conforming.
Required: Probability that drawing only one item, it will be either non-conforming or conforming.
Solution: $P_{ges} = 3/50 + 47/50 = 1 = 100\%$.

3.3 Multiplication Formula (AND Formula)

On the assumption that the results A_i, for which the individual probabilities are $P_i = P_i(A_i)$, are independent of each other, the total probability is that the results will occur as additive

$$P_{ges} = \prod P_i = P_1 P_2 \cdots P_k.$$

Note. Results are then always independent of each other, if the probability of the one result does not depend on whether the other result occurs or not.

Exercise 3.3

Given: In one lot of $N = 50$ items, $d = 3$ items are non-conforming.
Required. Probability that, drawing two items, first one non-conforming AND then one conforming item will be drawn.
Solution: $P_{ges} = 3/50 \cdot 47/49 = 0.0576 = 5.76\%$.

Note. If the question (is altered to: probability in the sample $n = 2$ of finding one non-conforming and one conforming item, it must be borne in mind that the two items may be drawn in

this sequence ● ○

or

in this sequence ● ○.

Then:

$$P_{ges} = 3/50 \cdot 47/49 + 47/50 \cdot 3/49 = 2 \cdot 0.0576 = 11.51\%.$$

Exercise 3.4

Given: Two dice are thrown.
Required: Probability of a total of six.
Solution: Note that the total of six can occur either by

1 + 5
or 5 + 1
or 2 + 4
or 4 + 2
or 3 + 3

$$P_{ges} = 5/36 = 0.13\overline{8} = 13.89\%.$$

Exercise 3.5

Given: In one lot of $N = 60$ items, $d = 9$ items are non-conforming.
Required: Probability of drawing (without putting back) $n = 4$ items and finding $x = 2$ non-conforming items; the solution follows the rules of probability calculation.
Solution: The probability of drawing first two non-conforming and then two conforming items is

$$P = \frac{9}{60} \cdot \frac{8}{59} \cdot \frac{51}{58} \cdot \frac{50}{57} = 0.015688.$$

For the favourable sequence to occur there are 6 possibilities

● ● ○ ○
or ● ○ ● ○
or ● ○ ○ ●
or ○ ● ● ○
or ○ ● ○ ●
or ○ ○ ● ●

so that

$$P(x = 2) = 6 \cdot 0.015688 = 0.09413 = 9.413\%.$$

4 Probability Distributions of Quantitative Characteristics

4.1 General

A population is the total quantity of all relevant items. A true population always has a finite range, for example a lot with a total N. An assumed population can have an infinite or finite range.

True populations with a quite definite probability distribution (model distributions) can be simulated using an assumed population with a finite range. This simulation is only possible if the always continuous distributions of quantitative characteristics are made discrete.

The most important parameters of finite, discrete probability distributions independent of the distribution form are:

— range N, not a genuine parameter, but included for the better clarity of the following text with the parameters,

—mean value

$$\mu = \frac{\sum x}{N}, \tag{4.1}$$

—variance

$$\sigma_N^2 = \frac{1}{N} \sum (x - \mu)^2, \tag{4.2}$$

—standard deviation

$$\sigma_N = +\sqrt{\sigma^2}. \tag{4.3}$$

Lines (4.1) to (4.3) have the advantage of approximately corresponding with the formulae given in Chap. 5 for the statistical characteristics used in evaluating samples. Generally the formulae for the parameters of discrete probability distributions are:

—mean value

$$\mu = \sum x g(x), \tag{4.4}$$

—standard deviation

$$\sigma = +\sqrt{\sum (x - \mu)^2 g(x)}. \tag{4.5}$$

4.2 Normal Distribution

The normal distribution is the most important probability distribution. Even if several or even many limiting quantities cause random deviations, a normal distribution can occur.

This is described by a probability density function

$$g(x) = \frac{1}{\sqrt{2\pi}\sigma} e^{-1/2\cdot(x-\mu)^2/\sigma^2} \tag{4.6}$$

and through its distribution function

$$G(x) = \int_{-\infty}^{x} g(x)\,dx. \tag{4.7}$$

Both functions are shown in Fig. 4.1.

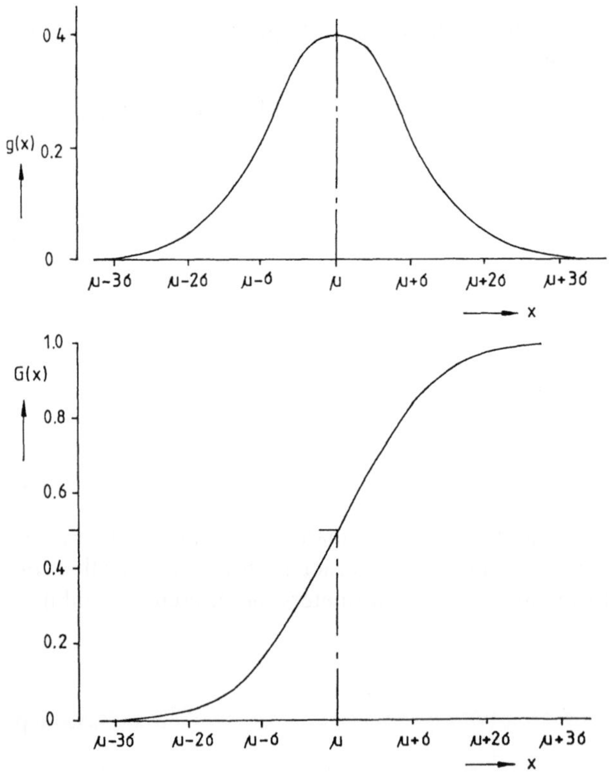

Fig. 4.1. Probability density function $g(x)$ and distribution function $G(x)$ of the normal distribution

Since the parameters μ and σ can take various values, there can be as many normal distributions as you wish. By the transformation

$$u = \frac{x - \mu}{\sigma}$$

every possible normal distribution can be adjusted relative to the origin and according to its width to some extent stretched or compressed, so that all normal distributions are the same, that is they retain a standardised form or shape. A diagrammatic explanation is given in Fig. 4.2.

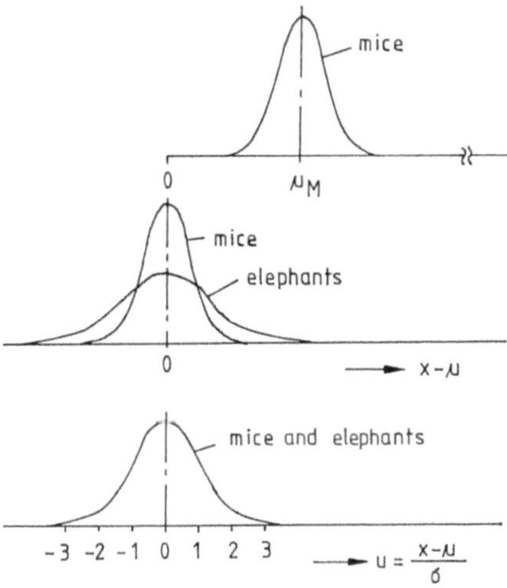

Fig. 4.2. Transformation of variable normal distribution into standardised form (u-distribution)

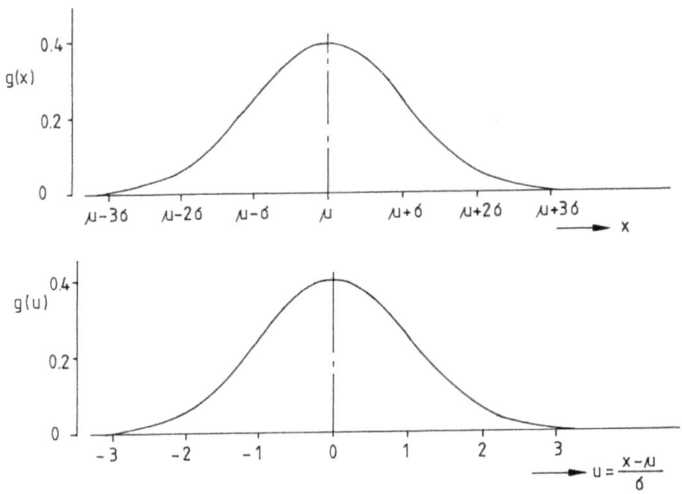

Fig. 4.3. Transition from $g(x; \mu, \sigma)$ to $g(u; 0, 1)$ by standard transformation

Figure 4.3 also shows the transition from a variable normal distribution to a standardised form.

The parameters of the standardised ND are $\mu = 0$ and $\sigma = 1$.

The standard normal distribution can be simply tabulated (Table A.1 in Appendix).

The ND Table A.1 (u-table) can be derived from the following values:

For $u \geqq 0$:	$G(u)$	section of the distribution in the range from $-\infty$ to u,
	$Q(u)$	section of the distribution in the range from u to $+\infty$,
	$G(u) - Q(u)$	section of the distribution in the range from $-u$ to $+u$.
For $u < 0$:	$G(-u)$	section of the distribution in the range from $-\infty$ to $-u$. Reference: $G(-u) = Q(u)$.

Exercise 4.1

Given: On an automatic lathe, shafts are produced according to the specified design size $d = 20^{+0.05}_{-0.05}$.
Once production is completed it is established that the lot is normally distributed with the parameters $\mu = 20.01$ mm and $\sigma = 0.03$ mm.

Required: Proportion of non-conforming items (proportion under $G_u = 19.95$ mm and over $G_o = 20.05$ mm).

Solution: With Figure 4.4: interval between μ and G_o in σ items.

$$u_o = \frac{G_o - \mu}{\sigma} = \frac{20.05 - 20.01}{0.03} = +1.33.$$

Interval between μ and G_u is in σ items

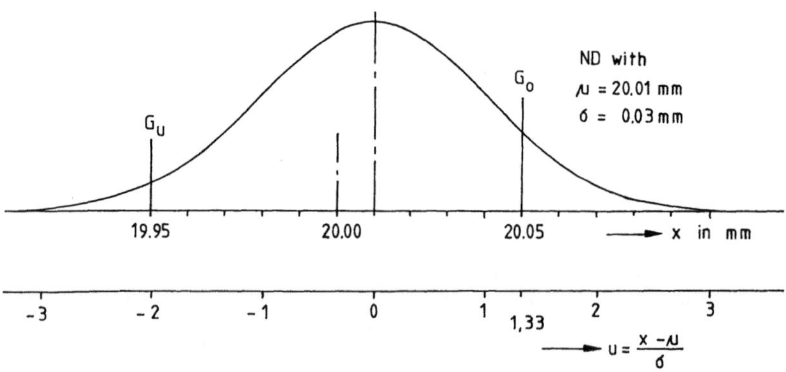

Fig. 4.4. Normally distributed shaft diameter dimensions and tolerance zone

$$u_u = \frac{G_u - \mu}{\sigma} = \frac{19.95 - 20.01}{0.03} = -2.$$

Thus the non-conforming proportions according to the ND table are:

above G_o	p_o	$= 0.091\,76 = 9.2\%$
below G_u	p_u	$= 0.022\,75 = 2.3\%$
in total	p	$= 11.5\%$

Proportion in the tolerance zone $1 - p = 88.5\%$.

The Probability Grid

The relationship between a normally distributed variable x and the standard normal variable u is given by

$$u = \frac{x - \mu}{\sigma}.$$

This is a linear equation which, in the coordinate system $u(x)$ runs through the point $u(x = \mu) = 0$ with the slope $1/\sigma$. This straight line is shown in Fig. 4.5.

Further points on the line are, for example,

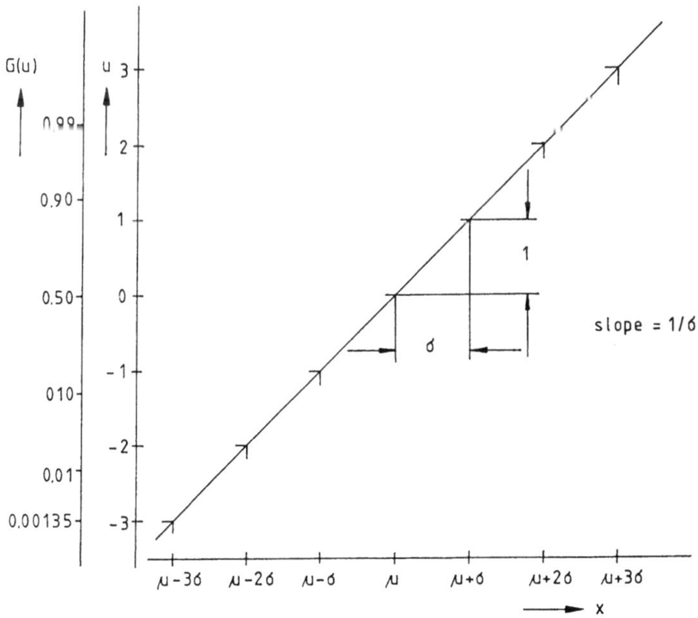

Fig. 4.5. Representation of the linear equation $u = (x - \mu)/\sigma$; with the addition of a $G(u)$ scale (probability grid)

$$u = -3 \quad \text{for} \quad x - \mu = -3\sigma$$

or

$$u = 3 \quad \text{for} \quad x - \mu = 3\sigma.$$

If with the help of the ND table a scale for the probability total $G(u)$ is inserted into this coordinate system, then the probability grid (PG) is constructed. In this PG the distribution function (total function) of each possible normal distribution is linearized.

The significance of the PG is not that normally distributed populations can be represented in it but that

—abnormally distributed probability distributions are inserted and the deviations can be estimated in comparison with the ND with the same parameters of shape and dimension (those sections outside certain limits).

—sample results (frequency totals) can be inserted into the PG in order to estimate the characteristics $\bar{x}$ and s graphically (Chap. 5). Also the extent of the confirmity between the sample and the ND shape can be subjectively estimated.

ND and Random Spread

If a population's parameters and distribution form is known, then random spread zones (RSZ) can be given within whose limits certain proportions—e.g. 95 or 99%—lay.

In the case of the ND, this is possible using the **u**-table or the PG.

Exercise 4.2

Given: Normally distributed shafts with the parameters $\mu = 20.01$ mm and $\sigma = 0.03$ mm.

Required:
1. RSZ by calculation in which 95% of the shafts lie.
2. RSZ by graph in which 95% of the shafts lie.
3. Possible evidence of the RSZ.

Solution:
1. With a normally distributed population the RSZ is

$$x = \mu \pm u\sigma$$

according to the u-table, for a proportion of 95% with bilateral demarcation $u = 1.9600$; so the RSZ is

$$x = 20.01 \pm 1.96 \cdot 0.03,$$

$$19.9512\,\text{mm} \leqq x \leqq 20.0688\,\text{mm}.$$

2. The graphical solution is shown in Fig. 4.6.
3. Possible evidence:

—in the RSZ lie 95% of all values, 2.5% lies below this and 2.5% over this RSZ,

—if one shaft is taken from the population, the probability that this
lies under the RSZ is 2.5%
lies within the RSZ is 95.0%
lies above the RSZ is 2.5%.

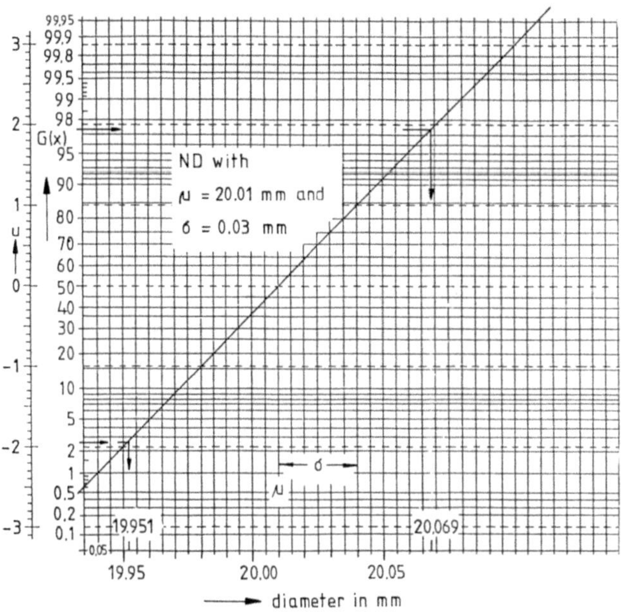

Fig. 4.6. Graphical solution to Exercise 4.2

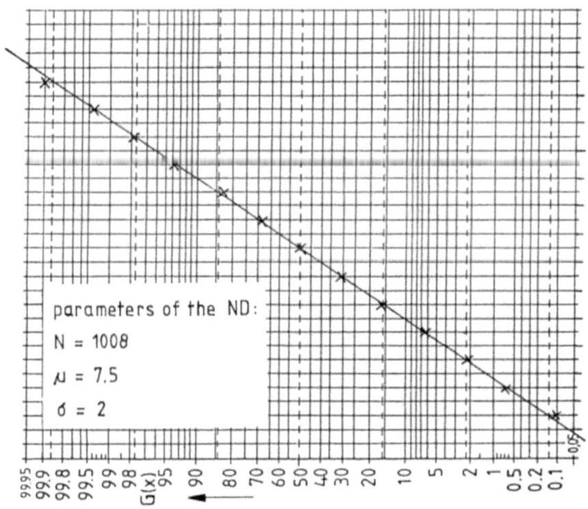

x_j	N_j	$\sum N_j$	G(x) in %
14	1	1008	100.0
13	5	1007	99.9
12	15	1002	99.4
11	40	987	97.9
10	103	947	93.9
9	150	844	83.7
8	190	694	68.8
7	190	504	50.0
6	150	314	31.2
5	103	164	16.3
4	40	61	6.1
3	15	21	2.1
2	5	6	0.7
1	1	1	0.1

Fig. 4.7. Model of a normal distribution with 14 classes

Models for Normal Distributions

True populations with a quite definite probability distribution can be simulated by assumed populations. This is however only the case for continuous distributions—and all distributions of quantitative characteristics are continuous—when the distribution is made discrete, i.e. is divided into classes.

In order to simulate closely the bell-shaped course of the ND, at least N = 1000 values are required.

Figure 4.7 contains the model of a ND which will be repeatedly used in the following sections for mixing and folding operations.

A further model of normal distribution is known by the name of counter box and is shown in Fig. 4.8. This model will also be used in the following sections.

x	g(x)	x
19	1	81
20	1	80
21	1	79
22	1	78
23	1	77
24	1	76
25	1	75
26	2	74
27	3	73
28	4	72
29	4	71
30	5	70
31	7	69
32	8	68
33	9	67
34	11	66
35	13	65
36	15	64
37	17	63
38	20	62
39	22	61
40	24	60
41	27	59
42	29	58
43	31	57
44	33	56
45	35	55
46	37	54
47	38	53
48	39	52
49	40	51
50	40	

Parameters of the ND:

$N = 1000$

$\mu = 50$

$\sigma_N = 9.9926 \approx 10$

Fig. 4.8. Model of a normal distribution with 62 classes (counter box)

Exercise 4.3

Given: ND model in Fig. 4.7.

Required: Determination of the parameters μ and σ with and without the aid of the statistical program on a calculator.

Solution: —With statistical program: $\mu = 7.5$ and $\sigma = 2$.

—Without statistical program: because of the symmetry of the distribution it is clear that

$$\mu = 7.5.$$

Calculation of the standard deviation follows with (4.5):

x	N_j	$g(x)$	$(x-\mu)$	$(x-\mu)^2 g(x)$
1	1		6.5	0.041 914 682
2	5		5.5	0.150 049 603
3	15		4.5	0.301 339 285
4	40		3.5	0.486 111 111
5	103	$N_j/1\,008$	2.5	0.638 640 873
6	150		1.5	0.334 821 428
7	190		0.5	0.047 123 015
8	190		0.5	
9	150		1.5	
10	103		2.5	
11	40		3.5	symmetrical
12	15		4.5	as above
13	5		5.5	
14	1		6.5	

$\sigma^2 = \sum (x-\mu)^2 g(x) = 4$

$\sigma = 2$

4.3 Abnormal Probability Distributions

Normal distributions can frequently be observed in nature; in production operations they only arise if the process is stable in terms of mean value and standard deviation. If either of these parameters changes over time, or both the whole production batch will have a mixed distribution. Mixed distributions can have various forms; they will be discussed in more detail in Chap. 6.

First type of mixed distribution (MD due to a continuous trend approaching the linear) is approximately symmetrical and narrower than the corresponding normal distribution (ND with similar parameters). This kind of mixed distribution is approximated on width (duration of the relevant trend) by the

—triangular distribution (TD),
—the trapezoid distribution (TPD) or at the simplest level by
—the rectangular distribution (RD).

In Fig. 4.9 these distributions are shown with information on the calculation of variances (from the range) and the range (from the variance).

A comparison of the distributions (Column 4) shows that the RD is the narrowest and the ND the widest distribution. There are also mixed distributions which are wider than the normal distribution; these are however rare (Chap. 6).

The formulae given in Fig. 4.9, Column 2 for the variance of the probability distributions is only valid for continuous distributions. If these are simulated by discrete distributions, the formulae in Column 5 should be used.

While it is necessary to have a large lot quantity in the normal distribution to simulate the bell-shape, the other symmetrical probability distributions need only small lot quantities; hence the following exercises:

1 Distribution	2 Variance	3 Range	4 Range proportion R/R_{RD}	5 Variance for simulation by discrete distribution with whole-numbered x-values
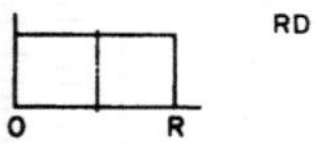	$\sigma^2 = \frac{R^2}{12}$	$R = 2\sqrt{3}\cdot\sigma$ $= 2\cdot 1.7321\cdot\sigma$	1	$\sigma^2 = \frac{R^2 - 1}{12}$
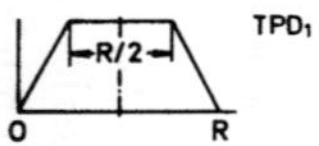	$\sigma^2 = \frac{10R^2}{192}$	$R = 2\sqrt{\frac{48}{10}}\cdot\sigma$ $= 2\cdot 2.1909\cdot\sigma$	1.2649	$\sigma^2 = \frac{10R^2 - 32}{192}$
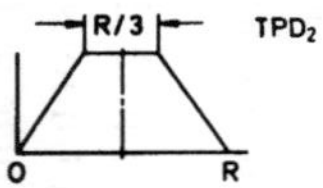	$\sigma^2 = \frac{5R^2}{108}$	$R = 2\sqrt{\frac{27}{5}}\cdot\sigma$ $= 2\cdot 2.3238\cdot\sigma$	1.3416	$\sigma^2 = \frac{5R^2 - 18}{108}$
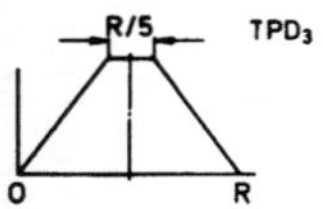	$\sigma^2 = \frac{13R^2}{300}$	$R = 2\sqrt{\frac{75}{13}}\cdot\sigma$ $= 2\cdot 2.4019\cdot\sigma$	1.3868	$\sigma^2 = \frac{13R^2 - 50}{300}$
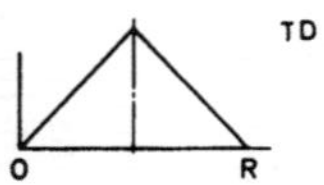	$\sigma^2 = \frac{R^2}{24}$	$R = 2\sqrt{6}\cdot\sigma$ $= 2\cdot 2.4495\cdot\sigma$	1.4142	$\sigma^2 = \frac{R^2 - 4}{24}$
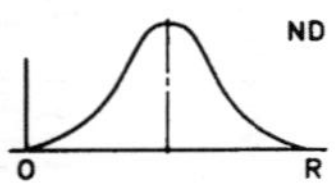	$\sigma^2 = \frac{R^2}{36}$	$R = 2\cdot 3\cdot\sigma$ with a proportion of $p = 0.27\%$ outside the R-zone	1.7321	ND-model from k folded RD; $R = k\cdot R_{RD}$ k: 4 | 8 | 16 $\sigma^2 = \frac{R^2 - 16}{48}$ | $\sigma^2 = \frac{R^2 - 64}{96}$ | $\sigma^2 = \frac{R^2 - 256}{192}$

Fig. 4.9. Information on different, symmetrical probability distributions

Exercise 4.4

Given: Simplest similar distribution (RD) with 2 classes

$N_1 = N_2 = 1.$

Required: Parameters of this probability distribution with and without the statistical program of a calculator.

Solution: —With the statistical program: $N = 2$, $\mu = 1.5$, $\sigma_N = 0.5$, $\sigma_{N-1} = 0.707\,1$.
Direction: for $N_i \to \infty$, $\sigma_{N-1} \to \sigma_N$.
Example: if $N_1 = N_2 = 1\,000$, then the parameters are $N = 2\,000$, $\mu = 1.5$, $\sigma_N = 0.5$, $\sigma_{N-1} = 0.500\,1$.
—Without statistical program: $\mu = 1.5$ on symmetry grounds

$$\sigma^2 = (R^2 - 1)/12 \quad \text{with} \quad R = 2.5 - 0.5 = 2$$

$$= \frac{4-1}{12} = 0.25,$$

$$\sigma = 0.5$$

or calculated with (4.5) $\sigma^2 = \sum (x - \mu)^2 g(x)$

$$= (1 - 1.5)^2 \cdot 0.5 + (2 - 1.5)^2 \cdot 0.5 = 0.25,$$

$$\sigma = 0.5.$$

Exercise 4.5

Given: Similar distribution (RD) with 7 classes.
Required: Parameters.
Solution: —With statistical program: $N = 7$, $\mu = 4$, $\sigma_N = 2$.
—Without statistical program: $\mu = 4$ on symmetry grounds

$$= \frac{49-1}{12} = 4,$$

$$= 2.$$

Exercise 4.6

Given: Trapezoid distribution of the type TPD_2 in Fig. 4.9.

x_j	N_j
1	1
2	2
3	3
4	3
5	3
6	3
7	2
8	1
N = 18	

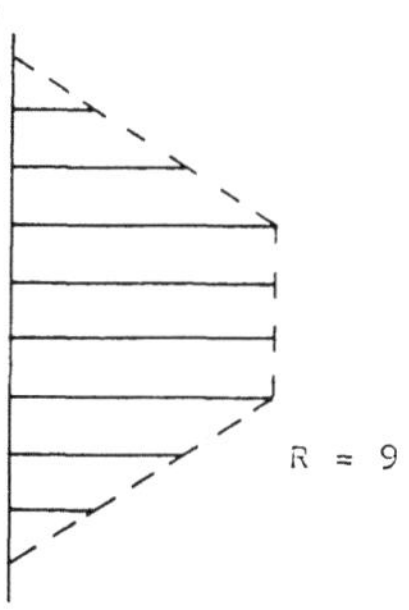

Required: Parameters.
Solution: —With statistical program: $N = 18$, $\mu = 4.5$, $\sigma = 1.8930$ (1.892969449).
—Without statistical program: $\mu = 4.5$ on symmetry grounds

$$\sigma^2 = \frac{5R^2 - 18}{108} = 3.5833,$$

$$\sigma = 1.8930 \quad (1.892969449).$$

Exercise 4.7

Given: Triangular distribution.

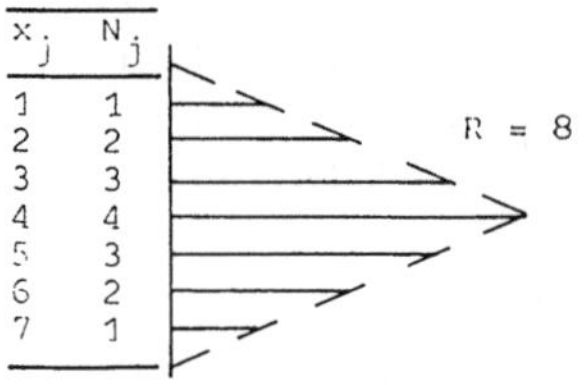

Required: Parameters.
Solution: —With statistical program: $N = 16$, $\mu = 4$, $\sigma = 1.5911$.

—Without statistical program: $\sigma^2 = \frac{8^2 - 4}{24} = 2.5,$

$$\sigma = 1.5811.$$

Exercise 4.8

Given: $N = 1000$ shafts with pivot pins are produced for which the specification is a diameter of $d = 60f7$ $(-60;\ -30)$. After production all the pins are automatically measured; the test results are stored in the computer and analysed; the frequency total function is drawn with a plotter. There is a normal distribution with $\mu = 59.960$ mm and $\sigma = 0.006$ mm.
Required: Estimate of the proportion of non-conforming shafts.
Solution: The drawing shows that only the upper limit value (machining side) is exceeded.

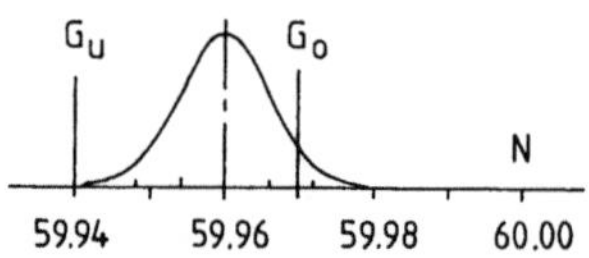

The interval from the mean value to the upper limit value in σ items is

$$u_{1-p} = \frac{G_O - \mu}{\sigma} = \frac{59.970 - 59.960}{0.006} = 1.\bar{6}.$$

According to the u-table, the proportion exceeding the limit is $p = 4.812\%$.
Note: this is only an estimate, which is however substantially exact because $n = N = 1000$.

Exercise 4.9

Given: Trapezoid distribution of the type TPD_1.

x_j	N_j
1	1
2	2
3	3
4	3
5	3
6	3
7	3
8	3
9	3
10	2
11	1

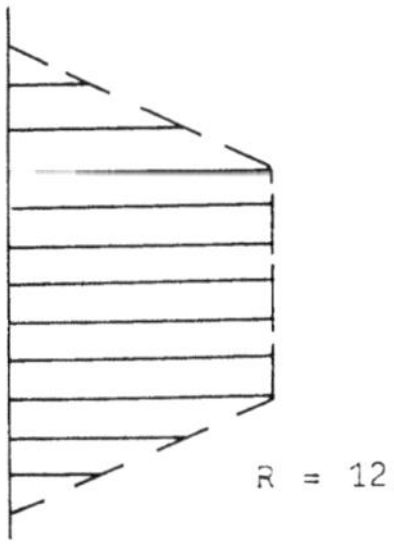

Required: Parameters.

Solution: 1. $\mu = 6$ on grounds of symmetry

$$\sigma^2 = \frac{10R^2 - 32}{192} \quad \text{as Fig. 4.9}$$

$$= \frac{10 \cdot 144 - 32}{192} = 7.\bar{3},$$

$$= 2.708\,0.\ \sigma$$

2. With automatic calculator:

$$N - 27,$$

$$\mu = 6,$$

$$\sigma_N = 1.708\,0.$$

Exercise 4.10

Given: Trapezoid distribution of the type TPD_2, as in Fig. 4.9.

x_j	N_j
1	1
2	2
3	3
4	4
5	4
6	4
7	3
8	2
9	1

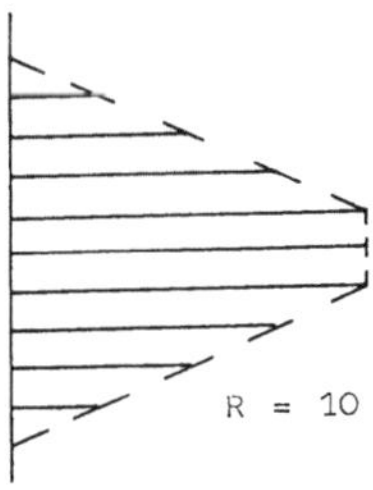

Required: Parameters.

Solution: 1. $\mu = 5$ on grounds of symmetry.

$$= \frac{13R^2 - 50}{300} \quad \text{as Fig. 4.9}$$

$$= \frac{1\,300 - 50}{300} = 4.167,$$

$$= 2.0412.$$

2. With automatic calculator:

$N = 24,$

$\mu = 5,$

$\sigma_N = 2.041\,2.$

Exercise 4.11

Given: Sloping triangular distribution; for this type the variance is:

—continuous distribution $\sigma^2 = \dfrac{R^2}{18},$

—discrete distribution $\sigma^2 = \dfrac{R^2 - 2.25}{18}.$

x_j	N_j
1	4
2	3
3	2
4	1

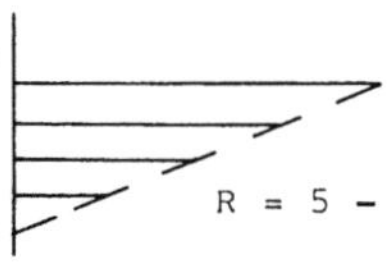

R = 5 - 0.5 = 4.5

Required: Parameters.

Solution: 1. With the above formula and the moment principle

$\mu = 2$ because $4 \cdot 1 = 2 \cdot 1 + 1 \cdot 2,$

$$\sigma^2 = \frac{4.5^2 - 2.25}{18} = 1,$$

$\sigma = 1.$

2. With automatic calculator:

$N = 10,$

$\mu = 2,$

$\sigma_N = 1.$

Note: Width of the continuous distribution

$$R = \sqrt{18} \cdot \sigma = 2 \cdot 2.121\,3 \cdot \sigma.$$

Exercise 4.12

Given: V-shaped distribution (VSD); for this type, the variance is

—continuous distribution $\sigma^2 = \dfrac{R^2}{8},$

—discrete distribution $\sigma^2 = \dfrac{R^2 - 1}{8}.$

x_j	N_j
1	4
2	3
3	2
4	1
5	0
6	1
7	2
8	3
9	4

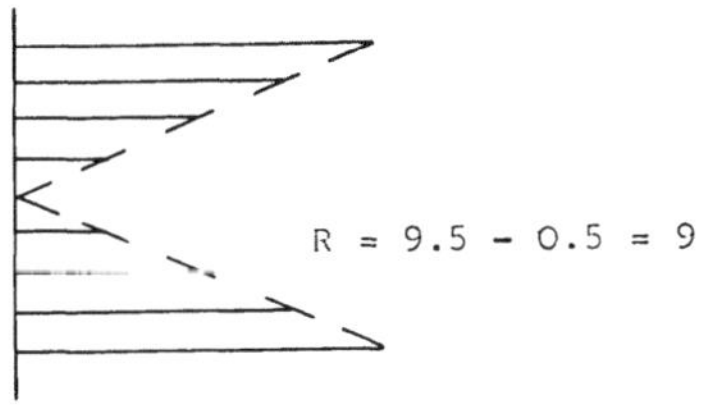

Required: Parameters.

Solution: 1. $\mu = 5$ on grounds of symmetry

$$\sigma^2 = \frac{9^2 - 1}{8} = 10$$

$$\sigma = 3.162\,277\,66.$$

2. With automatic calculator:

$$N = 20$$

$$\mu = 5$$

$$\sigma_N = 3.162\,277\,66.$$

Note: The V-distribution occurs operationally very seldom; it will be quoted in Exercise 7.4 to show that with four-fold folding of the VSD an ND is produced.

Note: Width of the continuous distribution

$$R = \sqrt{8} \cdot \sigma = 2\sqrt{2} \cdot \sigma = 2 \cdot 1.4142 \cdot \sigma,$$

i.e. less than half as wide as the ND with the same parameters.

4.4 Random Spread Zone for Mean Values

The Central Limiting Value Principle

If k values x_i from the same population or from different populations are added together, the sums $\sum x_i$ are normally distributed even when the population is not normally distributed; this holds from $k = 5$.

This principle is particularly significant in two cases:

- If linear dimensional chains are formed, the distribution of the closing values tend towards a normal distribution, even when the individual parts of the dimensional chains are not normally distributed.
- If the values x_i come from the same population and their sum is divided by the quantity, then the ND form does not alter; that means that the mean values

$$\bar{x} = \frac{\sum x_i}{n}$$

are continuously normally distributed even when the individual values x_i are not normally distributed.

The Deviation Progression Law

If k values x_i from the same or different populations are added, the variance of the total is the same as the total of the individual variances.

$$\sigma_{\text{ges}}^2 = \sum \sigma_i^2.$$

This law is particularly significant in two cases:

—If linear dimensional chains are formed, the variance of the closing values is exactly the same as the total of the variances of the k members of this dimensional chain, and indeed independent of their respective distributional forms.

—If the $n = k$ values x_i come from the same population, then the variance of the total is

$$\sigma^2 = n\sigma_x^2$$

and the standard deviation of the total is

$$\sigma = \sqrt{n}\sigma_x.$$

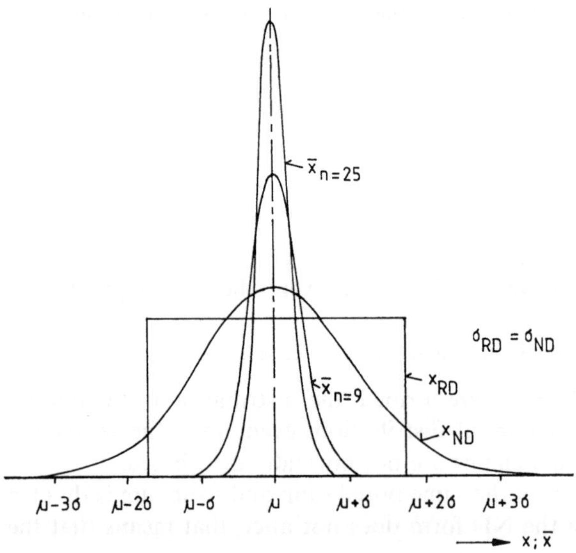

Fig. 4.10. Normal distribution and rectangular distribution with the parameters μ and σ and the distributions for mean values of samples with the quantities $n = 9$ and $n = 25$ from these probability distributions

If the totals are divided by n, the standard deviation is the (normally distributed) mean value.

$$\sigma_{\bar{x}} = \frac{\sqrt{n}\sigma_x}{n}$$

$$= \frac{\sigma_x}{\sqrt{n}}.$$

Within the random spread zone for individual values described in Sect. 4.2

$$x = \mu \pm u\sigma$$

is only valid if a normal distribution exists, so the analogue random spread for mean values is

$$\bar{x} = \mu \pm u\frac{\sigma}{\sqrt{n}}$$

independent of which distribution form is shown by the individual values, given that $n \geqq 5$. Figure 4.10 gives further information.

Exercise 4.13

Given: Possible, abnormally distributed population with the parameters $\mu = 7.5$ and $\sigma = 2$.

Required: Random spread zone in which 95% of the mean values from samples with a quantity of $n = 10$ are expected to fall.

Solution:
$$\bar{x} = \mu \pm u\frac{\sigma}{\sqrt{n}}$$

$$= 7.5 \pm 1.96\frac{2}{\sqrt{10}}$$

$$6.26 \leqq \bar{x} \leqq 8.74.$$

5 Mathematical and Graphical Evaluation of Dimensional Series

5.1 General

The populations mentioned in Chap. 4 are model laws. Actual populations occurring operationally (supplier batches, production batches) are usually unknown as regards position, spread and distribution form and must be estimated by sampling.

Since we are here only concerned with quantitative (measurable) characteristics, this always means dimensional series. A sample of quantity n components are taken at random from the batch. These n components are tested for one or more relevant dimensions or parameters. The results are a dimensional series with n values.

Either or both mathematical and graphical evaluation is used according to the particular process to decide between

—small samples ($n < 50$), these are evaluated without classification and
—large samples ($n \geqq 50$), these are usually evaluated after classification.

Classification means combining two or several possible neighbouring values to a total of 10 to 20 classes. Size values usually fall into

—0.1 mm	slide rule
—0.01 mm	micrometer
—0.001 mm	millimetric meter

Because of accuracy limits the size values classified in this way have already been classified, i.e. several of the n value results are each as large as each other. Thus classification is really a further classification of already classified existing values.

The results of the statistical evaluation of a dimensional series are statistical characteristics. The most important statistical characteristics are

—the mean value $\bar{x}$, estimated value for the parameter μ of the population taken from the sample and
—the standard deviation s, estimated value for the parameter σ of the population taken from the sample.

5.2 Mathmetical Evaluation Without Classification

If all the values from a dimensional series are used directly in the evaluation (for example individually inputted into the computer), then the mean value is

$$x = \frac{\sum x}{n}$$

and the standard deviation

$$s = +\sqrt{\frac{1}{n-1}\sum (x - \bar{x})^2}.$$

Instead of this definition formula it is preferable to use the identity formula

$$s = +\sqrt{\frac{1}{n-1}\left[\sum x^2 - \frac{1}{n}(\sum x^2)\right]}.$$

Exercise 5.1

Given: Dimensional series

i	x_i in mm
1	37.005
2	37.002
3	37.001
4	37.000
5	37.006
6	37.003
7	36.998

Required: Characteristics $\bar{x}$ and s

Solution: Using the statistical program of a calculator, $\bar{x} = 37.002\,14\,\text{mm}$ and $s = 0.002\,80\,\text{mm}$.

Also using a calculator, it may be appropriate to take the transformation

$$z = \frac{x - A}{B}$$

to determine the characteristics for z

$$\bar{z} = \frac{\sum z}{n}$$

and

$$s_z = +\sqrt{\frac{1}{n-1}\left[\sum z^2 - \frac{1}{n}(\sum z)^2\right]}$$

and to transform this back to x

$$\bar{x} = B\bar{z} + A$$

and

$$s_x = Bs_z.$$

Exercise 5.2

Given: Dimensional series from Exercise 5.1
Required: Characteristics $\bar{x}$ and s by proper transformation.
Solution: The transformation

$$z = \frac{x_i - 37.000}{10^{-3}}$$

amounting to a repetition of the deviations in 0.001 mm of $d = 37.000$ mm.

x_i	z_i	z_i^2
37.005	5	25
37.002	2	4
37.001	1	1
37.000	0	0
37,006	6	36
37,003	3	9
36.998	−2	4
$\sum$	15	79

—Without statistical program

$$\bar{z} = \frac{15}{7} = 2.1429$$

$$\bar{x} = 37.00214\,\text{mm}$$

$$s_z = +\sqrt{\frac{1}{n-1}\left[79 - \frac{1}{7}\cdot 15^2\right]} = 2.7946$$

$$s_x = 0.00279\,\text{mm}.$$

—With statistical program

$$\bar{z} = 2.1429$$

$$s_z = 2.7946.$$

Note: $z_4 = 0$ must be input into the calculator!

5.3 Graphical Evaluation Without Classification

The values of the dimensional series are sorted by size and inserted in the probability grid according to the probability totals of the expected values in Table A3.

The order of values can also mean that the values are inserted densely as points about the x-axis of the PG.

After entering the frequency totals in the PG, a balancing line is drawn through the points.

The point at which the straight line intersects the 50% line is the mean value $\bar{x}$ on the x-axis; half the difference of the intersection with the 84% line or the 16% line is the standard deviation s.

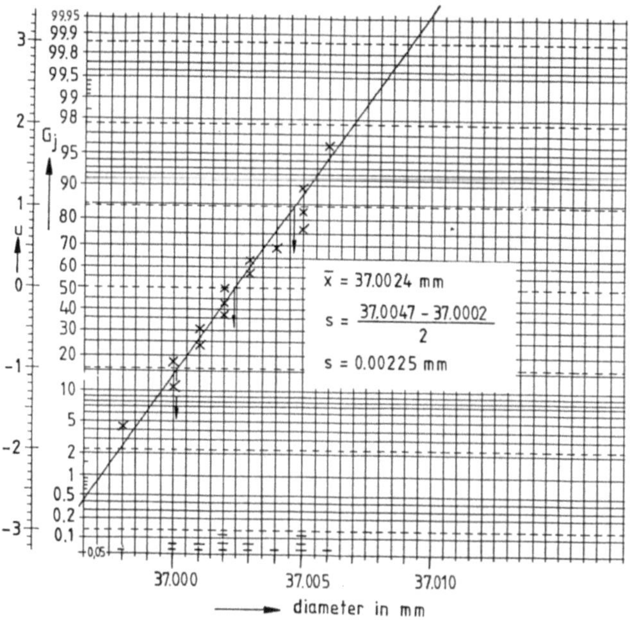

Fig. 5.1. Graphical evaluation of the dimensional series in Exercise 5.3

Exercise 5.3

Given: Dimensional series with $n = 15$ values

37.005	37.003	37.001
37.002	36.998	37.000
37.001	37.003	37.005
37.000	37.005	37.004
37.006	37.002	37.002

Required: $\bar{x}$ and s graphically.

Solution: Fig. 5.1

Mathematical control with the statistical program on the calculator.

$n = 15$

$\bar{x} = 37.002\,4\bar{6}\,\text{mm}$

$s = 0.002\,263\,\text{mm}.$

Note. The statistical characteristics and parameters always have the same unit of measurement as the individual values (e.g. in mm).

5.4 Mathematical Evaluation of Classified Dimensional Series

Concepts of Classification (Further Classification)

Classification. Allocation of the value range of a characteristic in part ranges (classes), which are mutually exclusive and complete the value range.

Evaluation Format	
characteristic: diameter, exercise 5.4 and 5.5	
lot size: sample size: 100	N = 37 mm G_o = 37.011 G_u = 36.995

class boundary lower	class boundary upper	class centre x_j	occupation number: check list	n_j	x'_j [1)]	$x'_j \cdot n_j$	$x'^2_j \cdot n_j$	Σn_j	$\frac{\Sigma n_j - 0.5}{n} \cdot 100$
1		2	3	4	5	6	7	8	9
−6.5	−5.5	−6	||	2	−7	−14	98	2	1.5
		−5	||	2	−6	−12	72	4	3.5
		−4	|||	3	−5	−15	75	7	6.5
		−3	卌	5	−4	−20	80	12	11.5
		−2	卌||	7	−3	−21	63	19	18.5
		−1	卌||||	9	−2	−18	36	28	27.5
		0	卌|||	8	−1	−8	8	36	35.5
		1	卌|||	8	0	0	0	44	43.5
		2	卌卌|	11	1	11	11	55	54.5
		3	卌卌|	11	2	22	44	66	65.5
		4	卌卌|	11	3	33	99	77	76.5
		5	卌|||	8	4	32	128	85	84.5
		6	卌	5	5	25	125	90	89.5
		7	||||	4	6	24	144	94	93.5
		8	|||	3	7	21	147	97	96.5
		9	||	2	8	16	128	99	98.5
9.5	10.5	10	|	1	9	9	81	100	99.5
class width w = 1 µm		auxiliary value from column 2 a = 1 µm	sum	100 = n		85 = Σ_1	1339 = Σ_2		1) $x'_j = \frac{x_j - a}{w}$

evaluation of mean $\bar{x}$

$$a + w \cdot \frac{\Sigma_1}{n} = 1 + 1 \cdot \frac{85}{100} = 0.85$$

$$= 1.85\ \text{um}$$

$$\bar{x} = \boxed{37.00185\,\text{mm}}$$

evaluation of standard deviation s

$$\Sigma_2 = 1339$$

$$-\frac{\Sigma_1^2}{n} = -\frac{85^2}{100} = -72.25$$

$$\text{Diff} = 1266.75$$

$$\frac{w^2 \cdot \text{Diff}}{n-1} = \frac{1^2 \cdot 1266.75}{99} = 12.7955 = s^2$$

$$s = +\sqrt{} = 3.5771 = \boxed{0.0036\,\text{mm}}$$

Fig. 5.2. Mathematical evaluation and preparation for graphical evaluation of the dimensional series in Exercise 5.4 ($n = 100$ stock pins from E-engine shafts). Bank form provided with kind approval of the Deutsche Gesellschaft für Qualität e.V., Frankfurt (German Quality Association)

Class. A section within a classification.

Class boundary. Value of the upper or lower limit of a class.

Nominal class boundary. Specified class boundary. Note: Use not recommended.

Genuine class boundary. Boundary between the actual value as rounded up and as rounded down to the middle of the class.

Class centre x_j. Arithemetical middle value between the genuine class boundaries.

Class width w. Genuine upper minus genuine lower class boundary; also class breadth.

Occupation number n_j. Number of individual values in a class.

Summed occupation number $\sum n_j$. Number of individual values which do not go beyond a characteristic value (class).

Frequency total $G_j = \sum n_j / n$. Summed occupation number divided by the total number of individual values (often with correction).

Pointers for classification

- —Class widths of similar size must be assigned; different class widths will not cause problems in graphical evaluation but will complicate mathematical evaluation.
- —Class boundaries must be chosen so that each value falls clearly into a class; this is the case if class boundaries = rounding boundaries.
- —Class number and class width must be reconciled so that 10 to 20 classes are produced.

After classification a slash list should be produced in which each value is ordered into the appropriate class (column 3 in the evaluation sheet, Fig. 5.2).

After producing the slash list, the occupation numbers n_j can be counted (Fig. 5.2, Column 4).

After this the temporary value a is chosen, it must be a class centre of a middle class.

Then follows the transformation

$$z_j = \frac{x_j - a}{w}.$$

This follows from the following formula: in column 5 the class with $x_j = a$ contains the value $z_j = 0$; thus the lowest class is negative, the highest class is numbered positively. The further mathematical evaluation follows according to the headings given on the evaluation sheet.

Exercise 5.4

Given: $n = 100$ diameter values of stock pins on E-engine shafts. The nominal size is $37\,j\,6$ or 37^{+11}_{5}.
The following values are provided by a precision measuring instrument with the zero set—with the aid of a master control—to the nominal size and give:

5	1	-2	4	-3	2	0	4	5	3
2	0	10	-5	1	8	7	-2	3	-1
1	5	-2	3	-1	6	-3	2	4	4
0	4	8	-2	3	-1	6	2	-4	5
6	2	-1	4	-4	9	-3	1	3	0
3	-6	0	9	-2	2	4	3	-1	6
-2	6	2	-1	3	-3	8	4	3	1
3	-1	7	0	4	1	-2	2	5	-6
5	4	-3	3	-1	7	2	0	7	1
2	-1	5	1	0	4	-5	5	2	-4

Required: Mathematical evaluation according to $\bar{x}$ and s.
Solution: Fig. 5.2

Notes.

—In this example a further classification is not necessary; the test results fall into an optimal classification.

—This example shows a special case of a two-fold transformation; the first follows the recording of the test results (deviations in μm of 37.000 mm), the second after evaluation.

—Calculators with statistical programs are often capable of producing $x_j n_j$; arithmetical check:

$$n = 100, \quad \bar{x} = 1.85\,\mu\text{m}, \quad s = 3.577\,1\,\mu\text{m}.$$

5.5 Graphical Evaluation of Classified Dimensional Series

Up to the stage of determination of the n_j values, the calculation is as for mathematical evaluation. After this stage the summed occupation numbers are derived $\sum n_j$ (Fig. 5.2, Column 8).

From these are derived the frequency totals $G_j = \sum n_j / n$; a sample range up to $n = 100$ will make the calculation of the frequency totals with correction somewhat more precise.

$$G_j = \frac{\sum n_j - 0.5}{n}.$$

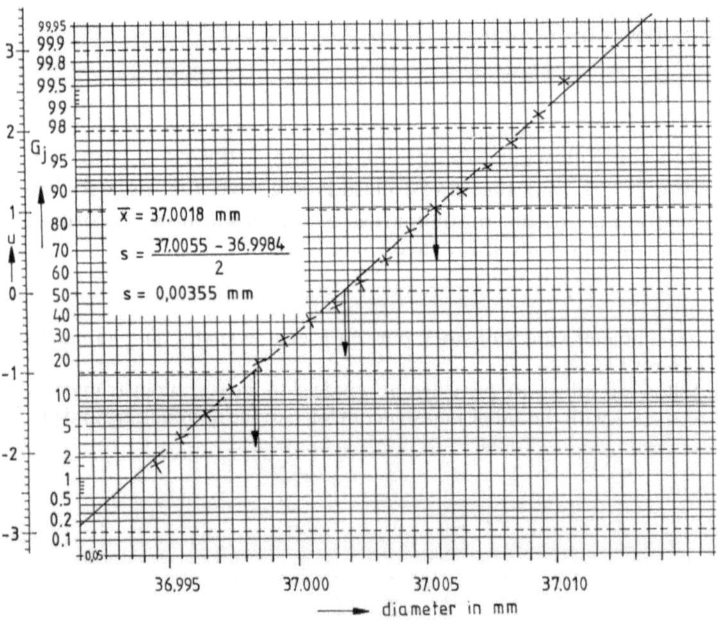

Fig. 5.3. Graphical evaluation of the dimensional series in Exercise 5.4 ($n = 100$ stock pins)

During the insertion of the G_j values into the PG, care should be taken that these are inserted above the upper genuine class limits.

When fitting the balancing line to these data G_j values up to 5% and above 95% should be omitted.

The further evaluation follows as for dimensional series without classification.

Exercise 5.5

Given: Dimensional series given in Exercise 5.4.
Required: Graphical evaluation according to $\bar{x}$ and s.
Solution: Figs. 5.2 and 5.3.

sample No	individual values x	$\bar{x}$	$\tilde{x}$	s^2	s	R	x_{min}	x_{max}
1	32 70 40 39 51	46.4	40	220	14.8	38	32	70
2	41 44 43 51 47	45.2	44	15	3.9	10	41	51
3	55 57 44 22 45	44.6	45	193	13.9	35	22	57
4	54 40 55 42 54	49.0	54	54	7.4	15	40	55
5	41 63 48 53 64	53.8	53	97	9.8	23	41	64
6	57 48 54 55 67	56.2	55	48	6.9	19	48	67
7	59 43 43 53 52	50.0	52	48	6.9	16	43	59
8	48 56 43 69 50	53.2	50	100	10.0	26	43	69
9	56 36 51 55 67	53.0	55	126	11.2	31	36	67
10	47 54 46 34 51	46.4	47	58	7.6	20	34	54
11	53 42 36 45 42	43.6	42	38	6.2	17	36	53
12	37 80 59 69 45	58.0	59	301	17.4	43	37	80
13	48 53 56 45 59	52.2	53	33	5.7	14	45	59
14	61 63 41 30 42	47.4	42	200	14.2	33	30	63
15	56 39 68 55 47	53.0	55	118	10.8	29	39	68
16	55 43 44 50 40	46.4	44	36	6.0	15	40	55
17	71 45 57 34 43	50.0	45	205	14.3	37	34	71
18	48 41 54 58 47	49.6	48	43	6.6	17	41	58
19	38 43 57 48 41	45.4	43	55	7.4	19	38	57
20	43 55 33 52 51	46.8	51	79	8.9	22	33	55
21	67 56 37 57 45	52.4	56	135	11.6	30	37	67
22	34 56 57 51 47	49.0	51	87	9.3	23	34	57
23	52 51 64 43 65	55.0	52	88	9.4	22	43	65
24	54 59 52 45 46	51.2	52	34	5.8	14	45	59
25	60 43 41 46 53	48.6	46	61	7.8	19	41	60
26	47 46 51 46 59	49.8	47	31	5.5	13	46	59
27	47 45 60 55 41	49.6	47	60	7.7	19	41	60
28	33 44 53 41 61	46.4	44	118	10.9	28	33	61
29	47 41 52 57 47	48.8	47	36	6.0	16	41	57
30	52 60 39 39 63	50.6	52	128	11.3	24	39	63
31	53 42 62 41 41	47.8	42	89	9.4	21	41	62
32	37 57 40 53 60	49.4	53	106	10.3	23	37	60
33	40 52 35 47 35	41.8	40	57	7.5	17	35	52
34	56 73 55 43 51	55.6	55	121	11.0	30	43	73
35	50 77 42 36 49	50.8	49	247	15.7	41	36	77
36	42 52 59 51 48	50.4	51	38	6.2	17	42	59
37	44 64 41 54 42	49.0	44	97	9.9	23	41	64
38	54 73 64 31 59	56.2	59	248	15.7	42	31	73
39	50 48 61 33 49	48.2	49	100	10.0	28	33	61
40	44 45 69 60 44	52.4	45	132	11.5	25	44	69
$\sum$	9966	1993.2	1958	4083	382.4	954	1536	2490
	mean values:	49.83	48.95	102.1	9.56	23.9	38.4	62.3

Fig. 5.4. $m = 40$ samples of the quantity $n = 5$ from the counter box (ND model with 62 classes and the parameters $\mu = 50$ and $\sigma = 10$). Applied Statistics 1, Springer 1970

For the insertion of the frequency totals in the PG it should be pointed out that—even with normally distributed populations—the random deviations from the line are greater, the smaller the sampler range. Hence the next exercise.

Exercise 5.6

Given: Model of a normal distribution with 62 classes, Fig. 4.8. From this $m = 40$ samples of $n = 5$ are taken, Fig. 5.4.

Required: Representation of the frequency totals in the PG for the first $n = 10$, $n = 20$, $n = 50$, $n = 100$ and for all $n = 200$ values.

Solution: Fig. 5.5.

5.6 Other Statistical Characteristics

Up to now only the most important statistical characteristics have been discussed: man value $\bar{x}$ and standard deviation s.

In addition there is the central value for the position of the distribution (median) $\tilde{x}$ that is the middle of the sizes of the ordered values x_i and for the width of the distribution, the range $R = x_{max} - x_{min}$.

The range R is often used to estimate the smoothed instant standard deviation for a production batch with a trend. To this end at regular intervals (e.g. hourly) a total of m samples of the same quantity n (preferably $n = 5$) are taken. It is accepted that the trend within the consecutively produced n items can be neglected. For all m samples R is calculated; the middle range

$$\bar{R} = \frac{\sum R}{m}$$

allows the estimation of the 'instant standard deviation'.

$$\hat{\sigma} = \frac{\bar{R}}{d_n}$$

with d_n in Table A2.

Note. The 'instant standard deviation' can also be estimated through the middle variance of m samples.

$$\hat{\sigma} = \sqrt{\bar{s^2}}.$$

Exercise 5.7

Given: For the diameter of a lathe, the measurement $d = 10.2 \pm 0.06\,\text{mm}$ is given. To estimate the 'instant spread' of the trend-based production, hourly $n = 5$ consecutively produced items are taken and tested.

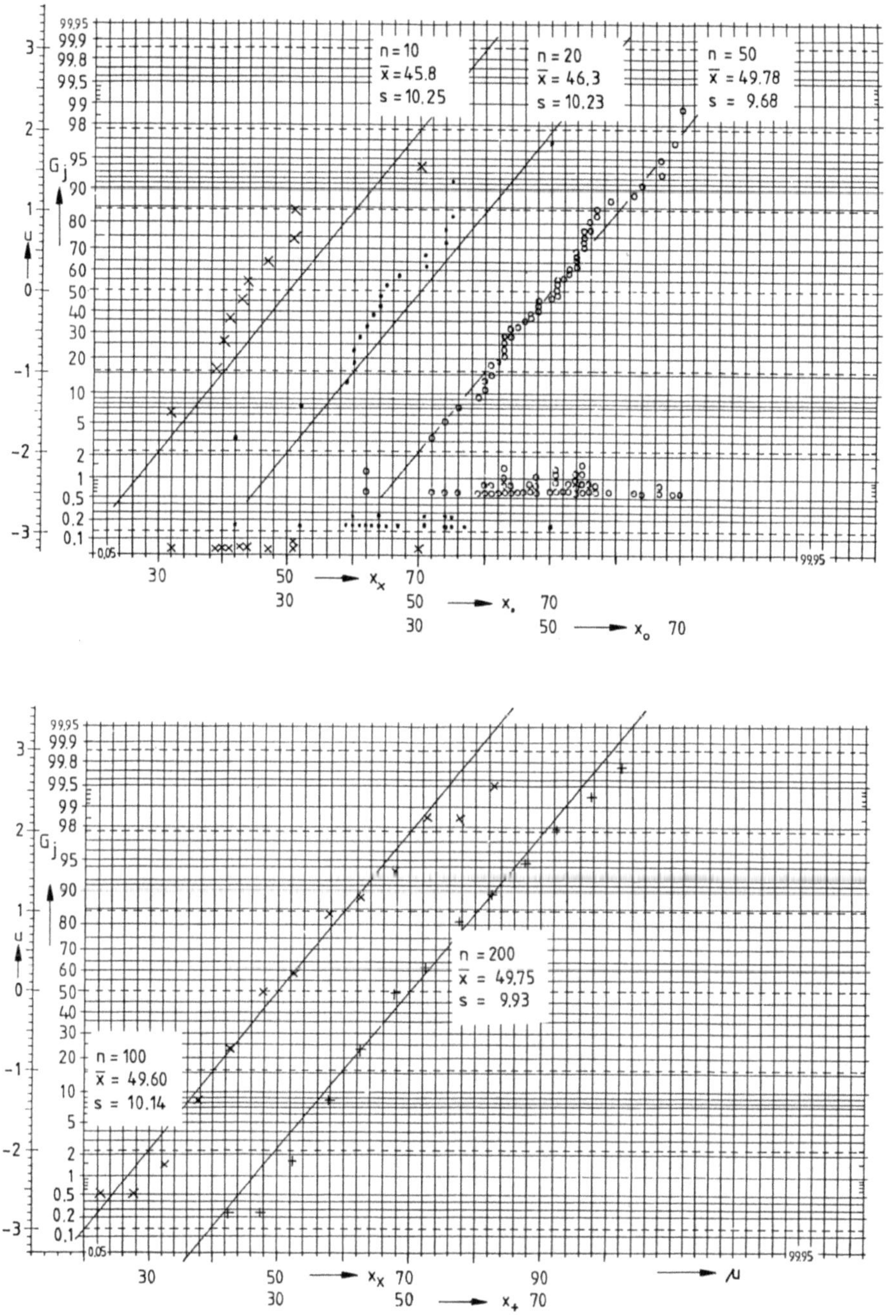

Fig. 5.5. Frequency totals in the PG of samples with the quantities $n = 10$ to $n = 200$ from the counter box; the lines drawn are the probability totals of the population with $\mu = 50$ and $\sigma = 10$

	10.173	10.161	10.175	10.198	10.200
	10.162	10.184	10.187	10.215	10.227
	10.182	10.167	10.170	10.197	10.192
	10.161	10.180	10.182	10.185	10.186
	10.161	10.197	10.170	10.193	10.199
R =	0.021	0.036	0.017	0.030	0.041
	10.200	10.209	10.227	10.230	10.232
	10.210	10.229	10.246	10.228	10.233
	10.217	10.206	10.237	10.241	10.257
	10.209	10.219	10.204	10.213	10.248
	10.206	10.207	10.232	10.229	10.232
R =	0.017	0.023	0.042	0.028	0.025

Required: Instant standard deviation.
Solution: Middle range

$$\bar{R} = \frac{\sum R}{m} = \frac{0.280}{10} = 0.028 \text{ mm.}$$

Instant standard deviation

$$\hat{\sigma} = \frac{\bar{R}}{d_n} = \frac{0.028}{2.326} = 0.012 \text{ mm.}$$

Note. If all $n \cdot m = 50$ values are evaluated, $s = 0.026$ mm; the total spread is more than twice as big as the instant spread.
Note. Above the middle variance:

$$\hat{\sigma} = \sqrt{\overline{s^2}} = 0.011\,54 \text{ mm.}$$

Exercise 5.8

Given: 120 bolt diameter sizes subdivided into groups of five.

x	$\bar{x}$ R s^2	x	$\bar{x}$ R s^2	x	$\bar{x}$ R s^2	x	$\bar{x}$ R s^2	x	$\bar{x}$ R s^2	x	$\bar{x}$ R s^2
6.54 6.44 6.53 6.46 6.45	6.484 0.10 0.00223	6.48 6.51 6.48 6.54 6.51	6.504 0.06 0.00063	6.44 6.48 6.39 6.55 6.48	6.468 0.16 0.00347	6.58 6.52 6.47 6.59 6.51	6.534 0.12 0.00253	6.53 6.51 6.59 6.47 6.52	6.524 0.12 0.00188	6.49 6.56 6.51 6.38 6.50	6.488 0.18 0.00437
6.54 6.52 6.54 6.51 6.49	6.520 0.05 0.00045	6.46 6.56 6.49 6.53 6.62	6.532 0.16 0.00367	6.48 6.51 6.47 6.58 6.63	6.534 0.16 0.00473	6.50 6.49 6.50 6.48 6.48	6.490 0.02 0.00010	6.48 6.49 6.49 6.44 6.47	6.474 0.05 0.00043	6.58 6.55 6.62 6.43 6.41	6.518 0.19 0.00867
6.59 6.52 6.53 6.49 6.47	6.520 0.12 0.00210	6.51 6.55 6.56 6.44 6.59	6.530 0.15 0.00335	6.59 6.54 6.54 6.48 6.43	6.516 0.16 0.00383	6.44 6.44 6.53 6.45 6.44	6.460 0.09 0.00155	6.48 6.60 6.48 6.62 6.52	6.540 0.14 0.00440	6.42 6.49 6.43 6.46 6.49	6.458 0.07 0.00107
6.44 6.50 6.45 6.50 6.50	6.478 0.06 0.00092	6.46 6.52 6.48 6.52 6.47	6.490 0.06 0.00080	6.45 6.56 6.52 6.50 6.52	6.510 0.11 0.00160	6.49 6.42 6.50 6.54 6.48	6.486 0.12 0.00188	6.51 6.44 6.49 6.57 6.49	6.500 0.13 0.00220	6.48 6.44 6.51 6.50 6.52	6.490 0.08 0.00100

Required: Evaluate without and with classification (further classification).

Solution: First all values are input into the calculator with $\bar{x} - s$ automatic facility:

$n \quad = 120$

$\bar{x} \quad = 6.502\,\text{mm}$

$s_{n-1} = 0.050\,503\,3\,\text{mm}.$

The total middle value is exact so

$$\bar{\bar{x}} = \frac{\sum \bar{x}}{m} = \frac{156.048}{24} = 6.502\,\text{mm}.$$

The standard deviation is also approximated

$$\hat{\sigma} = \frac{\bar{R}}{d_n} = \frac{0.110\,833}{2{,}326} = 0.047\,65\,\text{mm}$$

or

$$\hat{\sigma} = \sqrt{\overline{s^2}} = 0.049\,185\,\text{mm}.$$

Evaluation with further classification:

class boundary	class centre	slash list	n_j	z_j	$z_j n_j$	$z_j^2 n_j$
6.375/6.395	6.385	||	2	-5	-10	50
6.396/6.415	6.405	|	1	-4	- 4	16
6.415/6.435	6.425	~~||||~~	5	-3	-15	45
6.435/6.455	6.445	~~||||~~ ~~||||~~ ||||	14	-2	-28	56
6.455/6.475	6.465	~~||||~~ ~~||||~~	10	-1	-10	10
6.475/6.495	6.485	~~||||~~ ~~||||~~ ~~||||~~ ~~||||~~ ~~||||~~ |	26	0	0	0
6.495/6.515	6.505	~~||||~~ ~~||||~~ ~~||||~~ ||||	19	1	19	19
6.515/6.535	6.525	~~||||~~ ~~||||~~ ~~||||~~	15	2	30	60
6.535/6.555	6.545	~~||||~~ ~~||||~~	10	3	30	90
6.555/6.575	6.565	~~||||~~	5	4	20	80
6.575/6.595	6.585	~~||||~~ |||	8	5	40	200
6.595/6.615	6.605	|	1	6	6	36
6.615/6.635	6.625	||||	4	7	28	196
w = 0.02	a = 6.485		120		106 $=\sum_1$	858 $=\sum_2$

Calculation of the mean value $\bar{x}$

$$\bar{x} = a + w\frac{\sum_1}{n} = 6.485 + 0.02\cdot\frac{106}{120} = 6.502\,\bar{6} = 6.502\,7\,\text{mm}.$$

Calculation of the standard deviation s; first the sum of the quadratic deviations

$$SQ = \sum\nolimits_2 - \frac{\sum_1^2}{n} = 858 - \frac{106^2}{120} = 764.3\bar{6}$$

$$s^2 = \frac{w^2\cdot SQ}{n-1} = \frac{4\cdot 10^{-4}\cdot 764.3\bar{6}}{119} = 25.693\,0\cdot 10^{-4}$$

$$s = 0.050\,69\,\text{mm}.$$

Note on Exercise 5.8

This exercise has the main objective of practising the mathematical technique used in the evaluation of dimensional series.

In practice the nominal size and the variations of the bolt diameters must be given.

If the design instruction requires

$$d = 6.5^{+0.1}_{-0.1}$$

then the error proportion can be calculated.

$$u_{1-\hat{p}} = \frac{G_0 - \bar{x}}{s} = \frac{6.6 - 6.502}{0.0505} = 1.941$$

and therefore $\hat{p}_0 = 2.61\%$.

$$u_{1-\hat{p}} = \frac{G_u - \bar{x}}{s} = \frac{6.4 - 6.502}{0.0505} = -2.02$$

and therefore $\hat{p}_u = 2.17\%$.

Total error proportion: $\hat{p}_o + \hat{p}_u = \hat{p} = 4.78\%$.

5.7 The Significance of Statistical Characteristics

If an individual item is tested in relation to a relevant characteristic, then the test result only demonstrates a "valid degree of conformance" if a comparison of the test result with a standard value is carried out.

This standard value can be

—normal value,
—ideal value,
—desired value,
—limit value

or an appropriate value, for example one of the named values of a derived value.

Exercise 5.9

Given: An $x = 1.80$ m tall man stands on the scales in the morning and determines his weight $m = 90$ kg.

Required: Judgement.

Solution: The ideal weight is known to be the numerical value of the body length in cm minus 100 cm in kg, less 10%. For a body height of 180 cm

$$m_{\text{ideal}} = 80 \cdot (1 - 0.1) = 72\,\text{kg}.$$

The actual weight lies 18 kg over the ideal weight. After the introduction of a diet and regular weight checks, the standard value is the previous weight;

it is not so much the overweight as the removal of weight as a result of the diet which is of interest.

Exercise 5.10

Given: A stranger tells that he has a sister who weighs $m = 60$ kg.
Required: Judgement
Solution: The listener accepts that the sister is of average weight and concludes that her weight is 'normal'. This assumption is reasonable because most women are average weight.
This judgement must be corrected if the stranger adds that his sister is 135 cm (195 cm) tall.

If the characteristic is a tolerated size, then the limit values are the standard values, through which the judgement of the actual value follows in the final test. Actual values which lie in the tolerance field—within the limit values—are good; actual values outside this range are non-conforming.

The judgement of an individual item should in a similar way to be considered on the judgement of a number of individual items, of batches.

A sample is taken from a batch and evaluated according to $\bar{x}$ and s. For the clarification of the judgement of this characteristic it is generally accepted that the standard deviation s is not necessary, because the standard deviation σ of the population, of the batch, is known. This case rarely occurs—especially with strange batches—but does simplify the explanation and the understanding of the philosophy behind the judgement of batches.

In the case where the standard deviation σ of the population is known, there are two methods for judgement of the mean value $\bar{x}$.

— From the mean value $\bar{x}$ and the known standard deviation σ a confidence interval for the mean value μ of the population is calculated.

$$\mu = \bar{x} \pm u \frac{\sigma}{\sqrt{n}}.$$

If the desired value is within these limits it is assumed that the mean value $\bar{x}$ conforms with the desired value or only deviates from it by chance.

This method is meaningless in the judgement of tolerated sizes; therefore we will not go further into the confidence interval at this point.

— If the limit values are given, acceptable limits (operating limits) are established which lie within tolerance zones, which with a definite small error proportion p_u or p_o in normally distributed individual values will have a high acceptable probability P_a.

For example, the acceptable limits are such that with an actual non-conforming proportion above the upper limit value of $p_o = 1\%$, the acceptable probability $P_a = 90\%$, Fig. 5.6. If on the other hand the actual non-conforming proportion *was* $p_o = 7.3\%$, then the acceptable probability is only $P_a = 10\%$, see bottom of Fig. 5.6.

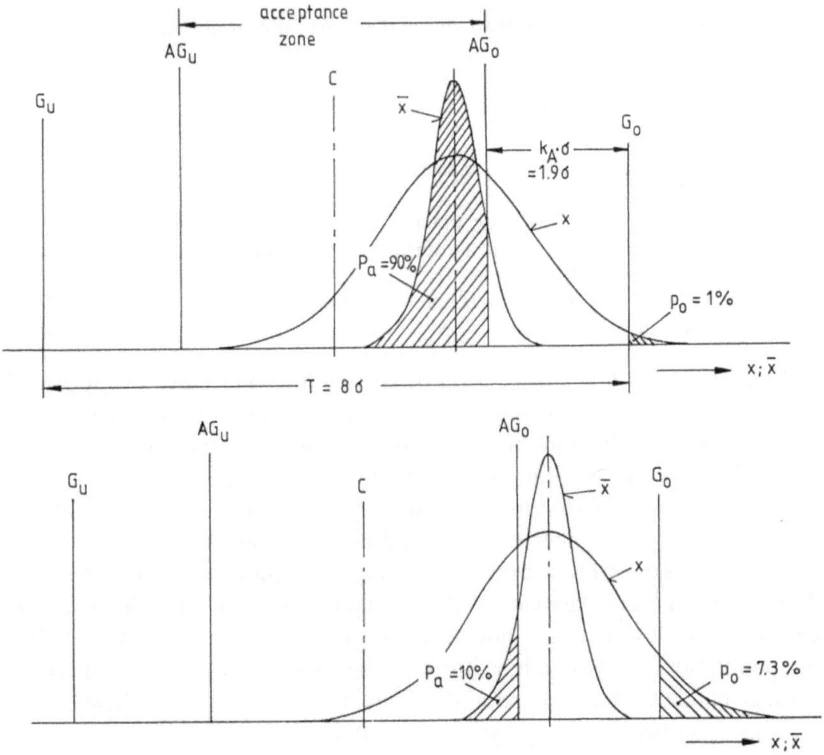

Fig. 5.6. Judgement of a batch on the basis of the mean value $\bar{x}$; standard deviation σ of the population is known; sample range $n = 9$; the acceptable limits (AG) lie such that the batch is accepted by $P_a = 90\%$ a nonconforming proportion of $p = 1\%$; if the nonconforming proportion is $p = 7.3\%$, then $P_a = 10\%$

This philosophy for the judgement of mean values $\bar{x}$ finds application in

—input testing (acceptance testing) and
—production testing (quality control chart techniques).

Four specific data are important for the effectiveness of this judgement method

—sample range n,
—acceptance factor k_A
—a point of the operation characteristic OC, P_a (p)
—a further point of the OC.

Through two of these specific data, the other two can be clearly established. Further explanation in Chap. 13.

6 Mixed Distributions

6.1 Types of Mixed Distributions

With production processes in which the parameters μ and σ remain constant, production batches are usually normally distributed despite the many sources of variability.

If in the course of production one or both of the parameters μ or σ change, mixed distributions result. Two can be distinguished:

- mixed distributions of the first type: these arise through a constant change, a trend whereby they can become linear, progressive, degressive or mixed, and
- mixed distributions of the second type: these arise from the sudden erratic change of one or both of the parameters. This type of mixed distribution includes the case of two or more batches which arise from different processes becoming mixed. This case should not occur, however, if this can be avoided and if the rule is followed that batches from different processes are either not mixed at all or only mixed when each has been checked and cleared.

6.2 Mixed Distributions Type 1

As long as the trend is linear, a mixed distribution will result which is symmetrical and narrower than a normal distribution having the same parameters, Fig. 6.1.

If the linear trend is steeper or runs longer without intervention, then a roughly trapezoid shaped distribution results, Fig. 6.2.

In Fig. 6.3 and in Fig. 6.4 initially stable production with a subsequent linear trend is simulated, which finally becomes progressive; in Fig. 6.4 this is rather more marked than in Fig. 3. More sloping distributions arise which are noticeably narrower than the relevant normal distribution.

A constant change to the standard deviation can only be simulated at great cost; and is therefore not shown here. It should however be noted that a constant change of standard deviation is not so noticeable in effect as an erratic change, which will be described in Sect. 6.3.

→ time

x_j	1	2	3	4	5	6	7	8	9	10	11	12	13	14	15	N_j	$\sum N_j$	G(x) in %
1	1															1	1	0.0066
2	5	1														6	7	0.0463
3	15	5	1													21	28	0.1852
4	40	15	5	1												61	89	0.5886
5	103	40	15	5	1											164	253	1.67
6	150	103	40	15	5	1										314	567	3.75
7	190	150	103	40	15	5	1									504	1071	7.08
8	190	190	150	103	40	15	5	1								694	1765	11.7
9	150	190	190	150	103	40	15	5	1							844	2609	17.3
10	103	150	190	190	150	103	40	15	5	1						947	3556	23.5
11	40	103	150	190	190	150	103	40	15	5	1					987	4543	30.0
12	15	40	103	150	190	190	150	103	40	15	5	1				1002	5545	36.7
13	5	15	40	103	150	190	190	150	103	40	15	5	1			1007	6552	43.3
14	1	5	15	40	103	150	190	190	150	103	40	15	5	1		1008	7560	50.0
15		1	5	15	40	103	150	190	190	150	103	40	15	5	1	1008	8568	56.7
16			1	5	15	40	103	150	190	190	150	103	40	15	5	1007	9575	63.3
17				1	5	15	40	103	150	190	190	150	103	40	15	1002	10577	70.0
18					1	5	15	40	103	150	190	190	150	103	40	987	11564	76.5
19						1	5	15	40	103	150	190	190	150	103	947	12511	82.7
20							1	5	15	40	103	150	190	190	150	844	13355	88.3
21								1	5	15	40	103	150	190	190	694	14049	92.92
22									1	5	15	40	103	150	190	504	14553	96.25
23										1	5	15	40	103	150	314	14867	98.33
24											1	5	15	40	103	164	15031	99.1414
25												1	5	15	40	61	15092	99.8148
26													1	5	15	21	15113	99.9537
27														1	5	6	15119	99.9934
28															1	1	15120	100.0000

parameters of the instant ND:

N = 1008

σ = 2

parameters of the mixed distribution:

N = 15120

μ = 14.5

σ = 4.7610; $6\cdot\sigma$–zone from 0.22 to 28.78

a

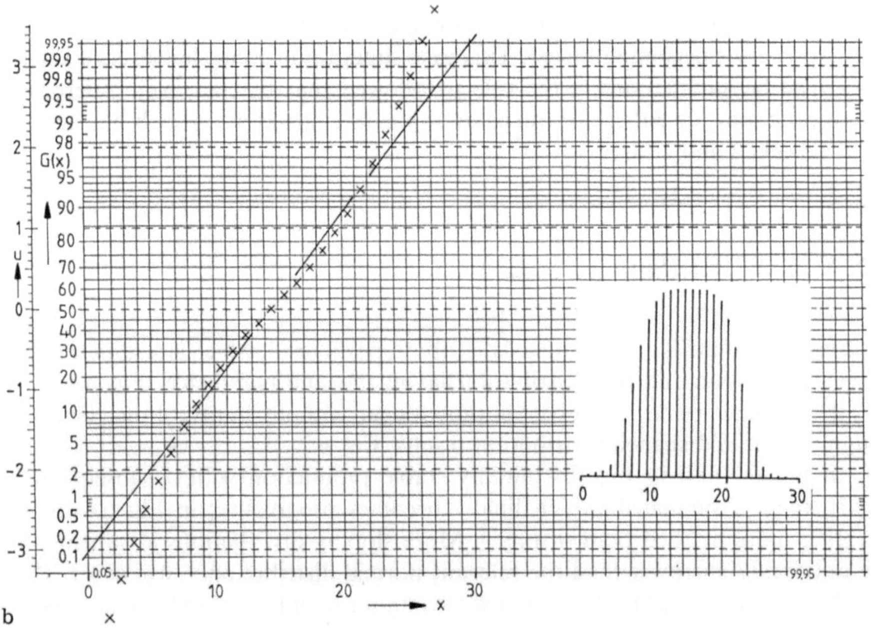

b

Fig. 6.1 a, b. Formation **a** and representation **b** of a mixed distribution by simulation of a linear trend, $\Delta\mu = 0.5\sigma$ between two time points

x_j	1	2	3	4	5	6	7	8	9	10	11	12	13	14	15	N_j	$\sum N_j$	G(x) in%
											→ time							
1	1															1	1	0.0066
2	5															5	6	0.0397
3	15	1														16	22	0.1455
4	40	5														45	67	0.4431
5	103	15	1													119	186	1.23
6	150	40	5													195	381	2.52
7	190	103	15	1												309	690	4.56
8	190	150	40	5												385	1075	7.11
9	150	190	103	15	1											459	1534	10.1
10	103	190	150	40	5											488	2022	13.4
11	40	150	190	103	15	1										499	2521	16.7
12	15	103	190	150	40	5										503	3024	20.0
13	5	40	150	190	103	15	1									504	3528	23.3
14	1	15	103	190	150	40	5									504	4032	26.7
15		5	40	150	190	103	15	1								504	4536	30.0
16		1	15	103	190	150	40	5								504	5040	33.3
17			5	40	150	190	103	15	1							504	5544	36.7
18			1	15	103	190	150	40	5							504	6048	40.0
19				5	40	150	190	103	15	1						504	6552	43.3
20				1	15	103	190	150	40	5						504	7056	46.7
21					5	40	150	190	103	15	1					504	7560	50.0
22					1	15	103	190	150	40	5					504	8064	53.3
23						5	40	150	190	103	15	1				504	8568	56.7
24						1	15	103	190	150	4C	5				504	9072	60.0
25							5	40	150	190	103	15	1			504	9576	63.3
26							1	15	103	190	150	40	5			504	1oo80	66.7
27								5	40	150	190	103	15	1		504	10584	70.0
28								1	15	103	190	150	40	5		504	11088	73.3
29									5	40	150	190	103	15	1	504	11592	76.7
30									1	15	103	190	150	40	5	504	12096	80.0
31										5	40	150	190	103	15	503	12599	83.3
32										1	15	103	190	150	40	499	13098	86.6
33											5	40	150	190	103	488	13586	89.9
34											1	15	103	190	150	459	14045	92.99
35												5	40	150	190	385	14430	95.44
36												1	15	103	190	309	14739	97.48
37													5	40	150	195	14934	98.77
38													1	15	103	119	15053	99.5569
39														5	40	45	15098	99.8545
40														1	15	16	15114	99.9603
41															5	5	15119	99.9934
42															1	1	15120	100.0000

parameters of the instant ND:

N = 1008

σ = 2

parameters of the mixed distribution:

N = 15120

μ = 21.5

σ = 8.8694

6 σ zone from -5.11 to 48.11

a

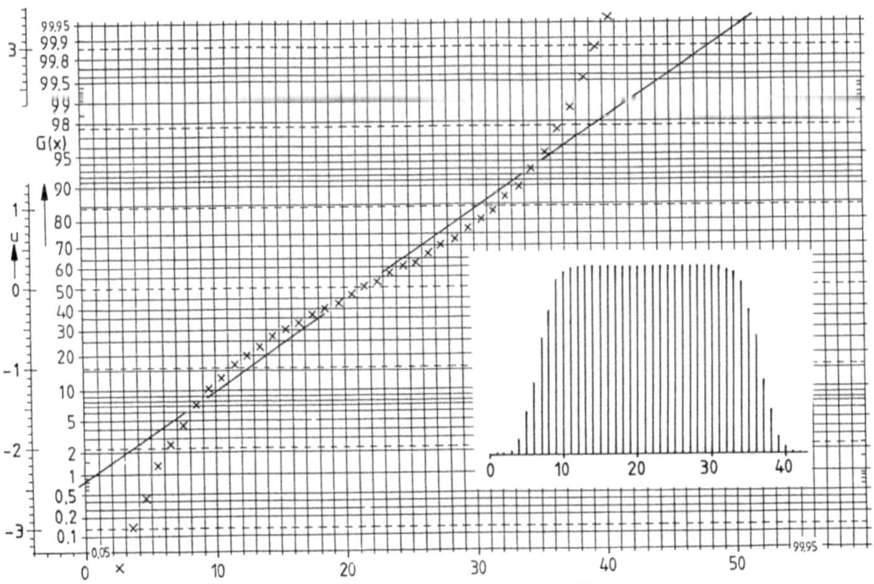

b

Fig. 6.2 a, b. Formation **a** and representation **b** of a mixed distribution of Type 1 by simulation of a linear trend with $\Delta\mu = \sigma$ between two time points

x_j	1	2	3	4	5	6	7	8	9	10	N_j	$\sum N_j$	G(x) in %
1	1	1									2	2	0.0198
2	5	5	1								11	13	0.1290
3	15	15	5								35	48	0.4762
4	40	40	15	1							96	144	1.43
5	103	103	40	5							251	395	3.92
6	150	150	103	15	1						419	814	8.07
7	190	190	150	40	5						575	1389	13.8
8	190	190	190	103	15	1					689	2078	20.6
9	150	150	190	150	40	5					685	2763	27.4
10	103	103	150	190	103	15	1				665	3428	34.0
11	40	40	103	190	150	40	5				568	3996	39.6
12	15	15	40	150	190	103	15	1			529	4525	44.9
13	5	5	15	103	190	150	40	5			513	5038	50.0
14	1	1	5	40	150	190	103	15			505	5543	55.0
15			1	15	103	190	150	40			499	6042	60.0
16				5	40	150	190	103	1		489	6531	64.8
17				1	15	103	190	150	5		464	6995	69.4
18					5	40	150	190	15		400	7395	73.4
19					1	15	103	190	40		349	7744	76.8
20						5	40	150	103	1	299	8043	79.8
21						1	15	103	150	5	274	8317	82.5
22							5	40	190	15	250	8567	85.0
23							1	15	190	40	246	8813	87.4
24								5	150	103	258	9071	89.99
25								1	103	150	254	9325	92.51
26									40	190	230	9555	94.79
27									15	190	205	9760	96.83
28									5	150	155	9915	98.36
29									1	103	104	10019	99.3948
30										40	40	10059	99.7917
31										15	15	10074	99.9405
32										5	5	10079	99.9901
33										1	1	10080	100.0000

(Columns 1–10: → t)

parameters of the instant ND: $N = 1008$; $\sigma = 2$

parameters of the mixed distribution:

$N = 10080$

$\mu = 14.5$

$\sigma = 6.4962$

$6 \cdot \sigma$ - zone from −4.99 to 33.99

a

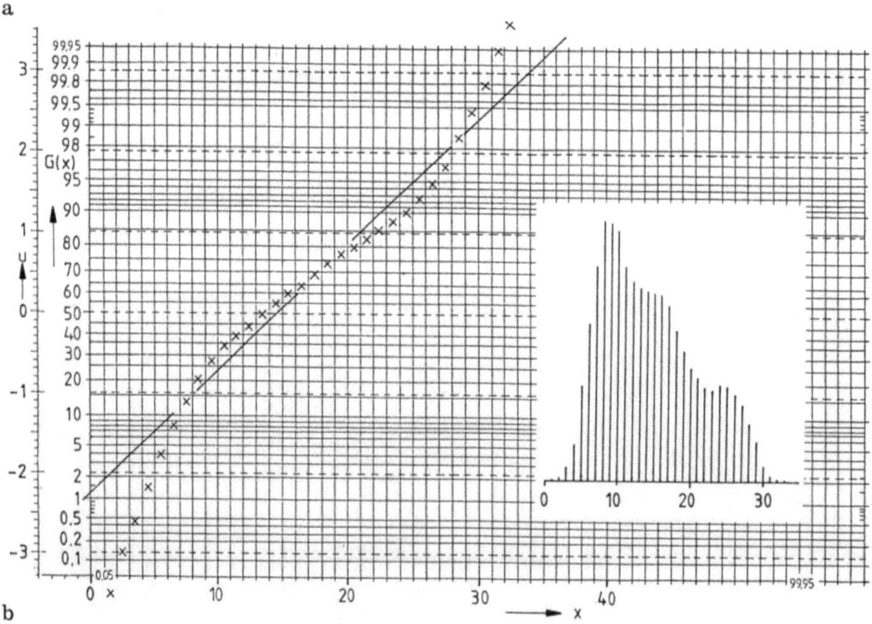

b

Fig. 6.3 a, b. Formation **a** and representation **b** of a mixed distribution of Type 1 by simulation of an only partly linear trend

x_i	1	2	3	4	5	6	7	8	9	10	11	12	13	14	15	N_i	$\sum N_i$	G(x) in%
													→ time					
1	1	1	1	1	1											5	5	0.0331
2	5	5	5	5	5	1	1									27	32	0.2116
3	15	15	15	15	15	5	5	1	1							87	119	0.7870
4	40	40	40	40	40	15	15	5	5	1	1					242	361	2.39
5	103	103	103	103	103	40	40	15	15	5	5	1	1			637	998	6.60
6	150	150	150	150	150	103	103	40	40	15	15	5	5			1076	2074	13.7
7	190	190	190	190	190	150	150	103	103	40	40	15	15			1566	3640	24.1
8	190	190	190	190	190	190	190	150	150	103	103	40	40			1916	5556	36.7
9	150	150	150	150	150	190	190	190	190	150	150	103	103			2016	7572	50.1
10	103	103	103	103	103	150	150	190	190	190	190	150	150			1875	9447	62.5
11	40	40	40	40	40	103	103	150	150	190	190	190	190	1		1467	10914	72.2
12	15	15	15	15	15	40	40	103	103	150	150	190	190	5		1046	11960	79.1
13	5	5	5	5	5	15	15	40	40	103	103	150	150	15		656	12616	83.4
14	1	1	1	1	1	5	5	15	15	40	40	103	103	40		371	12987	85.9
15						1	1	5	5	15	15	40	40	103		225	13212	87.4
16								1	1	5	5	15	15	150		192	13404	88.7
17										1	1	5	5	190	1	203	13607	89.99
18												1	1	190	5	197	13804	91.30
19														150	15	165	13969	92.39
20														103	40	143	14112	93.33
21														40	103	143	14255	94.28
22														15	150	165	14420	95.37
23														5	190	195	14615	96.66
24														1	190	191	14806	97.92
25															150	150	14956	98.92
26															103	103	15059	99.5966
27															40	40	15099	99.8611
28															15	15	15114	99.9603
29															5	5	15119	99.9934
30															1	1	15120	100.0000

parameters of the mixed distribution:

N = 15120

μ = 10.5667

σ = 4.7253

$6 \cdot \sigma$ - zone from -3.61 to 24.74

a

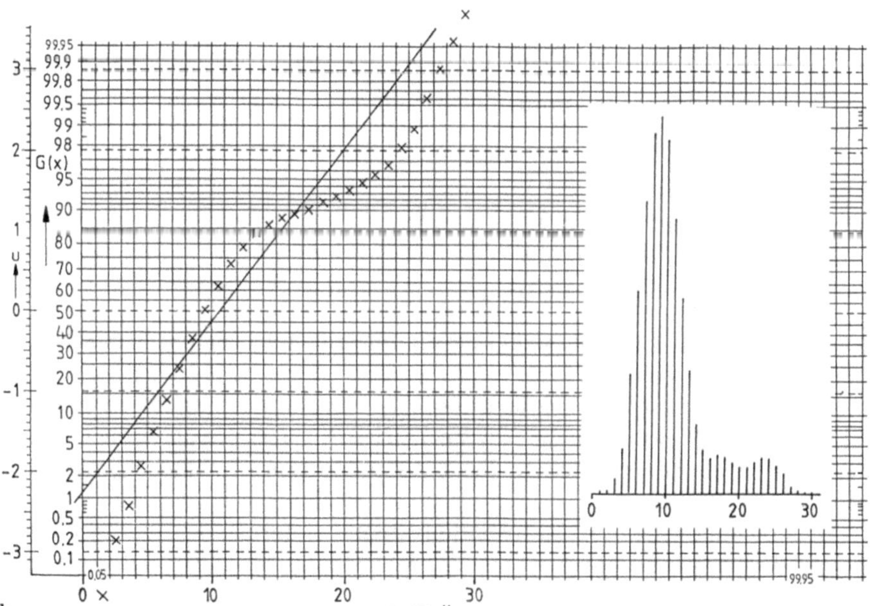

b

Fig. 6.4 a, b. Formation **a** and representation **b** of a mixed distribution of Type 1 by simulation of an initially stable production batch with a subsequent trend

Summary

The mixed distributions arising from a trend with constant spread are narrower than normal distributions having the same parameters. This should not however be described as a favourable deviaton from the shape of the normal distribution, since, in a dimensional chain, each possible distribution acts as if it were normal, Chap. 7.

6.3 Mixed Distributions Type 2

It should first be accepted that for an erratic change of process parameters, only the angle is altered, change of position remains constant through the standard deviation. Moreover it should be accepted that only one alteration occurs, there are only two batches mixed.

In Fig. 6.5, two normal distributions of the same form are mixed. This produces two-pointed distributions, both of which points are narrower than an ND which leads to the result that the probability totals in the PG:

—are steeper at the outset,
—are flatter in the middle,
—and at the end are again steeper

than the plotted line for the normal distribution.

In the cases

$N_2 = 0.5\,N_1$ — Fig. 6.6,
$N_2 = 0.2\,N_1$ — Fig. 6.7,
$N_2 = 0.1\,N_1$ — Fig. 6.8 or
$N_2 = 0.05\,N_1$ — Fig. 6.9

the total probability curve in the PG moves upwards to the right, which is how the non-symmetry arises.

In all cases almost all values lie within the—shifted to the right—6σ range. Only if $N_2 < 0.1\,N_1$ will have the very small extreme values not all register, also after the shift of the 6σ range. In Fig. 6.9 a proportion of about 3% remain outside the 6σ range.

With the following mixed distributions from two normal distributions it is assumed that the production position remains constant, during which the spread changes erratically.

In Fig. 6.10 two production batches of the same size are mixed, where the standard deviation of the second batch is $\sigma_2 = 3\sigma_1$. The mixed distribution is wider than the relevant normal distribution (straight in the PG); the proportion outside the 6σ range is $p = 2‰$.

Investigations with $\sigma_2 = 2\sigma_1$ and with $\sigma_2 = 4\sigma_1$ (not shown in figures here) produce a smaller proportion outside the 6σ range.

x_j	N_{j1}	N_{j2}	N_{j3}	N_{j4}
1	1			
2	5			
3	15			
4	40			
5	103	1		
6	150	5		
7	190	15		
8	190	40		
9	150	103	1	
10	103	150	5	
11	40	190	15	
12	15	190	40	
13	5	150	103	1
14	1	103	150	5
15		40	190	15
16		15	190	40
17		5	150	103
18		1	103	150
19			40	190
20			15	190
21			5	150
22			1	103
23				40
24				15
25				5
26				1

Parameters and random spread zones of the mixed distributions:

	ND_1+ND_2	ND_1+ND_3	ND_1+ND_4
lot of size N	2016	2016	2016
mean μ	9.5	11.5	13.5
standard deviation σ	2.2884	4.4721	6.3246
$6\cdot\sigma$ - zone from to	1.01....17,98	-1.92....24,92	-5.47....32.47
proportion outside	0.10 %	0.00 %	0.00 %

a

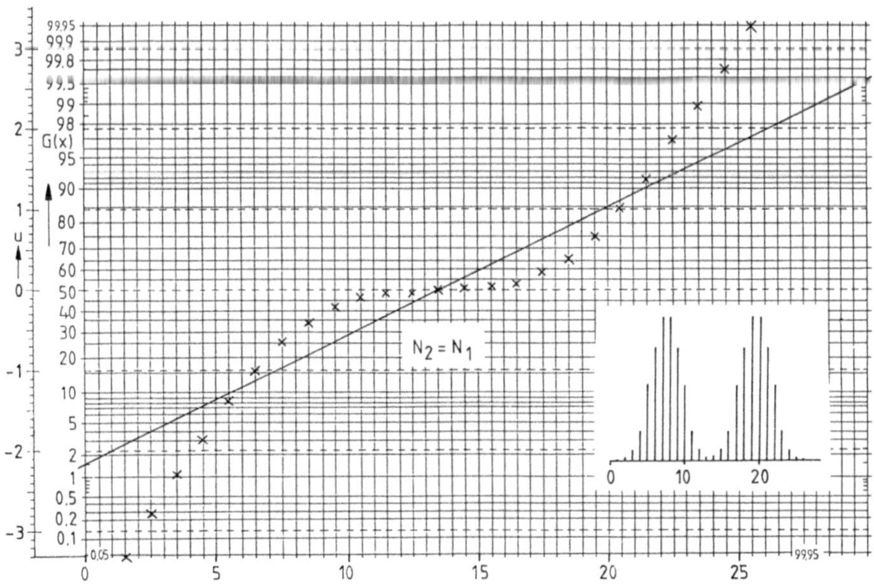

b

Fig. 6.5. a Formation of mixed distributions. A normal distribution ND_1 with the parameters $N = 1\,008$, $\mu_1 = 7.5$ and $\sigma_1 = 2$ is mixed with normal distribution of the same range, the same standard deviation but different mean values. **b** Representation of the mixed distribution arising from $ND_1 + ND_4$. This MD has the mean value $\mu = 13.5$

x_j	N_{j1}	N_{j2}	N_{j3}	N_{j4}
1	1			
2	5			
3	15			
4	40			
5	103	0.5		
6	150	2.5		
7	190	7.5		
8	190	20		
9	150	51.5	0.5	
10	103	75	2.5	
11	40	95	7.5	
12	15	95	20	
13	5	75	51.5	0.5
14	1	51.5	75	2.5
15		20	95	7.5
16		7.5	95	20
17		2.5	75	51.5
18		0.5	51.5	75
19			20	95
20			7.5	95
21			2.5	75
22			0.5	51.5
23				20
24				7.5
25				2.5
26				0.5

parameters and random spread zones of the mixed distributions:

	$ND_1 + ND_2$	$ND_1 + ND_3$	$ND_1 + ND_4$
lot size N	1512	1512	1512
mean μ	8.8333	10.1667	11.5000
standard deviation σ	2.7487	4.2687	6.0000
$6\cdot\sigma$ - zone from to	0.59....17.08	-2.64....22.97	-6.5....29.5
proportion outside	0.03%	0.00 %	0.00 %

a

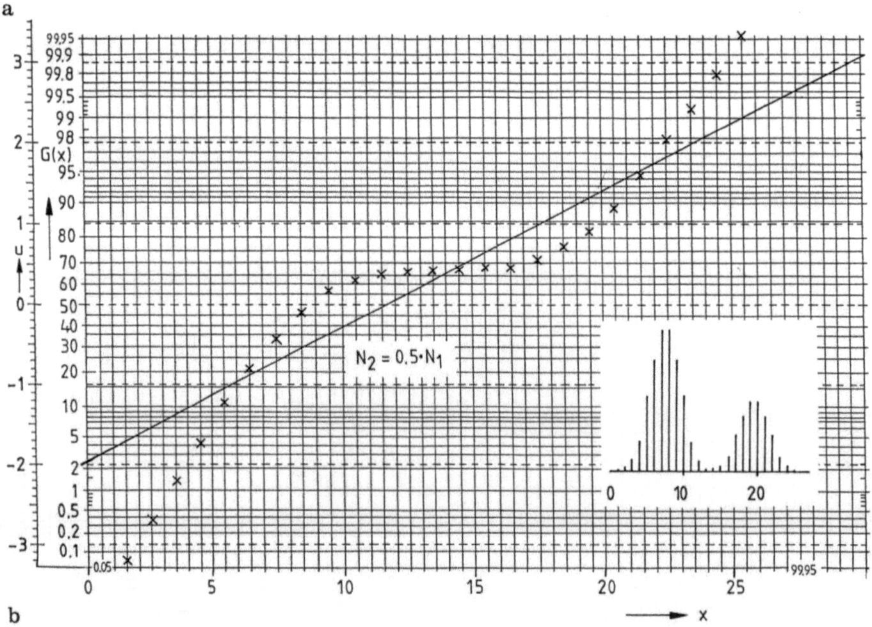

b

Fig. 6.6. a Formation of mixed distributions. A normal distribution ND_1 with the parameters $N = 1\,008$, $\mu_1 = 7.5$ and $\sigma_1 = 2$ is mixed with normal distribution of the same standard deviation, half the size and different mean values. **b** Representation of the MD with $\mu = 11.5$, arising from $ND_1 + ND_4$; $N_2 = 0.5\,N_1$; $\sigma_2 = \sigma_1$; $\Delta\mu = 6\sigma$

x_i	N_{j1}	N_{j2}	N_{j3}	N_{j4}
1	1			
2	5			
3	15			
4	40			
5	103	0.2		
6	150	1		
7	190	3		
8	190	8		
9	150	20.6	0.2	
10	103	30	1	
11	40	38	3	
12	15	38	8	
13	5	30	20.6	0.2
14	1	20.6	30	1
15		8	38	3
16		3	38	8
17		1	30	20.6
18		0.2	20.6	30
19			8	38
20			3	38
21			1	30
22			0.2	20.6
23				8
24				3
25				1
26				0.2

Parameters and random spread zones of the mixed distributions:

	$ND_1 + ND_2$	$ND_1 + ND_3$	$ND_1 + ND_4$
lot size N	1209.6	1209.6	1209.6
mean μ	8.1667	8.8333	9.5000
standard deviation σ	2.4944	3.5901	4.8990
$6\cdot\sigma$ - zone from to	0.68....15.65	-1.94....19.60	-5.20....24.20
proportion outside	0.35%	0.35%	0.10 %

b

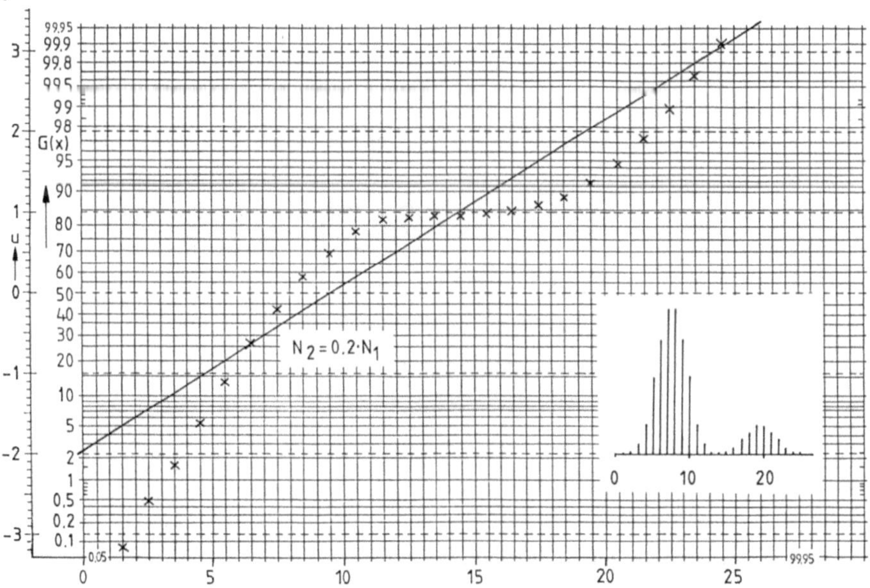

a

Fig. 6.7. a Formation of mixed distributions. A normal distribution ND_1 with the parameters $N_1 = 1008$, $\mu_1 = 7.5$ and $\sigma_1 = 2$ is mixed with normal distribution of the same standard deviation, with the range $N_2 = 0.2\,N_1$ and different mean values. **b** Representation of the MD with $\mu = 9.5$, arising from $ND_1 + ND_4$

x_j	N_{j1}	N_{j2}	N_{j3}	N_{j4}	mean values	distance to μ_1
1	1					
2	5					
3	15					
4	40					
5	103					
6	150					
7	190	0.1				
8	190	0.5			μ_1 = 7.5	
9	150	1.5				
10	103	4				
11	40	10.3				
12	15	15				
13	5	19	0.1			
14	1	19	0.5		μ_2 = 13.5	$3\cdot\sigma$
15		15	1.5			
16		10.3	4			
17		4	10.3			
18		1.5	15			
19		0.5	19			
20		0.1	19		μ_3 = 19.5	$6\cdot\sigma$
21			15			
22			10.3			
23			4			
24			1.5			
25			0.5			
26			0.1			
27						
28						
29						
30						
31				0.1		
32				0.5		
33				1.5		
34				4		
35				10.3		
36				15		
37				19		
38				19	μ_4 = 37.5	$15\cdot\sigma$
39				15		
40				10.3		
41				4		
42				1.5		
43				0.5		
44				0.1		

Parameters of the MD:

	$ND_1 + ND_2$	$ND_1 + ND_3$	$ND_1 + ND_4$
lot size N	1108.8	1108.8	1108.8
mean μ	8.0455	8.5909	10.2273
standard deviation σ	2.6411	3.9876	8.8533
$6\cdot\sigma$ – zone from to	0.12...15.96	-3.37...20.55	-16.33...36,79
proportion outside $6\cdot\sigma$ symmetrical to μ	1.48 %	2.83 %	6.26 %
proportion outside $6\cdot\sigma$ with x_{un} = 1,5	0.28 %	0.10 %	0.0 %

a

Fig. 6.8 a. Formation of mixed distributions. A normal distribution with the parameters $N_1 = 1\,008$, $\mu_1 = 7.5$ and $\sigma_1 = 2$ is mixed with normal distribution with the same standard deviation with the quantity $N = 0.1\,N_1$ and different mean values

x_j	N_{j1}	N_{j2}	N_{j3}	N_{j4}	N_{j5}	N_{j6}	N_{j7}
1	2						
2	10						
3	30						
4	80						
5	206	0.1					
6	300	0.5					
7	380	1.5					
8	380	4					
9	300	10.3	0.1				
10	206	15	0.5				
11	80	19	1.5				
12	30	19	4				
13	10	15	10.3	0.1			
14	2	10.3	15	0.5			
15		4	19	1.5			
16		1.5	19	4			
17		0.5	15	10.3	0.1		
18		0.1	10.3	15	0.5		
19			4	19	1.5		
20			1.5	19	4		
21			0.5	15	10.3	0.1	
22			0.1	10.3	15	0.5	
23				4	19	1.5	
24				1.5	19	4	
25				0.5	15	10.3	0.1
26				0.1	10.3	15	0.5
27					4	19	1.5
28					1.5	19	4
29					0.5	15	10.3
30					0.1	10.3	15
31						4	19
32						1.5	19
33						0.5	15
34						0.1	10.3
35							4
36							1.5
37							0.5
38							0.1

Parameters and random spread zones of the mixed distributions:

	ND_1+ND_2	ND_1+ND_3	ND_1+ND_4	ND_1+ND_5	ND_1+ND_6	ND_1+ND_7
lot size N	2116.8	2116.8	2116.8	2116.8	2116.8	2116.8
mean μ	7.6905	7.8810	8.0714	8.2619	8.4534	8.6429
standard deviation σ	2.1738	2.6273	3.2451	3.9509	4.7054	5.4884
$6 \cdot \sigma$ - zone from	1.17	0.00	-1.66	-3.69	-5.66	-7.82
to	...14.21	...15.76	...17.80	...20.11	...22.57	...25.11
proportion outside the $6 \cdot \sigma$ - zone	0.38 %	2.38 %	3.99 %	4.47 %	4.73 %	4.76 %

a

Fig. 6.9 a. Formation of mixed distributions. A normal distribution with the parameters $N_1 = 2016$, $\mu_1 = 7.5$ and $\sigma_1 = 2$ is mixed with normal distribution with the same standard deviation with the quantity $N = 0.05\,N_1$ and different mean values

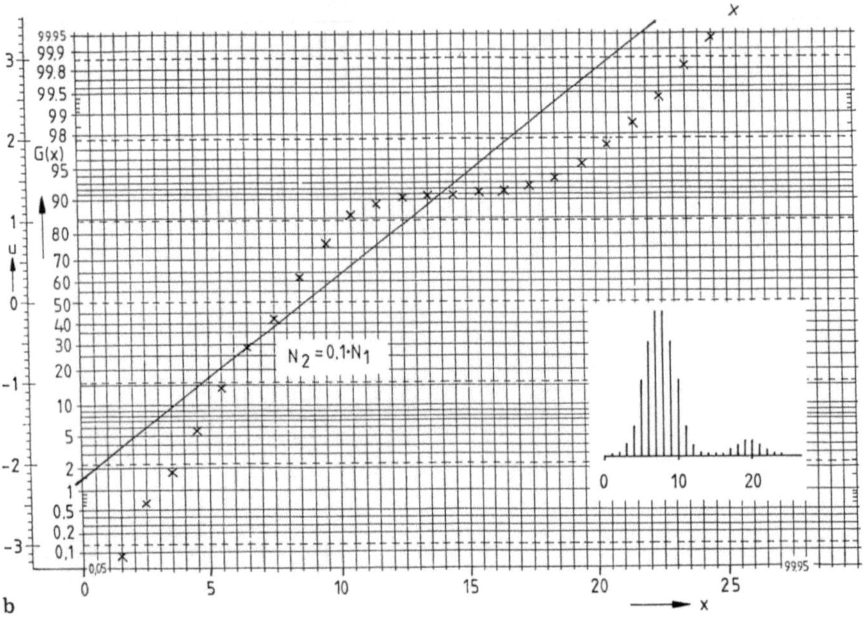

b

Fig. 6.8 b. Representation of the MD with $\mu = 8.59$ arising from $ND_1 + ND_3$; further details in Fig. 6.8a

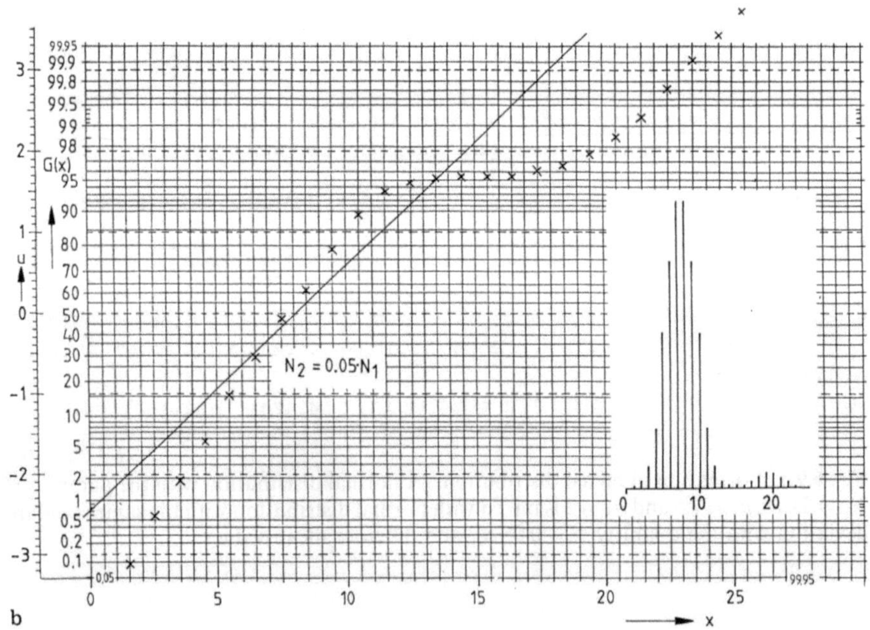

b

Fig. 6.9 b. Representation of the MD with $\mu = 8.07$ arising from $ND_1 + ND_4$; further details in Fig. 6.9a

	ND_1	ND_2	MD		
	$\mu_1 = 25.5$	$\mu_2 = 25.5$	$\mu = 25.5$		
	$\sigma_1 = 2$	$\sigma_2 = 6$	$\sigma = \sqrt{\frac{\sigma_1^2 + \sigma_2^2}{2}} = \sqrt{20} = 4.4721$		
x_j	N_j	N_j	N_j	$\sum N_j$	G(x) in %
6		1	1	1	0.0496
9		5	5	6	0.2976
12		15	15	21	1.0417
15		40	40	61	3.0258
18		103	103	164	8.1349
19	1		1	165	8.1845
20	5		5	170	8.4325
21	15	150	165	335	16.6171
22	40		40	375	18.6012
23	103		103	478	23.7103
24	150	190	340	818	40.5754
25	190		190	1008	50.0000
26	190		190	1198	59.4246
27	150	190	340	1538	76.2897
28	103		103	1641	81.3988
29	40		40	1681	83.3829
30	15	150	165	1846	91.5675
31	5		5	1851	91.8155
32	1		1	1852	91.8651
33		103	103	1955	96.9742
36		40	40	1995	98.9583
39		15	15	2010	99.7024
42		5	5	2015	99.9504
45		1	1	2016	100.0000

parameters of the MD calculated by calculator:

N = 2016

$\mu = 25.5$

$\sigma = 4.4721$

$6\cdot\sigma$ - zone: 12.08.....38.92

proportion outside: p = 2.0834 %

a

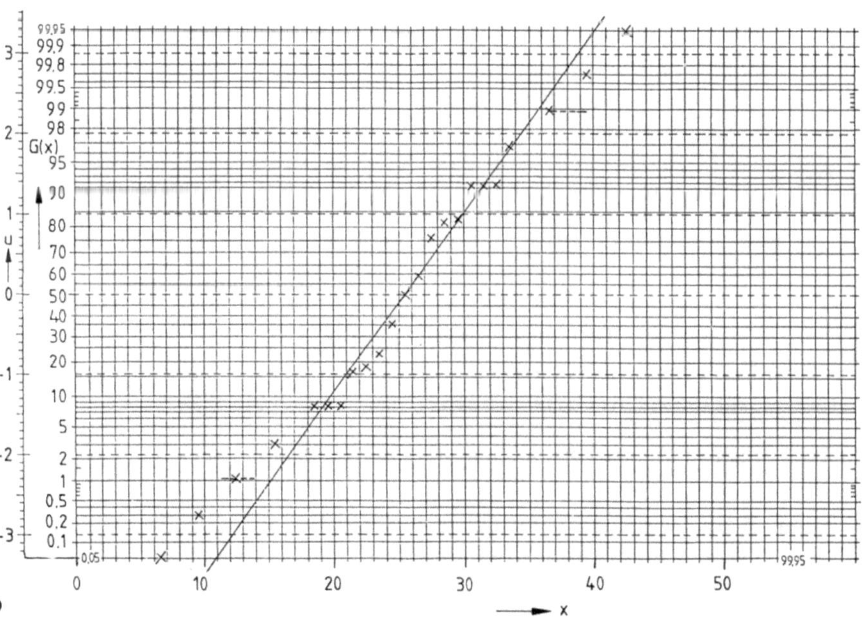

b

Fig. 6.10. Mixing two normal distributions

	$N_1 = 1008$	$N_2 = 504$	$N_3 = 504$	$N_4 = 1008$
	$\mu_1 = 25.5$	$\mu_2 = 25.5$	$\mu_3 = 25.5$	$\mu_4 = 25.5$
	$\sigma_1 = 2$	$\sigma_2 = 6$	$\sigma_3 = 2$	$\sigma_4 = 6$

x_j	N_{j1}	N_{j2}	N_{j3}	N_{j4}
6		0.5		1
9		2.5		5
12		7.5		15
15		20		40
18		51.5		103
19	1		0.5	
20	5		2.5	
21	15	75	7.5	150
22	40		20	
23	103		51.5	
24	150	95	75	190
25	190		95	
26	190		95	
27	150	95	75	190
28	103		51.5	
29	40		20	
30	15	75	7.5	150
31	5		2.5	
32	1		0.5	
33		51.5		103
36		20		40
39		7.5		15
42		2.5		5
45		0.5		1

Parameters and random spread zones of the mixed distributions:

	$ND_1 + ND_2$	$ND_3 + ND_4$
lot size N	1512	1512
mean μ	25.5	25.5
standard deviation σ	3.829708 $= \sqrt{\frac{2 \cdot 2^2 + 6^2}{3}}$	5.033223 $= \sqrt{\frac{2^2 + 2 \cdot 6^2}{3}}$
	(weighted average of variances)	
$6 \cdot \sigma$ - zone from to	14.01....36.99	10.10....40.50
proportion outside	21/1512 = 1.39 %	12/1512 = 0.79 %

Fig. 6.11. Formation of mixed distributions from normal distributions

Similar characteristics are shown by mixed distributions with different batch sizes, Figs. 6.11 and 6.12 (without PG diagram).

A further possibility in the formation of mixed distributions from two normal distributions is the simulation of the case where not only the position but also the spread changes.

In Fig. 6.13 it is accepted that both normally distributed batches have the same lot size; then the mixed distributions are always narrower than the relevant normal distributions.

	ND_1	ND_2	ND_3	ND_4
	$N_1 = 1008$	$N_2 = 100.8$	$N_3 = 100.8$	$N_4 = 1008$
	$\mu_1 = 25.5$	$\mu_2 = 25.5$	$\mu_3 = 25.5$	$\mu_4 = 25.5$
	$\sigma_1 = 2$	$\sigma_2 = 6$	$\sigma_3 = 2$	$\sigma_4 = 6$
x_j	N_{j1}	N_{j2}	N_{j3}	N_{j4}
6		0.1		1
9		0.5		5
12		1.5		15
15		4		40
18		10.3		103
19	1		0.1	
20	5		0.5	
21	15	15	1.5	150
22	40		4	
23	103		10.3	
24	150	19	15	190
25	190		19	
26	190		19	
27	150	19	15	190
28	103		10.3	
29	40		4	
30	15	15	1.5	150
31	5		0.5	
32	1		0.1	
33		10.3		103
36		4		40
39		1.5		15
42		0.5		5
45		0.1		1

Parameters and random spread zones of the mixed distributions:

	$ND_1 + ND_2$	$ND_3 + ND_4$
lot size N	1108.8	1108.8
mean μ	25.5	25.5
standard deviation σ	2.6285 $= \sqrt{\frac{10 \cdot 2^2 + 6^2}{11}}$	5.7525 $= \sqrt{\frac{2^2 + 10 \cdot 6^2}{11}}$
$6 \cdot \sigma$ - zone from to	17.61....33.39	8.24....42.75
proportion outside	1.10 %	0.18 %

Fig. 6.12. Formation of mixed distributions from normal distributions

In Fig. 6.14, two batches are mixed where $N_2 = 0.1\,N_1$. With the $MD = ND_1 + ND_s$ 4.55% lie above the 6σ range situated symmetrically in relation to the mean value; if this is shifted to higher values, however, than the proportion outside this range is nil, Fig. 6.14b.

If mixed distributions are formed from three normal distributions, then the case is simulated where a process is shifted erratically twice. In Fig. 6.15 it

x_j	N_{j1}	N_{j2}	N_{j3}	N_{j4}
1	1	1		
2	5			
3	15	5		
4	40			
5	103	15	1	
6	150			
7	190	40	5	
8	190			
9	150	103	15	1
10	103			
11	40	150	40	5
12	15			
13	5	190	103	15
14	1			
15		190	150	40
16				
17		150	190	103
18				
19		103	190	150
20				
21		40	150	190
22				
23		15	103	190
24				
25		5	40	150
26				
27		1	15	103
28				
29			5	40
30				
31			1	15
32				
33				5
34				
35				1

Parameters and random spread zones of the mixed distributions:

	$ND_1 + ND_2$	$ND_1 + ND_3$	$ND_1 + ND_4$
lot size N	2016	2016	2016
mean μ	10.75	12.75	14.75
standard deviation σ	4.5346	6.1288	7.9096
$6 \cdot \sigma$ - zone from to	-2.85...24.35	-5.64...31.14	-8.98...38.48
proportion outside	0.30 %	0.00 %	0.00 %

Fig. 6.13. Formation of mixed distributions. A normal distribution ND_1 with the parameters $N_1 = 1\,008$, $\mu_1 = 7.5$ and $\sigma_1 = 2$ is mixed with normal distribution of the same range, with doubled standard deviation and with different mean values

is assumed that the (symmetrically) ordered outside distributions are of equal size and together have the same range as the middle distribution. Independent of the distance between the two outside distributions, the mixed distributions are narrower than the relevant normal distribution.

Similar characteristics are shown by mixed distributions where the range of the outside distributions together account for 20% of the range of the middle distribution, Fig. 6.16.

x_j	N_{j1}	N_{j2}	N_{j3}	N_{j4}	N_{j5}
1	1	0.1			
2	5				
3	15	0.5			
4	40				
5	103	1.5	0.1		
6	150				
7	190	4	0.5		
8	190				
9	150	10.3	1.5	0.1	
10	103				
11	40	15	4	0.5	
12	15				
13	5	19	10.3	1.5	0.1
14	1				
15		19	15	4	0.5
16					
17		15	19	10.3	1.5
18					
19		10.3	19	15	4
20					
21		4	15	19	10.3
22					
23		1.5	10.3	19	15
24					
25		0.5	4	15	19
26					
27		0.1	1.5	10.3	19
28					
29			0.5	4	15
30					
31			0.1	1.5	10.3
32					
33				0.5	4
34					
35				0.1	1.5
36					
37					0.5
38					
39					0.1

Parameters and random spread zones of the mixed distributions:

	$ND_1 + ND_2$	$ND_1 + ND_3$	$ND_1 + ND_4$	$ND_1 + ND_5$
lot size N	1108.8	1108.8	1108.8	1108.8
mean μ	8.0909	8.4545	8.81818	9.1818
standard deviation σ	2.9296	3.7680	4.7399	5.7772
$6\cdot\sigma$ - zone from to	-0.70...16.88	-2.85...19.76	-5.40...23.04	-8.15...26.51
proportion outside	2.83 %	2.83 %	2.83 %	4.55 %

a

Fig. 6.14a. Formation of mixed distributions. A normal distribution ND_1 with the parameters $N_1 = 1\,008$, $\mu_1 = 7.5$ and $\sigma_1 = 2$ are mixed with normal distribution of the range $0.1\,N_1$, with doubled standard deviation and with different mean values

x_j	N_{j1}	N_{j2}	N_{j3}	N_{j4}	N_{j5}	N_{j6}	N_{j7}	mean values
6						0.5		
7						2.5		
8						7.5		
9						20		
10						51.5		
11						75		
12				0.5		95		
13				2.5		95		$\mu_6 = 12.5$
14				7.5		75		
15				20		51.5		
16				51.5		20		
17				75		7.5		
18		0.5		95		2.5		
19		2.5		95		0.5		$\mu_4 = 18.5$
20		7.5		75				
21		20		51.5				
22		51.5		20				
23		75		7.5				
24	1	95		2.5				
25	5	95		0.5				$\mu_2 = 24.5$
26	15	75						
27	40	51.5						
28	103	20						
29	150	7.5						
30	190	2.5	0.5					
31	190	0.5	2.5					$\mu_1 = 30.5$
32	150		7.5					
33	103		20					
34	40		51.5					
35	15		75					
36	5		95		0.5			
37	1		95		2.5			$\mu_3 = 36.5$
38			75		7.5			
39			51.5		20			
40			20		51.5			
41			7.5		75			
42			2.5		95		0.5	
43			0.5		95		2.5	$\mu_5 = 42.5$
44					75		7.5	
45					51.5		20	
46					20		51.5	
47					7.5		75	
48					2.5		95	
49					0.5		95	$\mu_7 = 48.5$
50							75	
51							51.5	
52							20	
53							7.5	
54							2.5	
55							0.5	

Parameters and random spread zones of the mixed distributions:

	$ND_1 + ND_2 + ND_3$	$ND_1 + ND_4 + ND_5$	$ND_1 + ND_6 + ND_7$
lot size N	2016	2016	2016
mean μ	30.5	30.5	30.5
standard deviation σ	4.6904	8.7178	12.8841
$6\cdot\sigma$ - zone from to	16.43...44.57	4.35...56.65	-8.15...69.15
proportion outside	0.00 %	0.00 %	0.00 %

a

Fig. 6.15 a. Formation of mixed distributions. A normal distribution ND_1 with the parameters $N_1 = 1\,008$, $\mu_1 = 30.5$ and $\sigma_1 = 2$ are mixed with two normal distributions with the parameters $N = 504$, $\sigma = 2$ and mean values lying symmetrically about μ_1

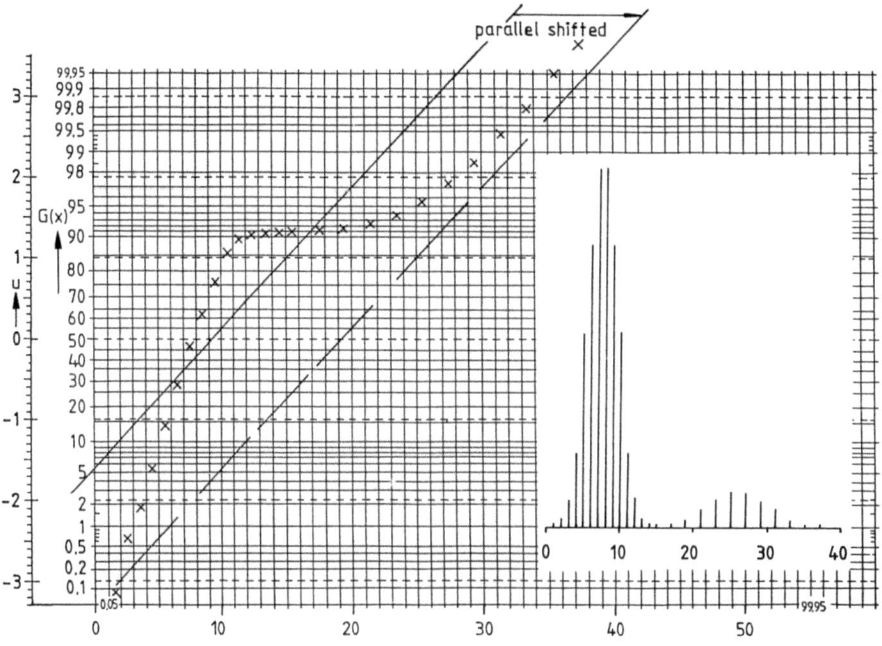

b

Fig. 6.14 b. Representation of the MD = $ND_1 + ND_5$ with $\mu = 9.18$; further details in Fig. 6.14a

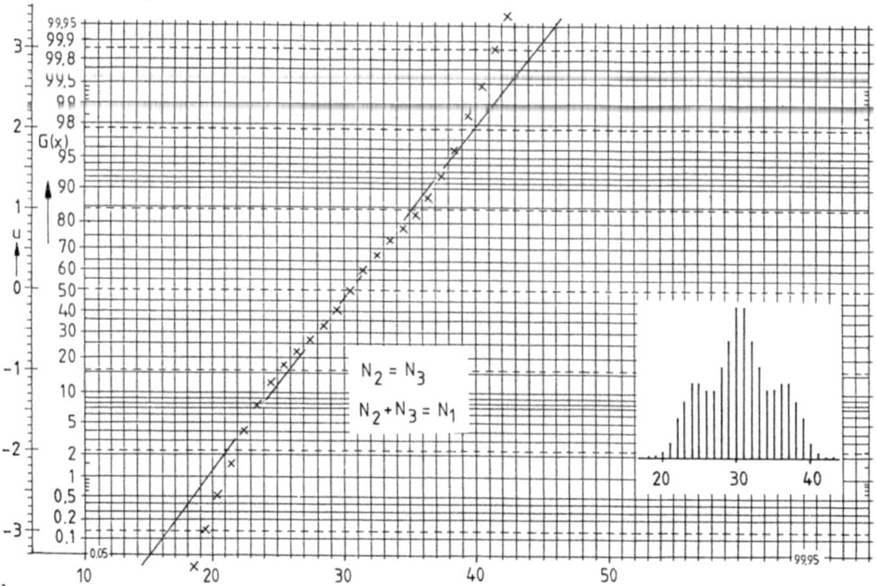

b

Fig. 6.15 b. Representation of the MD = $ND_1 + ND_2 + ND_3$; further details in Fig. 6.15a

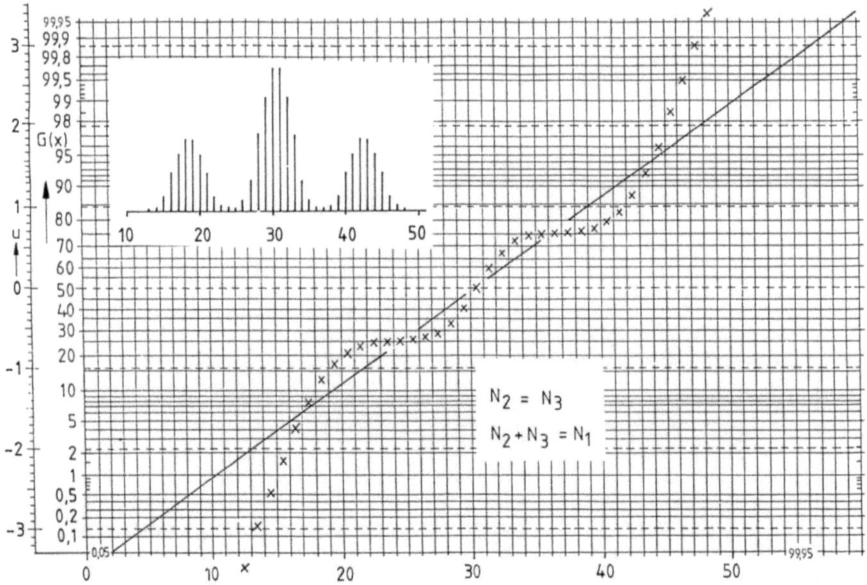

c

Fig. 6.15 c. Representation of the $MD = ND_1 + ND_4 + ND_5$; further details in Fig. 6.15a

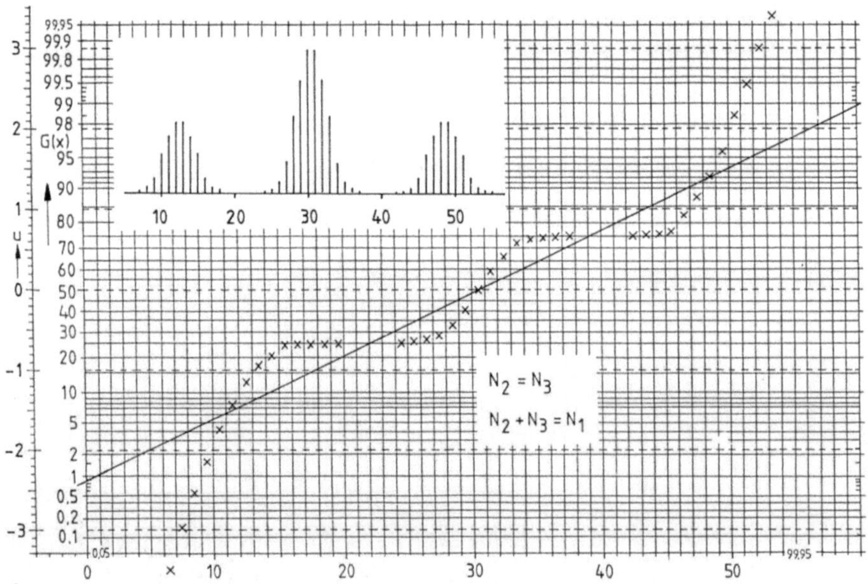

d

Fig. 6.15 d. Representation of the $MD = ND_1 + ND_6 + ND_7$; further details in Fig. 6.15a

x_i	N_{j1}	N_{j2}	N_{j3}	N_{j4}	N_{j5}	N_{j6}	N_{j7}	mean values
6						0.1		
7						0.5		
8						1.5		
9						4		
10						10.3		
11						15		
12				0.1		19		μ_6 = 12.5
13				0.5		19		
14				1.5		15		
15				4		10.3		
16				10.3		4		
17				15		1.5		
18		0.1		19		0.5		μ_4 = 18.5
19		0.5		19		0.1		
20		1.5		15				
21		4		10.3				
22		10.3		4				
23		15		1.5				
24	1	19		0.5				μ_2 = 24.5
25	5	19		0.1				
26	15	15						
27	40	10.3						
28	103	4						
29	150	1.5						
30	190	0.5	0.1					μ_1 = 30.5
31	190	0.1	0.5					
32	150		1.5					
33	103		4					
34	40		10.3					
35	15		15					
36	5		19		0.1			μ_3 = 36.5
37	1		19		0.5			
38			15		1.5			
39			10.3		4			
40			4		10.3			
41			1.5		15			
42			0.5		19		0.1	μ_5 = 42.5
43			0.1		19		0.5	
44					15		1.5	
45					10.3		4	
46					4		10.3	
47					1.5		15	
48					0.5		19	μ_7 = 48.5
49					0.1		19	
50							15	
51							10.3	
52							4	
53							1.5	
54							0.5	
55							0.1	

Parameters and random spread zones of the mixed distributions:

	$ND_1+ND_2+ND_3$	$ND_1+ND_4+ND_5$	$ND_1+ND_6+ND_7$
lot size N	1209.6	1209.6	1209.6
mean μ	30.5	30.5	30.5
standard deviation σ	3.1623	5.2915	7.6158
$6\cdot\sigma$ – zone from to	21.01...39.99	14.63...46.37	7.65...53.35
proportion outside	$\frac{12.2}{1209.6}$ = 1.01 %	$\frac{4.2}{1209.6}$ = 0.35 %	$\frac{1.2}{1209.6}$ = 0.1%

Fig. 6.16 a. Formation of mixed distributions. A normal distribution ND_1 with the parameters $N_2 = 1\,008$, $\mu_1 = 30.5$ and $\sigma_1 = 2$ is mixed with two normal distributions with the parameters $N = 1\,008$, $\sigma = 2$ and mean values which lie symmetrically about μ_1

x_i	N_{j1}	N_{j2}	N_{j3}	N_{j4}	N_{j5}	N_{j6}	N_{j7}	mean values
6						0.1		
7						0.5		
8						1.5		
9						4		
10						10.3		
11						15		
12				0.1		19		$\mu_6 = 12.5$
13				0.5		19		
14				1.5		15		
15				4		10.3		
16				10.3		4		
17				15		1.5		
18		0.1		19		0.5		$\mu_4 = 18.5$
19		0.5		19		0.1		
20		1.5		15				
21		4		10.3				
22		10.3		4				
23		15		1.5				
24	2	19		0.5				$\mu_2 = 24.5$
25	10	19		0.1				
26	30	15						
27	80	10.3						
28	206	4						
29	300	1.5						
30	380	0.5	0.1					$\mu_1 = 30.5$
31	380	0.1	0.5					
32	300		1.5					
33	206		4					
34	80		10.3					
35	30		15					
36	10		19		0.1			$\mu_3 = 36.5$
37	2		19		0.5			
38			15		1.5			
39			10.3		4			
40			4		10.3			
41			1.5		15			
42			0.5		19		0.1	$\mu_5 = 42.5$
43			0.1		19		0.5	
44					15		1.5	
45					10.3		4	
46					4		10.3	
47					1.5		15	
48					0.5		19	$\mu_7 = 48.6$
49					0.1		19	
50							15	
51							10.3	
52							4	
53							1.5	
54							0.5	
55							0.1	

Parameters and random spread zones of the mixed distributions:

	$ND_1 + ND_2 + ND_3$	$ND_1 + ND_4 + ND_5$	$ND_1 + ND_6 + ND_7$
lot size N	2217.6	2217.6	2217.6
mean μ	30.5	30.5	30.5
standard deviation σ	2.6968	4.1341	5.7840
$6\cdot\sigma$ - zone from to	22.41...38.59	18.10...42.90	13.15...47.85
proportion outside	$\frac{32.8}{2217.6} = 1.48\,\%$	$\frac{100.8}{2217.6} = 4.55\,\%$	$\frac{138.8}{2217.6} = 6.26\,\%$

a

Fig. 6.17 a. Formation of mixed distribution. A normal distribution ND_1 with the parameters $N_1 = 2016$, $\mu = 30.5$ and $\sigma_1 = 2$ is mixed with two normal distributions with the parameters $N = 1008$, $\sigma = 2$ and mean values which lie symmetrically about μ

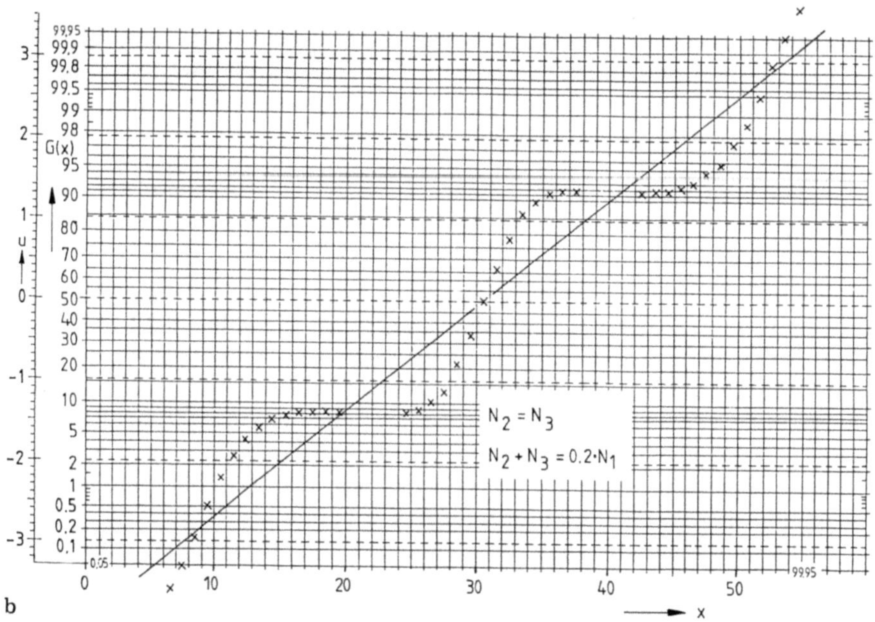

Fig. 6.16 b. Representation of the MD = $ND_1 + ND_6 + ND_7$; further details given in Fig. 6.16a

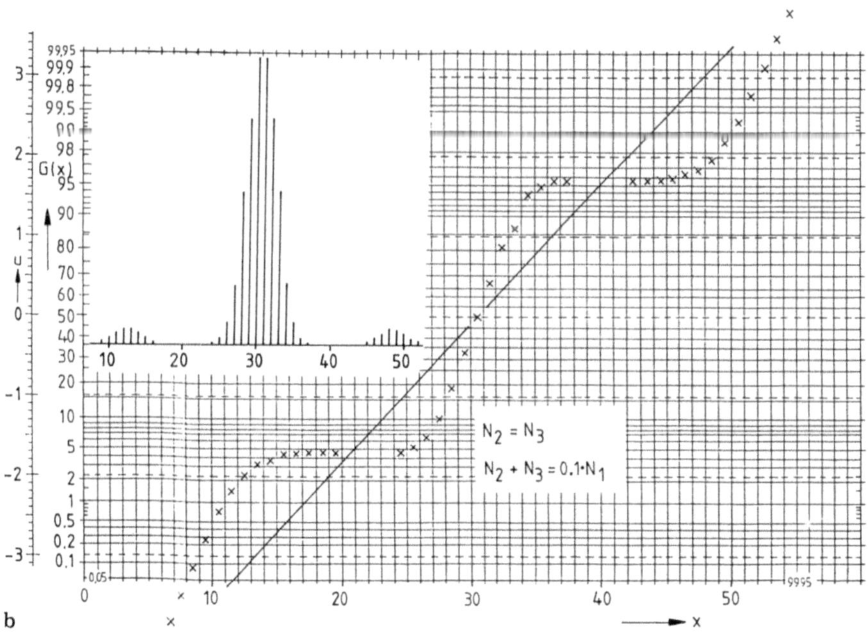

Fig. 6.17 b. Representation of the MD = $ND_1 + ND_6 + ND_7$; further details given in Fig. 6.17a

If the range of the outside distributions together account for only 10% of the range of the middle distribution, then the proportion outside the 6σ range constitute some percentage only when the outside distributions are more than 2σ in distance from the middle distribution, which in practice is very unlikely, Fig. 6.17.

Summary

The mixed distributions of Type 2 are also considerably narrower than their respective normal distributions. Almost all values lie within the 6σ range, at least with sloping distributions this is the case when the 6σ range is shifted.

Only when two batches of similar size with different standard deviations, for example $\sigma_2 = 3\sigma$, are mixed or when three distributions with small proportions in the symmetrically lying outer distributions are mixed can proportions of up to about 2% outside the 6σ range rise, provided these outer distributions are not further than 2σ from the middle distribution.

Mixed distributions which are wider than the relevant normal distributions should however not be described as unfavourably deviant, since each size distribution in a dimensional chain behaves as if it were normal, Chap. 7.

Exercise 6.1

Given: Two neighbouring V distributions as in Exercise 4.13 are mixed so that a three pointed mixed distribution results.

Required: Parameters of the mixed distribution.

Solution:

x_j	N_{j1}	N_{j2}	N_j
1	4		4
2	3		3
3	2		2
4	1		1
5	0		0
6	1		1
7	2		2
8	3		3
9	4		4
10		4	4
11		3	3
12		2	2
13		1	1
14		0	0
15		1	1
16		2	2
17		3	3
18		4	4

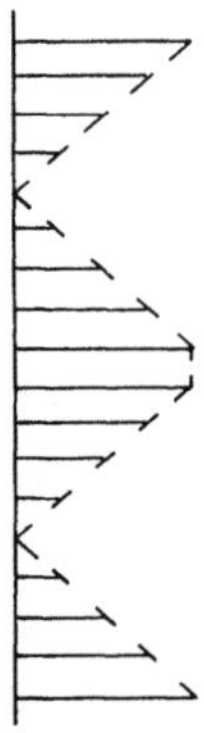

$N = 40$,
$\mu = 9.5$,
$\sigma_N = 5.5$

6σ range: $-7 \ldots 26$
proportion outside: 0%
4σ range: $-1.5 \ldots 20.5$
proportion outside: 0%

Note: The spread of this MD is $R = 18.5 - 0.5 = 18$ or $R = 2 \cdot 1.6364\sigma$; thus this distribution is little more than half as wide as the ND with the same parameters.

6.4 Tschebyscheff's Extreme Case

In almost every text book on technical statistics Tschebyscheff's inequable case is mentioned. In this possible distribution with the mean value μ and the standard deviation σ in the range of $\mu - u\sigma$ to $\mu + u\sigma$ there is a value proportion of at least

$$P \geqq 1 - \frac{1}{u^2}.$$

For $u = 3$ there is the least proportion

$$P = 1 - \frac{1}{9} = 0.8\bar{8} = 88.9\%$$

Consequently the proportion outside the range is

$$Q = \frac{1}{9} = 0.1\bar{1} = 11.1\%.$$

The formal derivation of this unusual case can be found in [10].

The highest proportion outside the 6σ range which had been found by mixing normal distributions was $p = 6.26\%$ in the $MD = ND_1 + ND_6 + ND_7$ in Fig. 6.17. In this case the outer distributions are about 9σ distant from the middle distribution.

If these outer distributions were further away from the middle, the proportion outside the 6σ range would be increased further, Fig. 6.18.

From this it can be projected that mixed distributions will only fulfill the extreme case according to Tschebyscheff if the spread was created by the mean values and not by the individual values.

This special case is illustrated in Fig. 6.19 by MD_1. If from $N = 10\,000$ values, 8 890 lie by 10 and the rest are symmetrically lying without spread in that half, for example with 9 and 11 (or equally with 9.9 and 10.1) then the outer values are not recorded in the the 6σ range and lie barely outside.

However as soon as the outer values lie not on the same side without spread but divided equally into five classes, then only about 4% will be outside the 6σ range, MD_2 in Fig. 6.19.

And it further the 8 890 middle values are divided into seven classes, then only about 2% will be outside the 6σ range, MD_3 in Fig. 6.19.

	MD_1	MD_2	MD_3	
N_{j2}	x_j	x_j	x_j	
0.1	24	18	6	
0.5	25	19	7	
1.5	26	20	8	
4	27	21	9	
10.3	28	22	10	
15	29	23	11	
19	30	24	12	
19	31	25	13	μ_2
15	32	26	14	
10.3	33	27	15	
4	34	28	16	
1.5	35	29	17	
0.5	36	30	18	
0.1	37	31	19	
N_{j1}				
2	54			
10	55			
30	56			
80	57			
206	58			
300	59			
380	60			
380	61			μ_1
300	62			
206	63			
80	64			
30	65			
10	66			
2	67			
N_{j3}				
0.1	84	90	102	
0.5	85	91	103	
1.5	86	92	104	
4	87	93	105	
10.3	88	94	106	
15	89	95	107	
19	90	96	108	
19	91	97	109	μ_3
15	92	98	110	
10.3	93	99	111	
4	94	100	112	
1.5	95	101	113	
0.5	96	102	114	
0.1	97	103	115	

Parameters and random spread zones of the mixed distributions:

	MD_1	MD_2	MD_3
lot size N	2217.6	2217.6	2217.6
mean μ	60.5	60.5	60.5
standard deviation σ	9.2638	11.0371	14.6101
$6\cdot\sigma$ - zone from to	32.17....88.29	27.39....93.61	16.67....104.33
proportion outside	$\frac{168.8}{2217.6}$ = 7.61 %	$\frac{189.4}{2217.6}$ = 8.54 %	$\frac{197.4}{2217.6}$ = 8.90 %

Fig. 6.18. Formation of mixed distribution. A normal distribution ND_1 with the parameters $N_1 = 2016$, $\mu_1 = 60.5$ and $\sigma_1 = 2$ are mixed with two normal distributions with the parameters $N = 100.8$, $\sigma = 2$ and mean values which lie symmetrically about μ_1

	MD_1	MD_2	MD_3
x_j	N_{j1}	N_{j2}	N_{j3}
8.8		111	111
8.9		111	111
9.0	555	111	111
9.1		111	111
9.2		111	111
9.7			1270
9.8			1270
9.9			1270
10.0	8890	8890	1270
10.1			1270
10.2			1270
10.3			1270
10.8		111	111
10.9		111	111
11.0	555	111	111
11.1		111	111
11.2		111	111
N	10000	10000	10000
μ	10	10	10
σ_N	0.333167	0.336482	0.385720
$6\cdot\sigma$ - zone from	9.0005	8.991	8.84
to	10.9995	11.009	11.16
proportion outside	11.1%	4.44%	2.22%

Fig. 6.19. Extreme mixed distribution

Summary

The limiting case according to Tschebyscheff is a theoretical special case and is unlikely to occur in practice. With real mixed distributions at most about 2% of the values lie outside the 6σ range.

Exercise 6.2

Given: $N = 10\,000$ values are ordered as a mixed distribution MD_1 so that Tschebyscheff's extreme case is shown.

Required: Parameters of this distribution MD_1 and the distribution MD_2 to MD_4 in which the distribution components are scattered.

Solution:

	MD_1	MD_2	MD_3	MD_4	
x_j	N_{j1}	N_{j2}	N_{j3}	N_{j4}	
9.3		111	111	111	
9.4		111	111	111	
9.5	555	111	111	111	
9.6		111	111	111	
9.7		111	111	111	1270
9.8			1778		1270
9.9			1778		1270
10.0	8890	8890	1778		1270
10.1			1778		1270
10.2			1778		1270
10.3		111	111	111	1270
10.4		111	111	111	
10.5	555	111	111	111	
10.6		111	111	111	
10.7		111	111	111	
N	10000	10000	10000	10000	
μ	10	10	10	10	
σ	0.166583	0.173118	0.218518	0.255988	
$6 \cdot \sigma$ - zone					
from	9.5003	9.4806	9.3444	9.2320	
to	10.4997	10.5194	10.6555	10.7680	
proportion outside	11.1 %	4.44 %	2.22%	0 %	

Note: If the distribution components overlap as in MD_4, the extreme values are "caught" by the 6σ range.

7 Distribution Folds

7.1 Description of Fold Methods

When distributions are mixed, as in Chap. 6, in theory for example two batches with the ranges N_1 and N_2 may be joined with a batch with the range $N = N_1 + N_2$.

With folding a quite different operation is simulated, namely the addition of two or more independent characteristics (batches) each with a quite distinct probability distribution in order to find the probability distribution (density function or sum function or both) of the closing value, the fold product.

Folding is a mathematical operation; in the case of continuous distributions the derivation of the fold integral value can be looked up in [10].

In the case in which two probability distributions are independent and discrete, or where two continuous distributions are simulated by discrete distributions, the probability for the closing value x is

$$P(x) = \sum_{i=-\infty}^{i=\infty} P1(i)P2(x-i). \tag{7.1}$$

The explanation of this formula follows from the example "the sum of two dice", which are physically symmetrical (not marked). Then the dice are also statistically symmetrical and thus show an equal distribution; the probability of each number on a die is as large as $P = 1/6$.

So the probability of throwing the sum $x = 5$ with two dice is:

$$\begin{aligned} P(x=5) = {} & P1(1)P2(4) && = 1/6 \cdot 1/6 \\ & + P1(2)P2(3) && + 1/6 \cdot 1/6 \\ & + P1(3)P2(2) && + 1/6 \cdot 1/6 \\ & + P1(4)P2(1) && + 1/6 \cdot 1/6 \\ \hline & \textstyle\sum = 4/36. \end{aligned}$$

In Fig. 7.1 the complete fold matrix for the folding of two rectangular distributions with the classes 1 to 6 (dice) is given. The fold product of two rectangular distributions is a triangular distribution with the classes (numbers) 2 to 12.

The range of the fold product is the same as the sum of the ranges of both folded distributions; in the example given in Fig. 7.1 this is $R_{TD} = 2R_{RD} = 12 = 13 - 1$.

Parameters
of the RD: N = 6; μ = 3,5; σ = 1,707825 = $\sqrt{\frac{6^2 - 1}{12}}$

x	g(x)	1 1	2 1	3 1	4 1	5 1	6 1·1/6
6	1	7 1	8 1	9 1	10 1	11 1	12 1·1/36
5	1	6 1	7 1	8 1	9 1	10 1	11 1
4	1	5 1	6 1	7 1	8 1	9 1	10 1
3	1	4 1	5 1	6 1	7 1	8 1	9 1
2	1	3 1	4 1	5 1	6 1	7 1	8 1
1	1 · 1/6	2 1	3 1	4 1	5 1	6 1	7 1

P(x=5) = 4/36

parameters of
the fold product:

N = 36

μ = 7

σ = 2,415229 = $\sqrt{2}\cdot 1,707825$ = $\sqrt{\frac{12^2 - 4}{24}}$

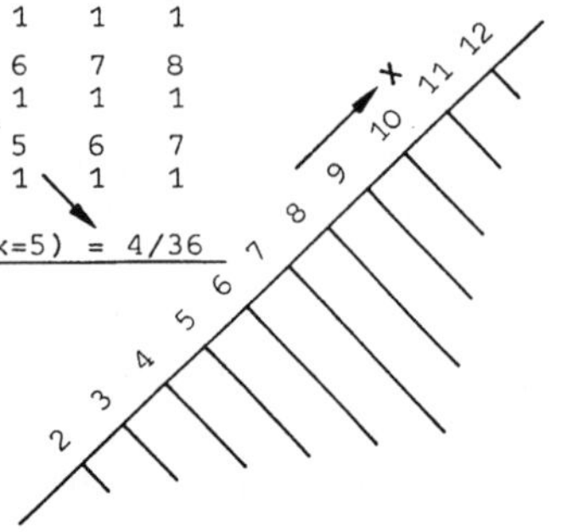

Fig. 7.1. Fold matrix for the folding of two similar rectangular distributions with the classes 1 to 6 (dice).

The folding of dimensional distributions simply means:

—formation of all possible dimension combinations and calculation of their probabilities with the multiplication formula,
—summation of the probabilities of all similar dimensional combinations by the addition formula.

If symmetrical distributions are folded, then the fold matrix is doubly symmetrical, so that only 1/4 of the matrix needs to be calculated; with sufficient practice the calculation of 1/8 of the matrix may be enough (examples in Sect. 7.2).

The purpose of folding operations is principally to learn and understand the fundamentals of the working of the basic laws using different probability distributions.

Law:	*Effects:*
Central limit value principle	Tendency towards normal distribution
Deviation progressive principle	The variances and not the standard deviations are added

Theoretically it is also possible to carry out folding operations with (actually given) frequency distributions.

Following from (7.1), computer programs can also be produced and applied [6]; no further details will however be given in this book since it is also possible to calculate a statistically tolerated dimensional chain quickly and reliably with just a calculator, see Sect. 11.2.

7.2 Examples of Folding Methods

In Fig. 7.2a two simple rectangular distributions are folded together.

Note. The term rectangular distribution is not quite accurate, since we are talking about discrete distributions; the term "equal distribution" would be more correct. However the term "rectangular distribution" will be used, because a (continuous) rectangular distribution should be simulated.

In Fig. 7.2a the fold operation is carried out a total of four times, so that in total 16 rectangular distributions are folded.

The fold product of the 16 rectangular distributions is shown in Fig. 7.2b. It happens that at the end of the probability sum function in the PG there are deviations from the ND line.

This is not—as would be assumed—a legacy of the abnormal derivation, but is a result of the fact that we are dealing here with discrete and not continuous distributions.

If 16 continuous rectangular distributions with the range $R = 2.5 - 0.5 = 2$ were to be folded, all values of the approximately normally distributed fold product would be in the value range of $16 \cdot 0.5 = 8$ to $16 \cdot 2.5 = 40$, or the range would be $R = 40 - 8 = 32 = 16 \cdot 2$.

However if a continuous rectangular distribution is simulated by a discrete distribution, then nominally the range is $R = 2$ (difference between the outer class limits), in fact then all the values of the approximately normal distributed fold product of 16 RD's in the value range of $16 \cdot 1 = 16$ to $16 \cdot 2 = 32$. This explains the deviations at the end of the probability sum function.

This does not alter if two normal distributions are folded. The fold product must be a normal distribution. If however two normal distributions which have been made discrete and therefore ultimately limited are folded, then deviations at the end of the probability sum function will occur as against the ND line, Fig. 7.3.

This is only recognisable when two ND's from the "counter box" are folded, Fig. 7.4. Although this model with 62 classes simulates the shape of the ND better than that ND model used in Fig. 7.3, the fold product has the deviations from the ND line arising from being made discrete.

In the Figs. 7.5 to 7.7, three probability distributions are folded as described in Fig. 4.9.

By the folding operations it can be seen that:

—fold products emulate the shape of a normal distribution, even when the output distributions deviate from a normal distribution, or
—by folding, deviations from the ND shape are ironed out, or

1. folding (2 RD are folded)

x	g(x)			
x		1	2	
	g(x)	1	1 · 1/2	
1	1	2	3	
		1	1	
2	1	3	4	
	· 1/2	1	1	1/4

2. folding (4 RD are folded)

x		2	3	4	
	g(x)	1	2	1 · 1/4	
2	1	4	5	6	
		1	2	1	
3	2	5	6	7	
		2	4	2 · 1/16	
4	1	6	7	8	
	· 1/4	1	2	1	

3. folding (8 RD are folded)

x		4	5	6	7	8
	g(x)	1	4	6	4	1 · 1/16
4	1	8	9	10	11	12
		1	4	6	4	1
5	4	9	10	11	12	13
		4	16	24	16	4 · 1/256
6	6	10	11	12	13	14
		6	24	36	24	6
7	4	11	12	13	14	15
		4	16	24	16	4
8	1	12	13	14	15	16
	· 1/16	1	4	6	4	1

4. folding (16 RD are folded)

x		8	9	10	11	12
	g(x)	1	8	28	56	70 · 1/256
8	1	16				
		1				
9	8	17	18	continuation		
		8	64	symmetrical		
10	28	18	19	20		
		28	224	784		
11	56	19	20	21	22	
		56	448	1568	3136 · 1/65536	
12	70	20	21	22	23	24
	· 1/256	70	560	1960	3920	4900
			continuation symmetrical			

parameters of the fold product:

N	μ	σ_N
4	3	$\sqrt{2}\cdot 0.5 = 0.7071$
16	6	$\sqrt{4}\cdot 0.5 = 1$
256	12	$\sqrt{8}\cdot 0.5 = 1.4142$
65536	24	$\sqrt{16}\cdot 0.5 = 2$

or by figure 4.9: $\sigma^2 = \dfrac{R^2 - 256}{192} = \dfrac{32^2 - 256}{192} = 4$

$\sigma = 2$

fold product of 16 folded RD

x	N_j	$\sum N_j$	G(x) in %	x	$\sum N_j$	G(x) in %
16	1	1	0.00153	32	65536	100.0000
17	16	17	0.0259	31	65535	99.9985
18	120	137	0.2090	30	65519	99.9741
19	560	697	1.0635	29	65399	99.7909
20	1820	2517	3.8406	28	64839	98.9365
21	4368	6885	10.5057	27	63019	96.1594
22	8008	14893	22.7249	26	58651	89.4943
23	11440	26333	40.1810	25	50643	77.2751
24	12870	39203	59.8190			

a

Fig. 7.2 a. Folds of two rectangular distributions with the parameters $N = 2$, $\mu = 1.5$ and $\sigma = 0.5$ with the continuation of the fold

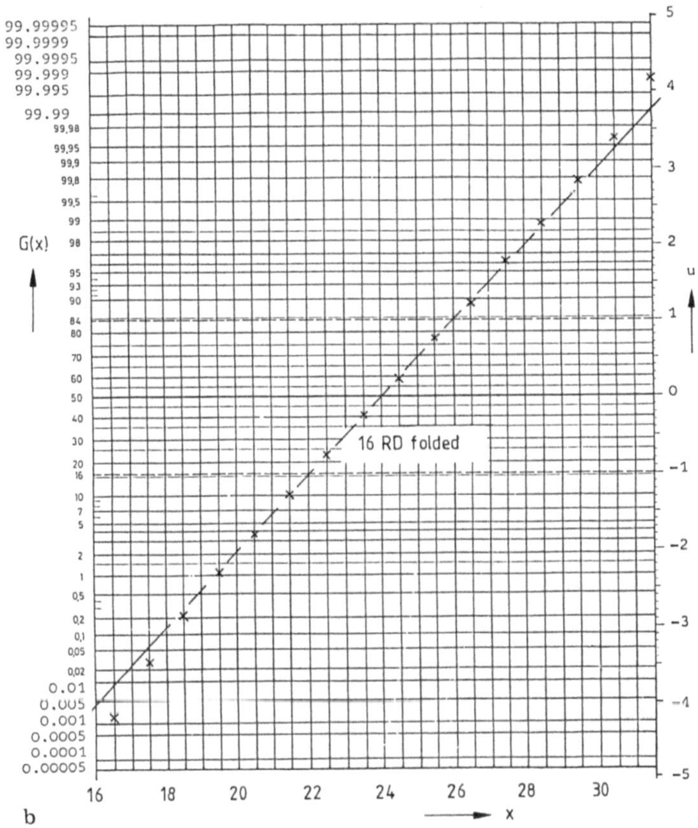

Fig. 7.2 b. Probability sum function of the fold product from 16 rectangular distributions

—each possible dimensional distribution, even if strongly deviant from the ND shape, will behave in the dimensional chain as if they were normal.

In Fig. 7.8, two mixed distributions of the type MD_1 as in Fig. 6.19 (Tschebyscheff's special case with a proportion of 11.1% out side the 6σ range) are folded after reduction of the occupation numbers (above right), two fold products are folded again.

The fold product of the fold products, Fig. 7.8b, is almost a normal distribution. 0.126% of the values lie outside the 6.1σ range; with an exact normal distribution this would be 0.114% according to the u-table.

Up to now only probability distributions with the same batch range N have been folded and the "batch size of the fold products" was N^2 respectively.

While the batch size in the mixing of probability distributions, Chap. 6, has considerable influence over the position and the shape of the mixed distribution, the batch size has absolutely no influence in the folding operations.

With folding, probabilities N_j/N are multiplied and their products added. If the batch sizes are omitted in denominators, then the sums $\sum N1_j N2_j$ are the parameters of the ND calculated by calculator:

N = 1008

μ = 7.5

σ_N = 2

distribution:

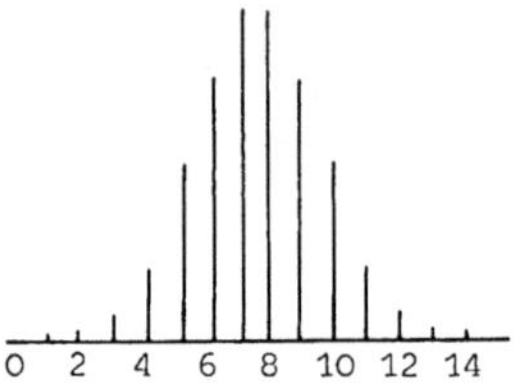

folding:

x g(x)	1 1	2 5	3 15	4 40	5 103	6 150	7 190 • 1/1008
1 1	2 1						
2 5	3 5	4 25					
3 15	4 15	5 75	6 225				
4 40	5 40	6 200	7 600	8 1600			
5 103	6 103	7 515	8 1545	9 4120	10 10609		
6 150	7 150	8 750	9 2250	10 6000	11 15450	12 22500	
7 190	8 190	9 950	10 2850	11 7600	12 19570	13 28500	14 36100
8 190	9 190	10 950	11 2850	12 7600	13 19570	14 28500	15 36100
9 150	10 150	11 750	12 2250	13 6000	14 15450	15 22500	
10 103	11 103	12 515	13 1545	14 4120	15 10609		
11 40	12 40	13 200	14 600	15 1650			
12 15	13 15	14 75	15 225				
13 5	14 5	15 25					
14 1 • 1/1008	15 1						

continuation symmetrical

parameters of the fold product:

N = 1016064

μ = 15

σ_N = 2.828427125 = $\sqrt{2} \cdot 2$

fold product:

x	N_j	$\sum N_j$	G(x) in %
2	1	1	0.000098
3	10	11	0.0011
4	55	66	0.0065
5	230	296	0.0291
6	831	1127	0.1109
7	2530	3657	0.3599
8	6570	10227	1.0065
9	15020	25247	2.4849
10	30509	55756	5.4875
11	53506	109262	10.7535
12	82450	191712	18.8681
13	111660	303372	29.8576
14	133600	436972	43.0064
15	142120	579092	56.9937
16	133600	712692	70.1425
17	111660	824352	81.1320
18	82450	906802	89.2466
19	53506	960308	94.5126
20	30509	990817	97.5153
21	15020	1005837	98.9936
22	6570	1012407	99.6402
23	2530	1014937	99.8892
24	831	1015768	99.9710
25	230	1015998	99.9935
26	55	1016053	99.9989
27	10	1016063	99.9999
28	1	1016064	100.0000

a

Fig. 7.3 a. Folds of two model normal distributions

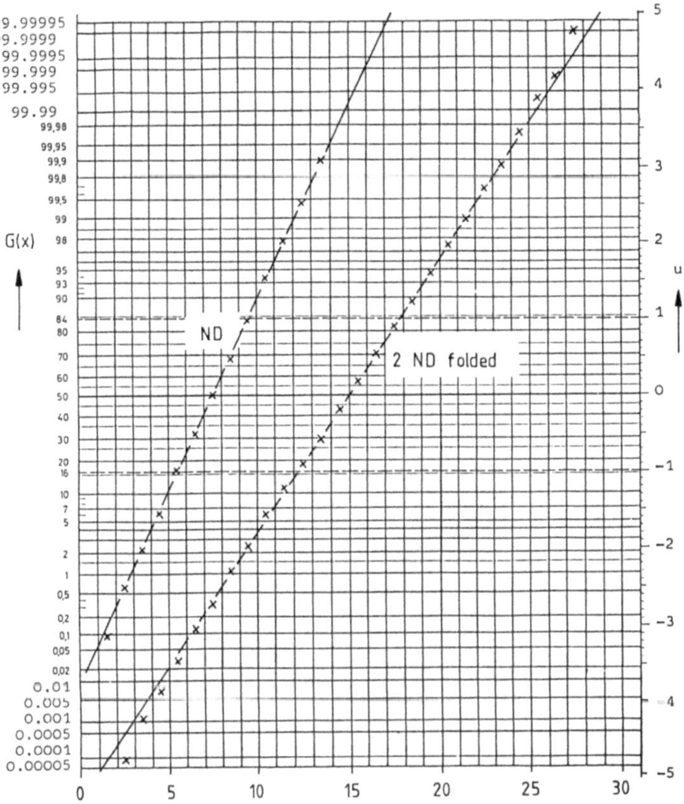

b

Fig. 7.3 b. Probability sum function of a model ND and the fold product from two of these ND's

"occupation numbers" in the individual classes of the fold products. These are fictitious and thus it is better to calculate the parameters of the fold product with the calculator. The "occupation numbers" of the fold product divided by the product of the batch quantities of the components gives the probabilities for the respective classes.

Example. If N shafts are mounted with N bushes, this results in exactly N constructed batches!

In Fig. 7.9a above a normal distribution is folded with a rectangular distribution. The probability sum function is shown in Fig. 7.9b. The same fold products results if the same normal distribution is folded with a rectangular distribution which has a batch range of "only" $N_{RD} = 7$, Fig. 7.9a below.

In Fig. 7.10a two normal distributions with different standard deviations are folded; the result is again a normal distribution, apart from the deviations in the end of the sum function caused by making then discrete, Fig. 7.10b.

Folding (initial stage)

x / g(x)	19 1	20 1	21 1	22 1	23 1	24 1	25 1	26 2	27 3	28 4	29 4	30 5	31 7 •1/1000
19 1	38 1												
20 1	39 1	40 1											
21 1	40 1	41 1	42 1										
22 1	41 1	42 1	43 1	44 1									
23 1	42 1	43 1	44 1	45 1	46 1		continuation symmetrical						
24 1	43 1	44 1	45 1	46 1	47 1	48 1							
25 1	44 1	45 1	46 1	47 1	48 1	49 1	50 1						
26 2	45 2	46 2	47 2	48 2	49 2	50 2	51 2	52 4 · 1/1000000					
27 3	46 3	47 3	48 3	49 3	50 3	51 3	52 3	53 6	54 9				
28 4	47 4	48 4	49 4	50 4	51 4	52 4	53 4	54 8	55 12	56 16			
29 4	48 4	49 4	50 4	51 4	52 4	53 4	54 4	55 8	56 12	57 16	58 16		
30 5	49 5	50 5	51 5	52 5	53 5	54 5	55 5	56 10	57 15	58 20	59 20	60 25	
31 7 •1/1000	50 7	51 7	52 7	53 7	54 7	55 7	56 7	57 14	58 21	59 28	60 28	61 35	62 49

Counter box

x	g(x) x •1000	
19	1	81
20	1	80
21	1	79
22	1	78
23	1	77
24	1	76
25	1	75
26	2	74
27	3	73
28	4	72
29	4	71
30	5	70
31	7	69
32	8	68
33	9	67
34	11	66
35	13	65
36	15	64
37	17	63
38	20	62
39	22	61
40	24	60
41	27	59
42	29	58
43	31	57
44	33	56
45	35	55
46	37	54
47	38	53
48	39	52
49	40	51
50	40	

Fold product (initial stage)

x	N_j	$\sum N_j$	G(x) in %	x	$\sum N_j$	G(x) in %
38	1	1	0.0001	162	1000000	100.0000
39	2	3	0.0003	161	999999	99.9999
40	3	6	0.0006	160	999997	99.9997
41	4	10	0.0010	159	999994	99.9994
42	5	15	0.0015	158	999990	99.9990
43	6	21	0.0021	157	999985	99.9985
44	7	28	0.0028	156	999979	99.9979
45	10	38	0.0038	155	999972	99.9972
46	15	53	0.0053	154	999962	99.9962
47	22	75	0.0075	153	999947	99.9947
48	29	104	0.0104	152	999925	99.9925
49	38	142	0.0142	151	999896	99.9896
50	51	193	0.0193	150	999858	99.9858
51		etc.				

parameters of the fold product:

N = 1000000

μ = 100

σ_N = 14,1317

a

Fig. 7.4 a. Extension for the folding of the ND model "counter box" with 62 classes and the parameters $N = 10000$, $\mu = 50$ and $\sigma = 9.9926 \approx 10$

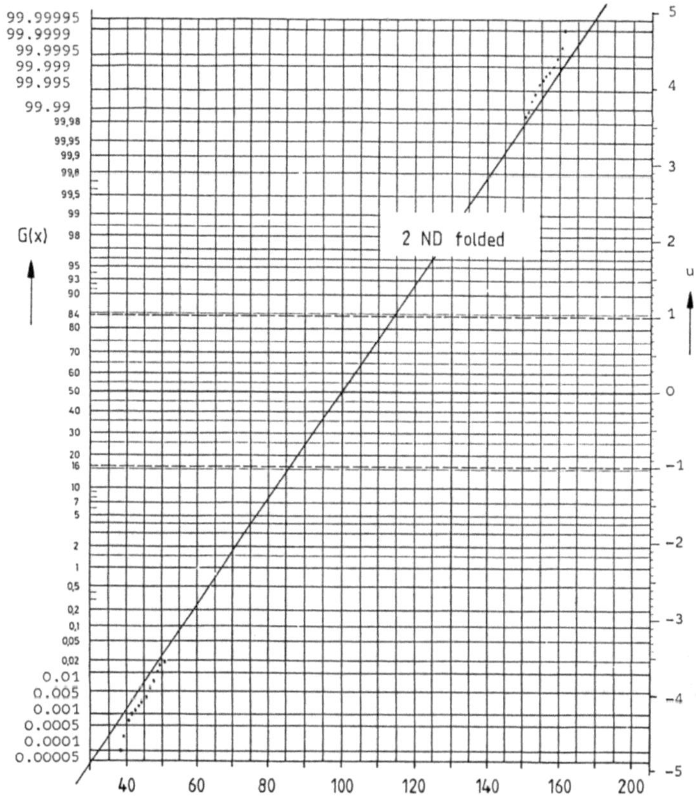

b

Fig. 7.4b Probabilty sum function of the fold products of 2 ND's from the "counter box"

distribution:

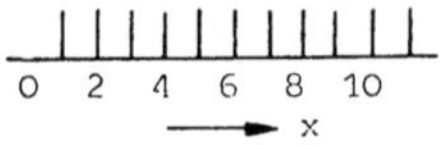

standard deviation: $\sigma = \sqrt{\frac{R^2 - 1}{12}} = \sqrt{\frac{121 - 1}{12}} = \sqrt{10} = 3.16227766$

parameters calculated
by calculator: $N = 11$; $\mu = 6$; $\sigma_N = 3.16227766$

Falten:

x	g(x)							
		1	2	3	4	5	6	
		1	1	1	1	1	1	• 1/11
1	1	2						
		1						
2	1	3	4		continuation			
		1	1					
3	1	4	5	6	symmetrical			
		1	1	1				
4	1	5	6	7	8			
		1	1	1	1			
5	1	6	7	8	9	10		
		1	1	1	1	1		
6	1	7	8	9	10	11	12	
	•	1	1	1	1	1	1	
	1/11	continuation symmetrical						

fold product:

x	N_j	$\sum N_j$	G(x) in %
2	1	1	0.8264
3	2	3	2.4793
4	3	6	4.9587
5	4	10	8.2645
6	5	15	12.3967
7	6	21	17.3554
8	7	28	23.1405
9	8	36	29.7521
10	9	45	37.1901
11	10	55	45.4545
12	11	66	54.5454
13	10	76	62.8099
14	9	85	70.2479
15	8	93	76.8595
16	7	100	82.6446
17	6	106	87.6033
18	5	111	91.7355
19	4	115	95.0413
20	3	118	97.5207
21	2	120	99.1735
22	1	121	100.0000

parameters of the fold product:

$N = 121$

$\mu = 12$

$\sigma_N = 4.472135955 = \sqrt{2} \cdot 3.16227766$

a

Fig. 7.5 a. Folding two rectangular distributions

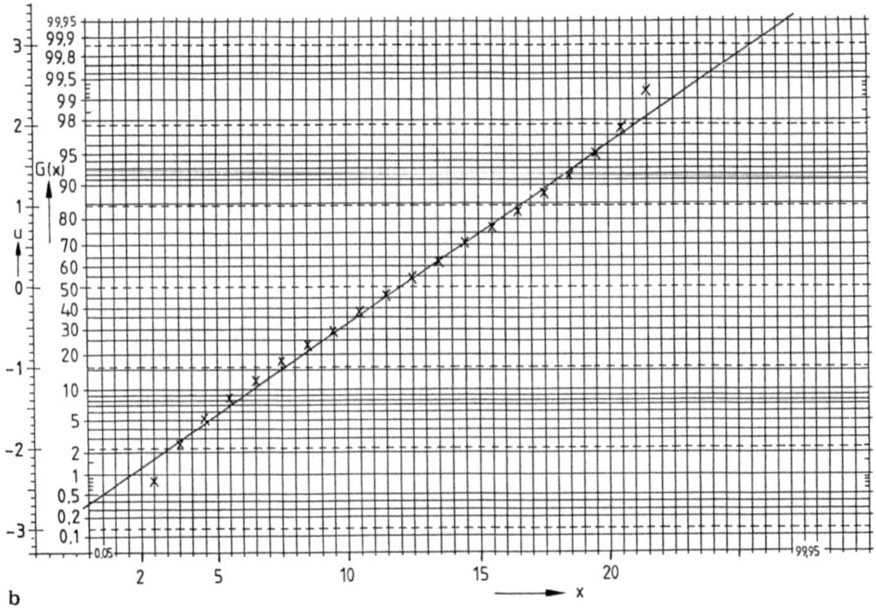

Fig. 7.5 b. Probability sum function of the fold product of two rectangular distributions

distribution:

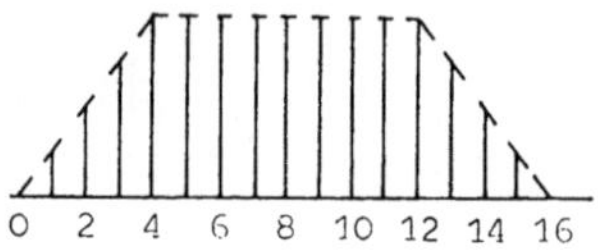

standard deviation: $\sigma = \sqrt{\frac{10R^2 - 32}{192}} = \sqrt{\frac{10 \cdot 16^2 - 32}{192}}$

$= 3.628590176$

parameters calculated
by calculator: $N = 48$; $\mu = 8$; $\sigma_N = 3.628590176$

folding:

x	g(x)								
x		1	2	3	4	5	6	7	8
	g(x)	1	2	3	4	4	4	4	4·1/48
1	1	2							
		1							
2	2	3	4						
		2	4						
3	3	4	5	6	continuation				
		3	6	9	symmetrical				
4	4	5	6	7	8				
		4	8	12	16				
5	4	6	7	8	9	10			
		4	8	12	16	16			
6	4	7	8	9	10	11	12		
		4	8	12	16	16	16		
7	4	8	9	10	11	12	13	14	
		4	8	12	16	16	16	16	
8	4	9	10	11	12	13	14	15	16
	•	4	8	12	16	16	16	16	16
	1/48	continuation symmetrical							

parameters of the fold product:

$N = 2304$

$\mu = 16$

$\sigma_N = 5.131601439$
$= \sqrt{2} \cdot 3.628590176$

fold product:

x	N_j	$\sum N_j$	G(x) in %
2	1	1	0.0434
3	4	5	0.2170
4	10	15	0.6510
5	20	35	1.6121
6	33	68	2.9514
7	48	116	5.0347
8	64	180	7.8125
9	80	260	11.2947
10	96	356	15.4514
11	112	468	20.3125
12	128	596	25.8681
13	144	740	32.1181
14	158	898	38.9757
15	168	1066	46.0069
16	172	1238	53.7326
17	168	1406	61.0243
18	158	1564	67.8819
19	144	1708	74.1319
20	128	1836	79.6875
21	112	1948	84.5486
22	96	2044	88.7153
23	80	2124	92.1875
24	64	2188	94.9653
25	48	2236	97.0486
26	33	2269	98.4809
27	20	2289	99.3490
28	10	2299	99.7830
29	4	2303	99.9566
30	1	2304	100.0000

Fig. 7.6 a. Folding two trapezoid distributions of the type TPD_1

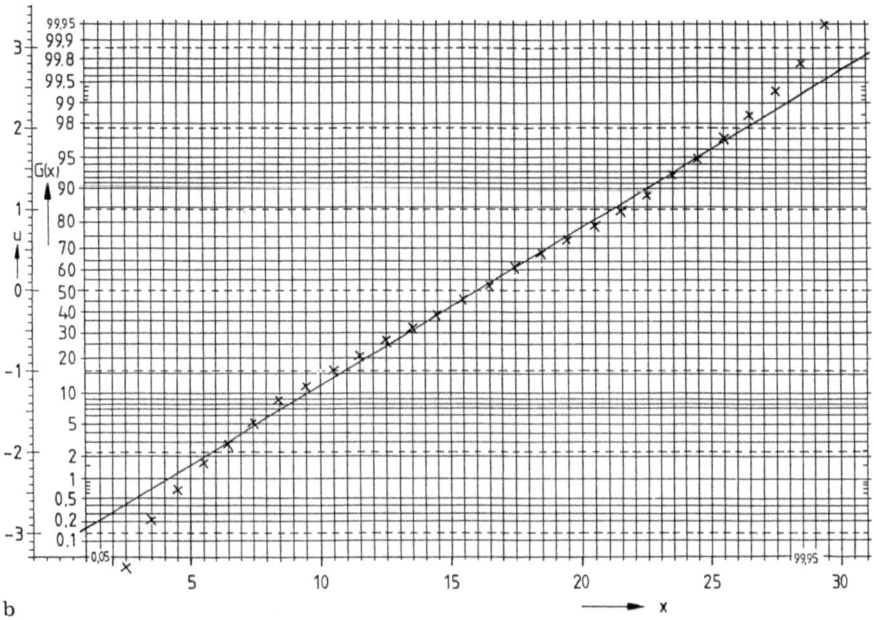

b

Fig. 7.6 b. Probability sum function of the fold product of two TPD_1's

distribution:

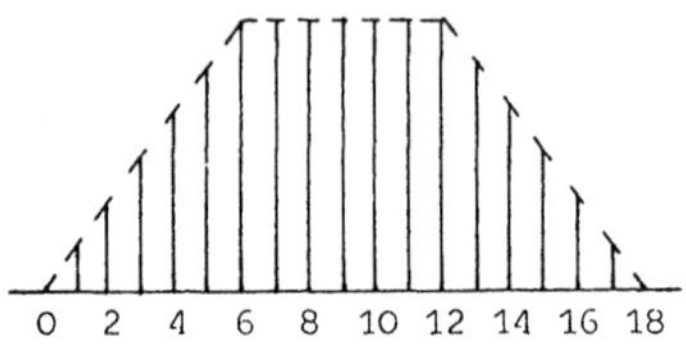

standard deviation: $\sigma = \sqrt{\dfrac{5R^2 - 18}{108}} = \sqrt{\dfrac{5 \cdot 18^2 - 18}{108}}$

$= 3.851406669$

parameters calculated by calculator: $N = 72;\ \mu = 9;$

$\sigma_N = 3.851406669$

folding:

x	g(x)	1	2	3	4	5	6	7	8	9
		1	2	3	4	5	6	6	6	6 • 1/72
1	1	2								
		1								
2	2	3	4							
		2	4							
3	3	4	5	6	continuation symmetrical					
		3	6	9						
4	4	5	6	7	8					
		4	8	12	16					
5	5	6	7	8	9	10				
		5	10	15	20	25				
6	6	7	8	9	10	11	12			
		6	12	18	24	3o	36			
7	6	8	9	10	11	12	13	14		
		6	12	18	24	30	36	36		
8	6	9	10	11	12	13	14	15	16	
		6	12	18	24	30	36	36	36	
9	6	10	11	12	13	14	15	16	17	18
	• 1/72	6	12	18	24	30	36	36	36	36
		continuation symmetrical								

parameters of the fold product:

$N = 5184$

$\mu = 18$

$\sigma_N = 5{,}446\,711\,546$

$= \sqrt{2} \cdot 3{,}851\,406\,669$

a

fold product:

x	N_j	$\sum N_j$	G(x) in %
2	1	1	0.0193
3	4	5	0.0965
4	10	15	0.2894
5	20	35	0.6752
6	35	70	1.3503
7	56	126	2.4306
8	82	208	4.0123
9	112	320	6.1728
10	145	465	8.9699
11	180	645	12.4421
12	216	861	16.6088
13	252	1113	21.4699
14	286	1399	26.9869
15	316	1715	33.0826
16	340	2055	39.6412
17	356	2411	46.5085
18	362	2773	53.4915
19	356	3129	60.3588
20	340	3469	66.9174
21	316	3785	73.0131
22	286	4071	78.0131
23	252	4323	83.8912
24	216	4539	87.5579
25	180	4719	91.0301
26	145	4864	93.8272
27	112	4976	95.9876
28	82	5058	97.5694
29	56	5114	98.6497
30	35	5149	99.3248
31	20	5169	99.7106
32	10	5179	99.9035
33	4	5183	99.9807
34	1	5184	100.0000

Fig. 7.7 a. Folding two trapezoid distributions of the type TPD_2

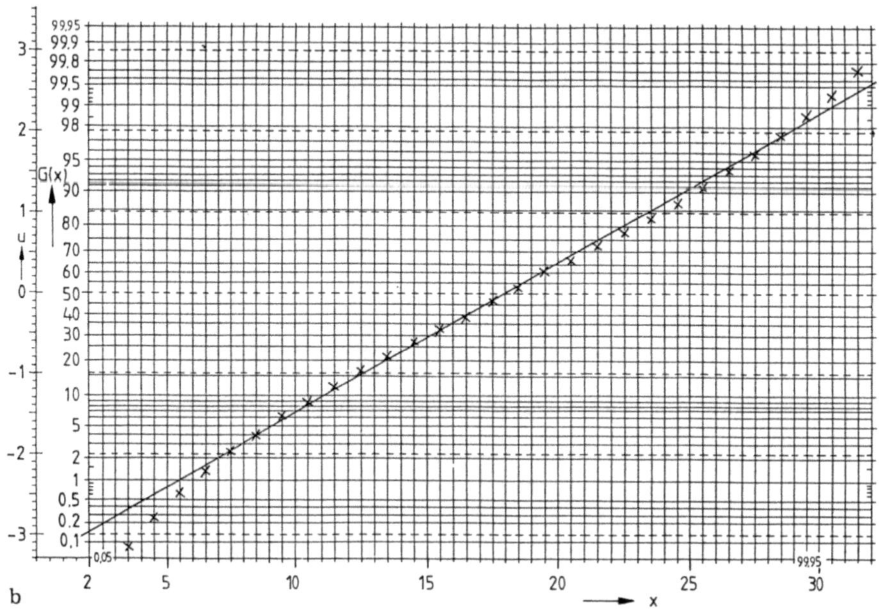

b

Fig. 7.7 b. Probability sum function of the fold product of two TPD_2's

x g(x)	9 555	10 8890	11 555·1/10000
9 555	18 308025	continuation symmetrical	
10 8890	19 4933950	20 79032100	
11 555 ·1/10000	20 308025	21 4933950	22 308025

fold product:

x_j	N_j	$\sum N_i$	G(x) in %	$N_{j,red.}$
18	308025	308025	0.3080	31
19	9867900	10175925	10.1759	987
20	79648150	89824075	89.8241	7964
21	9867900	99691975	99.6919	987
22	308025	100000000	100.0000	31

parameter of the fold product:

N = 100000000

μ = 20

σ_N = 0.471169 $\sigma_{N,red.}$ = 0.471381

folding of 2 fold products:

x g(x)	18 31	19 987	20 7964	21 987	22 31·1/10000
18 31	36 961				
19 987	37 30597	38 974169	continuation symmetrical		
20 7964	38 246884	39 7860468	40 63425296		
21 987	39 30597	40 974169	41 7860468	42 974169	
22 31 ·1/10000	40 961	41 30597	42 246884	43 30597	44 961

Fig. 7.8 a. Folding two extreme mixed distributions with a proportion of 11.1% outside the 6σ range

fold product of fold products:

x_j	N_i	$\sum N_i$	G(x) in %
36	961	961	0.000961
37	61194	62155	0.0621
38	1467937	1530092	1.53
39	15782130	17312222	17.31
40	65375556	82687778	82.69
41	15782130	98469908	98.47
42	1467937	99937845	99.9378
43	61194	99999039	99.9990
44	961	100000000	100.0000

parameter:

N = 100000000

μ = 40

σ = 0.666633

$6 \cdot \sigma$ - zone from 38.0001 to 41.9999
proportion outside: 3.06 %

$6.1 \cdot \sigma$ - zone from 37.97 to 42.03
proportion outside: 0.12 %

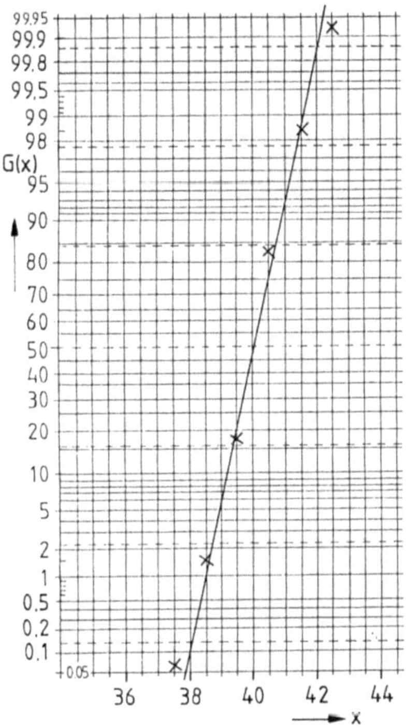

Fig. 7.8 b. Probability sum function of the fold product of four extreme mixed distributions with a proportion of 11.1% outside the 6σ range

Folding a normal distribution ND: N = 1008; μ = 7.5; σ_N = 2
with a rectangular distribution RD: N = 1008; μ = 4; σ_N = 2

folding:

x		1	2	3	4	5	6	7	8	9	10
	g(x)	1	5	15	40	103	150	190	190	150	103 • 1/1008
1	144	2	3	4	5	6	7	8	9	10	11
		144	720	2160	5760	14832	21600	27360	27360	21600	14832
2	144·	3	4	5	6	7	8	9	10	11	
		144	720	2160	5760	14832	21600	27360	27360	21600	
3	144	4	5	6	7	8	9	10	11		
		144	720	2160	5760	14832	21600	27360	27360		
4	144	5	6	7	8	9	10	11			
		144	720	2160	5760	14832	21600	27360			
5	144	6	7	8	9	10	11				
		144	720	2160	5760	14832	21600				
6	144	7	8	9	10	11		continuation			
		144	720	2160	5760	14832		symmetrical			
7	144	8	9	10	11						
	•	144	720	2160	5760						
	1/1008										

fold product:

x	N_j	$\sum N_j$	G(x) in %	x	$\sum N_j$	G(x) in %
2	144	144	0.0142	21	1016064	100.0000
3	864	1008	0.0992	20	1015920	99.9858
4	3024	4032	0.3968	19	1015056	99.9008
5	8784	12816	1.2613	18	1012032	99.6032
6	23616	36432	3.5856	17	1003248	98.7387
7	45216	81648	8.0357	16	979632	96.4144
8	72576	154224	15.1786	15	934416	91.9643
9	99792	254016	25.0000	14	861840	84.8214
10	120672	374688	36.8764	13	762048	75.0000
11	133344	508032	50.0000	12	641376	63.1236

Parameters of the fold product:

N = 1016064

μ = 11,5; $\sigma_N = 2{,}828427125 = \sqrt{2^2 + 2^2}$

folding a normal distribution ND: N = 1008; μ = 7,5; σ_N = 2
with a rectangular distribution RD: N = 7; μ = 4; σ_N = 2

folding: (initial stage)

x		1	2	3	4
	g(x)	1	5	15	40 · 1/1008
1	1	2	3	4	5
		1	5	15	40
2	1	3	4	5	
		1	5	15	
3	1	4	5		
		1	5		
4	1	5			
	•	1			
	1/7				

fold product:

x	N_j	$\sum N_j$	G(x) in %
2	1	1	0.0142
3	6	7	0.0992
4	21	28	0.3968
5	61	89	1.2613
			continuation as above

parameter of the fold product:

N = 7056

μ = 11.5

σ = 2.8284

a

Fig. 7.9 a. Folding a normal distribution with a rectangular distribution

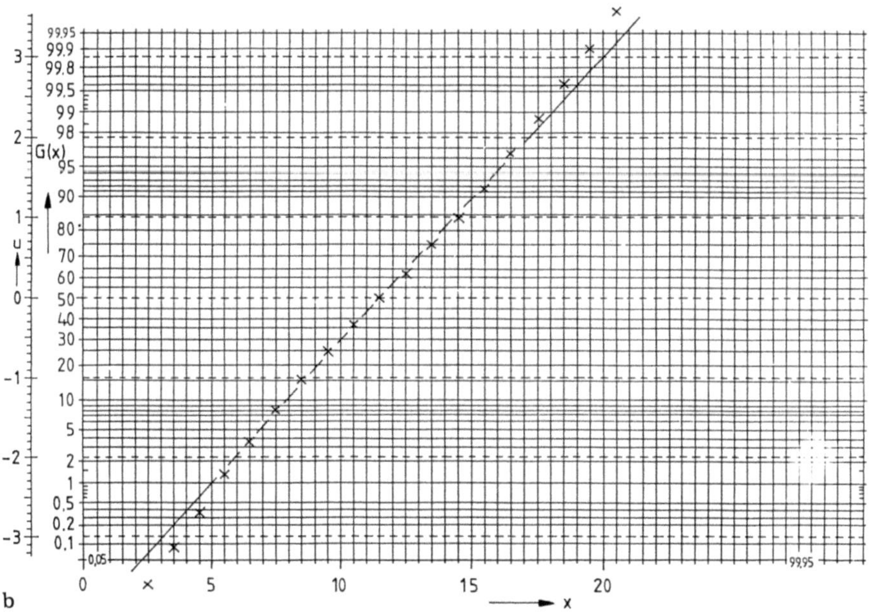

b

Fig. 7.9 b. Sum function of the fold product of a normal distribution and a rectangular distribution (both $\sigma = 2$)

folding of two normal distributions: ND_1: N_1 = 1008; μ_1 = 7,5; σ_{N_1} = 2

ND_2: N_2 = 1008; μ_2 = 27; σ_{N_2} = 8

folding:

x g(x)	1 1	2 5	3 15	4 40	5 103	6 150	7 190	8 190	9 150	10 103	11 40	12 15	13 5	14 1 • 1/1008
1 1	2 1	3 5	4 15	5 40	6 103	7 150	8 190	9 190	10 150	11 103	12 40	13 15	14 5	15 1
5 5	6 5	7 25	8 75	9 200	10 515	11 750	12 950	13 950	14 750	15 515	16 200	17 75	18 25	19 5
9 15	10 15	11 75	12 225	13 600	14 1545	15 2250	16 2850	17 2850	18 2250	19 1545	20 600	21 225	22 75	23 15
13 40	14 40	15 200	16 600	17 1600	18 4120	19 6000	20 7600	21 7600	22 6000	23 4120	24 1600	25 600	26 200	27 40
17 103	18 103	19 515	20 1545	21 4120	22 10609	23 15450	24 19570	25 19570	26 15450	27 10609	28 4120	29 1545	30 515	31 103
21 150	22 150	23 750	24 2250	25 6000	26 15450	27 22500	28 28500	29 28500	30 22500	31 15450	32 6000	33 2250	34 750	
25 190	26 190	27 950	28 2850	29 7600	30 19570	31 28500	32 36100	33 36100	34 28500					
29 190	30 190	31 950	32 2850	33 7600	34 19570									
33 150	34 150													
37 103														
41 40														
45 15														
49 5														
53 1 • 1/1008														

continuation symmetrical

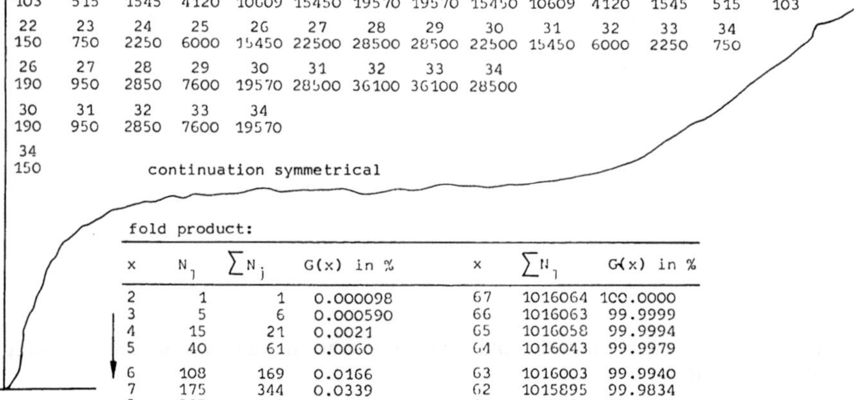

fold product:

x	N_j	$\sum N_j$	G(x) in %	x	$\sum N_j$	G(x) in %
2	1	1	0.000098	67	1016064	100.0000
3	5	6	0.000590	66	1016063	99.9999
4	15	21	0.0021	65	1016058	99.9994
5	40	61	0.0060	64	1016043	99.9979
6	108	169	0.0166	63	1016003	99.9940
7	175	344	0.0339	62	1015895	99.9834
8	265	609	0.0599	61	1015720	99.9661
9	390	999	0.0983	60	1015455	99.9401
10	680	1679	0.1652	59	1015065	99.9017
11	928	2607	0.2566	58	1014385	99.8348
12	1215	3822	0.3762	57	1013457	99.7434
13	1565	5387	0.5302	56	1012242	99.6238
14	2340	7727	0.7605	55	1010677	99.4698
15	2966	10693	1.0524	54	1008337	99.2395
16	3650	14343	1.4116	53	1005371	98.9476
17	4525	18868	1.8570	52	1001721	98.5884
18	6498	25366	2.4965	51	997196	98.1430
19	8065	33431	3.2902	50	990698	97.5035
20	9745	43176	4.2493	49	982633	96.7098
21	11945	55121	5.4249	48	972888	95.7507
22	16834	71955	7.0817	47	960943	94.5750
23	20335	92290	9.0831	46	944109	92.9183
24	23420	115710	11.3881	45	923774	90.9169
25	26170	141880	13.9637	44	900354	88.6119
26	31290	173170	17.0432	43	874184	86.0363
27	34099	207269	20.3992	42	842894	82.9568
28	35470.	242739	23.8901	41	808795	79.6008
29	37645	280384	27.5951	40	773325	76.1099
30	42775	323159	31.8050	39	735680	72.4049
31	45003	368162	36.2341	38	692905	68.1950
32	44950	413112	40.6580	37	637902	63.7659
33	45950	459062	45.1804	36	602952	59.3419
34	48970	508032	50.0000	35	557002	54.8196

parameters of the fold product:

N = 1016064

μ = 34.5

$\sigma_N = 8.246211251 = \sqrt{2^2 + 8^2}$

Fig. 7.10 a. Folding two normal distributions with different standard deviations

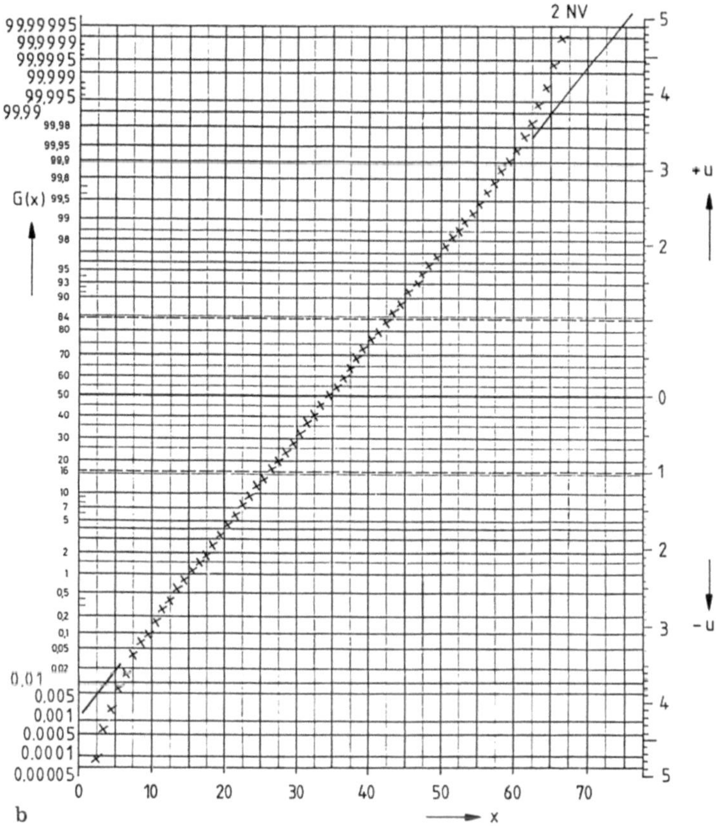

b

Fig. 7.10 b. Sum function of the fold product of two normal distributions of $\sigma_{N1} = 2$ and $\sigma_{N2} = 8$

Exercise 7.1

Given: Two rectangular distributions made discrete with the parameters $\mu = 4$ and $\sigma_N = 2$.

Required: Fold these distributions and their fold product.

Solution: Folding two rectangular distributions

with $N = 7$; $\mu = 4$; $\sigma_N = 2$

folding:

x	g(x)	1	2	3	4	5	6	7
		1	1	1	1	1	1	1 · 1/7
1	1	2 1	3 1	4 1	5 1	6 1	7 1	8 1
2	1	3 1	4 1	5 1	6 1	7 1	8 1	
3	1	4 1	5 1	6 1	7 1	8 1		
4	1	5 1	6 1	7 1	8 1			
5	1	6 1	7 1	8 1				
6	1	7 1	8 1					
7	1 · 1/7	8 1						

fold product:

x	N_j	$\sum N_j$	G(x) in %	x	N_j	$\sum N_j$	G(x) in %
2	1	1	2.0408	14	1	49	100.0000
3	2	3	6.1224	13	2	48	97.9592
4	3	6	12.2449	12	3	46	93.8776
5	4	10	20.4082	11	4	43	87.7551
6	5	15	30.6122	10	5	39	79.5918
7	6	21	42.8571	9	6	34	69.3878
8	7	28	57.1429				

parameters of the fold product:

$N = 49$

$\mu = 8$

$\sigma_N = 2.828427125 = \sqrt{8}$

Exercise 7.2

Given: Sloping triangular distribution with the parameters $\mu = 2$ and $\sigma = 1$.
Required: Fold product of four of these distributions with equal slopes.
Solution:

x / g(x)	1 4	2 3	3 2	4 1 · 1/10
1 4	2 16	3 12	4 8	5 4 · 1/100
2 3	3 12	4 9	5 6	6 3
3 2	4 8	5 6	6 4	7 2
4 1	5 4	6 3	7 2	8 1

fold product of two TD's:

x_i	N_j	$\sum N_j$	G(x) in %
2	16	16	16
3	24	40	40
4	25	65	65
5	20	85	85
6	10	95	95
7	4	99	99
8	1	100	100

folding the fold products:

x / g(x)	2 16	3 24	4 25	5 20	6 10	7 4	8 1
2 16	4 256	5	6	7	8	9	10
3 24	5 384	6 576	7	8	9	10	11
4 25	6 400	7 600	8 625	9	10	11	12
5 20	7 320	8 480	9 500	10 400	11	12	13
6 10	8 160	9 240	10 250	11 200	12 100	13	14
7 4	9 64	10 96	11 100	12 80	13 40	14 16	15
8 1	10 16	11 24	12 25	13 20	14 10	15 4	16 1

parameters:

2 TD: $N = 100$

$\mu = 5$

$\sigma_N = 1.4142 = \sqrt{2} \cdot 1$

4 TD: $N = 10000$

$\mu = 8$

$\sigma_N = 2 = \sqrt{4} \cdot 1$

G(x) is shown in figure 7.11

fold product of four triangular distributions:

x_i	N_j	$\sum N_j$	G(x) in %
4	256	256	2.56
5	768	1024	10.24
6	1376	2400	24.00
7	1840	4240	42.40
8	1905	6145	61.45
9	1608	7753	77.53
10	1124	8877	88.77
11	648	9525	95.25
12	310	9835	98.35
13	120	9955	99.55
14	36	9991	99.91
15	8	9999	99.99
16	1	10000	100.00

Exercise 7.3

Given: Sloping TD with the parameters $\mu = 2$ (3) and $\sigma = 1$.
Required: Fold product of four of these distributions, two of which have opposite slopes.
Solutions:

x q(x)		1 4	2 3	3 2	4 1·1/10
1	1	2 4	3 3	4 2	5 1 1/100
2	2	3 8	4 6	5 4	6 2
3	3	4 12	5 9	6 6	7 3
4	4	5 16	6 12	7 8	8 4

fold product of two TD's:

x_j	N_j	$\sum N_j$	G(x) in %
2	4	4	4
3	11	15	15
4	20	35	35
5	30	65	65
6	20	85	85
7	11	96	96
8	4	100	100

folding the fold products:

x q(x)		2 4	3 11	4 20	5 30	6 20	7 11	8 4
2	4	4 16						
3	11	5 44	6 121·1/10000					
4	20	6 80	7 220	8 400				
5	30	7 120	8 330	9 600	10 900			
6	20	8 80	9 220	10 400	11 600	12 400		
7	11	9 44	10 121	11 220	12 330	13 220	14 121	
8	4	10 16	11 44	12 80	13 120	14 80	15 44	16 16

parameters:

2 TD: $N = 100$

$\mu = 5$

$\sigma_N = 1.4142 = \sqrt{2} \cdot 1$

4 TD: $N = 10000$

$\mu = 10$

$\sigma_N = 2 = \sqrt{4} \cdot 1$

fold product of four triangular distributions:

x_j	N_j	$\sum N_j$	G(x) in %
4	16	16	0.16
5	88	104	1.04
6	281	385	3.85
7	680	1065	10.65
8	1220	2285	22.85
9	1728	4013	40.13
10	1974	5987	59.87
11	1728	7715	77.15
12	1220	8935	89.35
13	680	9615	96.15
14	281	9896	98.96
15	88	9984	99.84
16	16	10000	100.00

G(x) is shown in figure 7.11

Exercise 7.4

Given: V-shaped distribution with the parameters $\mu = 3$ and $\sigma = \sqrt{3}$.
Required: Fold product of four of these distributions.
Solution:

x / g(x)		1 2	2 1	3 0	4 1	5 2
1	2	2 4	3 2	4 0	5 2	6 4
2	1	3 2	4 1	5 0	6 1	7 2
3	0	4 0	5 0	6 0	7 0	8 0
4	1	5 2	6 1	7 0	8 1	9 2
5	2	6 4	7 2	8 0	9 2	10 4

fold product of two VD's:

x_j	N_j	$\sum N_j$	G(x) in %
2	4	4	11.11
3	4	8	22.22
4	1	9	25
5	4	13	36.11
6	10	23	63.89
7	4	27	75
8	1	28	77.78
9	4	32	88.89
10	4	36	100.00

$\mu = 6$ und $\sigma_N = 2.44949 = \sqrt{6}$

folding the fold products:

x / g(x)		2 4	3 4	4 1	5 4	6 10	7 4	8 1	9 4	10 4
2	4	4 16								
3	4	5 16	6 16							
4	1	6 4	7 4	8 1						
5	4	7 16	8 16	9 4	10 16					
6	10	8 40	9 40	10 10	11 40	12 100				
7	4	9 16	10 16	11 4	12 16	13 40	14 16			
8	1	10 4	11 4	12 1	13 4	14 10	15 4	16 1		
9	4	11 16	12 16	13 4	14 16	15 40	16 16	17 4	18 16	
10	4	12 16	13 16	14 4	15 16	16 40	17 16	18 4	19 16	20 16

fold product from 4 V-shaped distributions:

x_j	N_j	$\sum N_j$	G(x) in %
4	16	16	1.235
5	32	48	3.704
6	24	72	5.555
7	40	112	8.642
8	113	225	17.361
9	120	345	26.620
10	76	421	32.485
11	128	549	42.361
12	198	747	57.639
13	128	875	67.515
14	76	951	73.380
15	120	1071	82.639
16	113	1184	91.358
17	40	1224	94.444
18	24	1248	96.296
19	32	1280	98.843
20	16	1296	100.000

parameters:

$N = 1296$

$\mu = 12$

$\sigma_N = \sqrt{4} \cdot \sqrt{3} = 3.4641 = \sqrt{12}$

G(x) is shown in figure 7.12

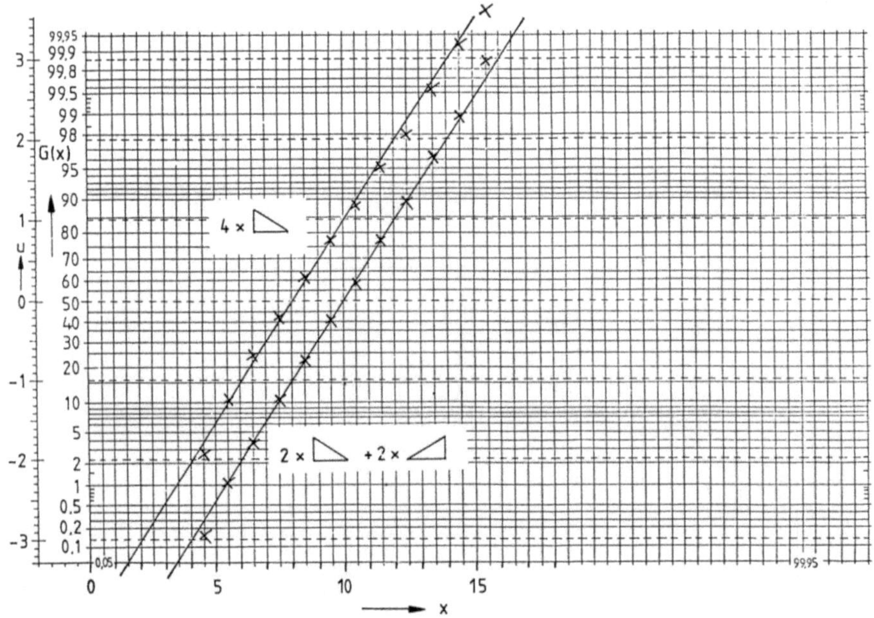

Fig. 7.11. Probability sum functions of the fold products of each four sloping triangular deviations of the parameters $\mu = 2(3)$ and $\sigma = 1$

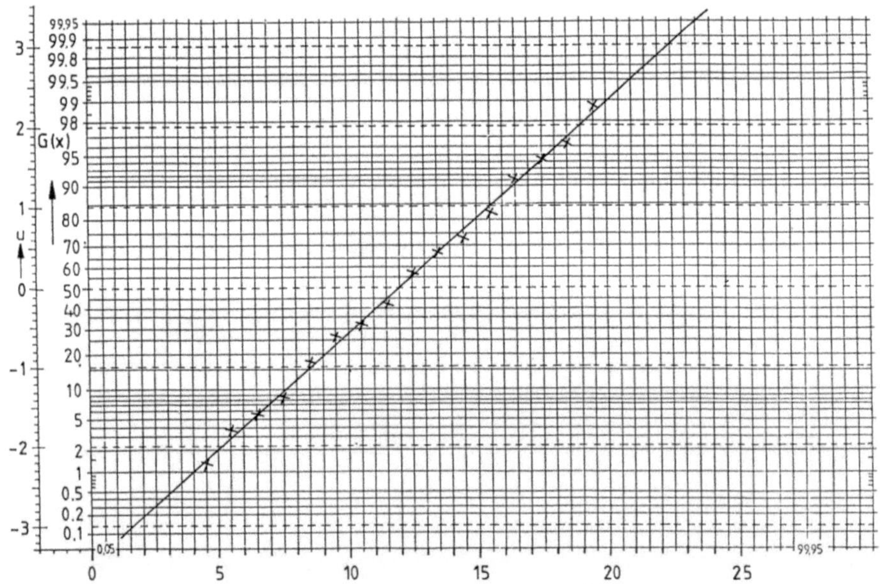

Fig. 7.12. Probability sum function of the fold product of four V-shaped distributions with the parameters $\mu = 3$ and $\sigma_N = \sqrt{3}$; Parameters of the fold product: $\mu = 12$ and $\sigma_N = \sqrt{4} \cdot \sqrt{3} = 3.4641$. The fold product is extraordinarily close to a normal distribution

8 Relationship Between Tolerances and Production Distributions

8.1 Direct and Indirect Function Characteristics

The ideas we have discussed so far:

—basic statistics for quantitative characteristics,
—mixed distributions, and
—folding operations,

should in this and the following chapters now be applied to tolerances, fits (two member dimensional chains) and multiple dimensional chains.

All characteristics given in production specifications (drawings) such as dimensions or material characteristics must usually fulfill a specific function for which they are intended.

—All characteristics are functional characteristics and therefore important for quality and reliability of technical products.

A distinction can be drawn between

—direct functional dimensions (characteristics) and
—indirect functional dimensions (characteristics).

This can be explained with the example of a gear shaft, which is positioned on both sides, Fig. 8.1. At the middle next to the pivot pins, 1 and 5, are the shaft recesses on both sides for the gear wheels, 2 and 4; the shaft section with the greatest diameter is in the middle, 3.

Direct functional dimensions are in Fig. 8.1 the diameters 1 and 5 or 2 and 4. They appear in direct function (in interaction) with the inner diameters of the pivots or gear wheels.

These direct functional characteristics are therefore especially important. The following points are valid for them:

—They must have narrow tolerances and therefore
—must be produced with high precision
—so careful quality control must be planned and carried out, so that
—a controlled process is ensured.
—Only small proportions of dimensions exceeding limit values (errors) can be accepted.

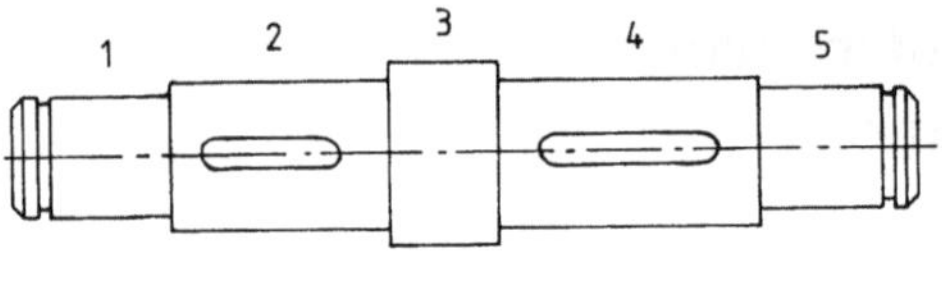

Fig. 8.1. Gear shaft with the Sects. 1 to 5.

—Direct functional characteristics are test values in the sense that they must be taken into account in test planning concerning the test sample quantity.

Test sizes can be shown in a special way in a drawing:

1. Example:

$25^{+0.03}_{-0.03}$.

Zeppelin shaped dimension; test dimension

2. Example:

$25^{+0.05}_{-0.05}$.

Bar dimension; Test dimension with which in addition the internal works specification for the actual dimension distribution in the tolerance field must be taken into account.

—Further examples for direct functional characteristics are
 —thread pitch diameter of a spindle,
 —pitch circle of a gear wheel,
 —piston diameter and cylinder diameter of an engine.

The diameter of the shaft Sect. 3 in the middle of the shaft in Fig. 8.1 belongs to the indirect functional characteristics. This diameter appears with no other dimension in a direct function. It fulfills its function indirectly in that it serves to ensure that the limit value of the flexing of the shaft or the limit value of the bending edge tension (permissible tension) is not exceeded.

Indirect functional characteristics are not especially important. For these the following points should be taken into account:

—They can have wide tolerances and therefore
—can be produced with average or low production precision and
—often need no quality control or quality testing.
—Certain proportions of dimensions which exceed limit values (errors) can be accepted, if these occur despite the wide tolerances. Non-comformity must only be rejected if as a result the part is considered unacceptable.

Note: If the diameter in the middle of the shaft in Fig. 8.1 for example with 60 ± 0.15 (general dimension with "fine" tolerance class) is not achieved, then it is not likely that shafts with the actual dimension $d = 59.80$ will adversely affect the function (not exceed the limit bend).

—Indirect functional characteristics are not test dimensions; they need not be particularly considered at the test planning stage with regard to the test sample quantity. For example, from one batch just one individual part is tested, the others are tested only visually.
—Indirect functional characteristics are usually general dimensions and are provided with general tolerances.

Note: The former term "free dimension" was to some extent clearer, because this expression shows that this dimension is "free" and is not constrained by other dimensions.

—Further examples for indirect functional characteristics are:
  —length of a screw
  —width of a gear
  —cast housing of a wall thickness
  —whole material characteristics.

Note. The classification given here for direct functional dimensions and indirect functional dimensions is very rough and unrestrictedly valid for the mentioned examples (pivot pins and middle diameters). In between there are also sliding scales. In these cases the above-mentioned classification is only applicable with restrictions.

8.2 Establishment and Observance of Tolerances

The statements in the introduction should be repeated and are summarised as follows:

Possible dimension	Establishment of tolerances
General dimension	there are norms for these; the decision of the designer is restricted to the choice of appropriate tolerance class.
Tolerance on fit	there are norms for these; the decision of the designer is restricted to the choice of tolerance class and tolerance zone position.
Dimensions which belong to multiple dimensional chains	there are no norms for these: the designer calculates the nominal dimensions and the required tolerances.

If norms exist, the choice of the appropriate tolerance class follows with consideration of

—the demands relative to fulfillment of function
—the production capabilities

The ideal case for the assurance of function fulfillment is in the designer's control, that all parts are at optimal dimension with no spread or deviate only as much from this as is normal in gauged manufacture, up to IT4. Then a high value is given to uniformity within the batch and the characteristics of, for example, friction bearings are uniformly optimal in terms of lubrication, warming, noise development or product life.

This ideal case is not practicable. The designer must give tolerances which can be maintained with economic methods.

The optimal case would be that—at the same time as fulfillment of the requirements of function—the tolerances in each individual case are reconciled with the capabilities of technical production.

This assumes that the capabilities of technical production are known to the designer. Against this:

—it is not known in advance which machine will be used
—the operating precision of the machine—according to [46]—is often not even known to the production people,
—the contact between design and production is often made difficult if the designers work centrally and the production operations are undertaken in distant locations.

The actual case is that the tolerances seldom correspond with the capabilities of technical production but are either too wide or too narrow, more often the latter.

If no norms are provided and the designers calculate according to their own judgement and decide on the establishment of tolerances, then this usually takes the worst case by arithmetical tolerance calculation, Sect. 8.5.

In these cases, the actual case can often appear intensified: tolerances seldom too wide, usually too narrow.

In this regard, the following historical-phychological points should be borne in mind:

—The production department believes that it is allowed to "exploit" tolerances. This claim can be traced back to norms which have been valid for decades, according to which the actual size lies "possibly" in the tolerance zone.
—The designers know this and therefore prefer to set tolerances rather too narrow (fear tolerances). They are ultimately responsible for the function of the item. If the function is not fulfilled, they are called to account. On the other hand, the designers are never held responsible if tolerances are too narrow. Because of this they are unhappy about widening tolerances,

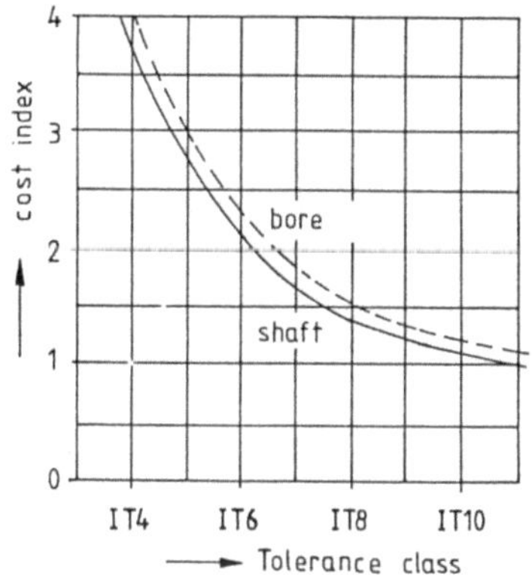

Fig. 8.2. Cost index for the machining of shafts and bores (from Zollikofer)

with the fear that then these widened tolerances would be "exploited" in the direction of lower conformity of the actual size.

In the sense of a current, total enterprise responsibility for quality, which applies to all sections involved, it should be precisely the other way round.

- As far as the design department is concerned, tolerances should be set as large as possible—from a statistical point of view too where multiple dimensional chains are concerned.
- As far as the production department is concerned, tolerances should not be "exploited"; even with sufficiently large tolerances production must be carried out to the highest possible conformity. Actual sizes should deviate as little as possible from the desired sizes. Limit values are included in the permissible value range; but this is only valid for individual low number values and only then when this is absolutely necessary. The middle of the actual size distribution should constantly be aimed at the desired value.
- The guarantee of functional fulfillment is not the sole preserve of the design department but the duty of both design and production departments.

So all departments involved need to get involved in a substantial re-think. It is the job of the company's management to change frequently inadequate cooperation into genuine cooperation between the design and production departments. Quality and efficiency can be fundamentally improved in this area.

8.3 Tolerances and Costs

Production costs depend on different factors. They include:

- production processes,
- output,

—materials,
—nominal size range,
—tolerance.

Exhaustive investigation into different materials and in different production processes (turning, boring, milling, grinding) all tend to lead to the same result.

As a rule of thumb it can be assumed that with small tolerances ($T < 0.01$ mm) halving the tolerance will result in four times higher production costs. Costs involved in producing internal fit surfaces are always higher than external fit surfaces.

In Fig. 8.2 an example is given of a cost index of tolerance classes for the machining of shafts and bores.

8.4 Errors and Waste

If actual sizes do not match the desired size but lie within the tolerance zone then we are talking about permissible deviations.

If actual sizes lie outside tolerance zones, then these are inadmissible deviations or non-conformities. A non-conformity says nothing about the efficiency of the component which shows the (or several) non-conformities.

Small non-conformity proportions are usually acceptable. Because of the constant slope of the probability density function in a normal distribution, at the end of the distribution non-conformities are likely near the limit values and become less frequent further away from the limit value, Fig. 8.3. This also goes for actual, abnormal distributions in principle.

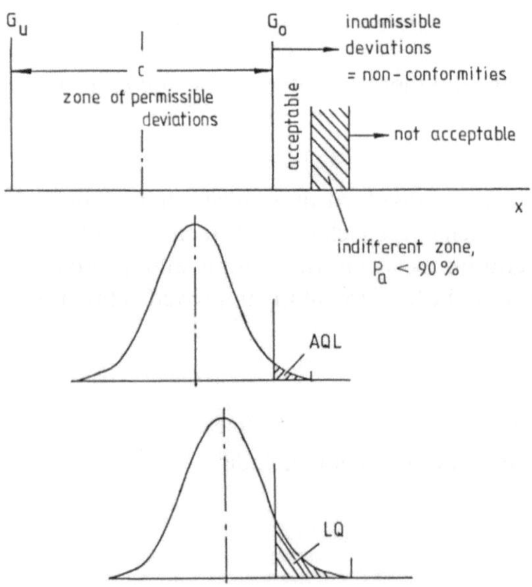

Fig. 8.3. Some concepts of quantitative sample testing

In sample testing, test plans are constructed so that small non-conformity proportions have a high acceptable probability ($P_a \gtrsim 90\%$). This non-conformity proportion is called AQL (acceptable quality level). With high non-conformity proportions where actual values are far from limit values there is a low acceptable probability ($P_a \lesssim 10\%$); this non-conformity proportion is called LQ (limiting quality). In summary:

— each inadmissible deviations is an non-conformity,
— batches with low non-conformity proportions are (usually) all useable,
— batches with high non-conformity proportions are (frequently) unuseable, sometimes just some of the actual values which lie outside the tolerance zone, sometimes all. Only unuseable items should be designated as waste.

8.5 Arithmetical Tolerance Calculation

In design (up to now) arithmetical tolerance calculation has been favoured. Thus with k members of a dimensional chain, the arithmetical closing tolerance—the tolerance of the closing value—is

$$T_a = \sum_{i=1}^{i=k} T_i.$$

Basic to the tolerance calculation is the assumption that the actual value in the batch belonging to the dimensional chain should "possibly" lie in the tolerance zone, so will often be near to the limit value. If this is the case with all the members of a dimensional chain, then it is highly probable that the actual values will be joined with the closing values which lie near or close by the arithmetically calculated limit values for the closing value. This case is the least favourable or "worst case".

The calculation of the closing values and their limit values (lower and upper closing values) can result from

— the deviations calculation or
— the limit value calculation.

Since the deviations calculation is more advantageous this will be used here.

The deviations calculation takes the following stages:

1. Preparation of the individual part drawing and the composite drawing.
2. Recording the individual sizes belonging to the dimensional chain with its nominal sizes and deviations.
3. Drafting a size plain; here individual sizes will be allocated to a
   — positive score direction or a
   — negative score direction. Care must be taken that
   — sizes which increase the closing value with their upper deviations are scored positive and
   — sizes which decrease the closing value with their upper deviations are scored negative.

The size plan need only be drawn diagrammatically (not to scale). It should be seen from this where the closing value M_0 is a clearance or an interference. This results first from the tolerance calculation but then is known from the design objective. For example a slide bearing with interference is absurd.

4. Calculation of the closing value with the aid of the format for calculation of tolerances.

Exercise 8.1

Given: Drawing with linear dimensional chain (section of a gear unit) with sizes M_1 to M_5 as given in Fig. 8.4.

Required: Arithmetically calculated closing value.

Solution: From the drawing and the composite of the sizes the following size plan can be deduced:

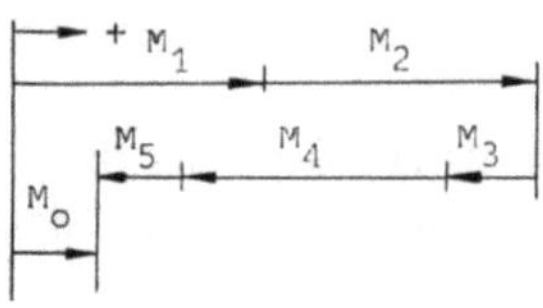

After insertion of all the nominal sizes and all deviations in the form, Fig. 8.5, their sums are found, row 14.

From the sums in the column "positive score direction", the sums in the column "negative score direction" are subtracted. Since the lower deviations of the positive sizes increase the closing value, they are added to the upper deviations of the positive sizes and inverted. In the form this is shown by a circled number.

The sums in line 16, columns 2, 3 and 4 are the nominal closing values and their deviations.

From line 16 the middle closing value can be calculated with its conforming deviations according to the total, line 17.

In column 8 all tolerances of the individual values

$$T_i = A_{oi} - A_{ui}$$

are calculated. The sum in line 14 must agree with the tolerance in line 16 or 17.

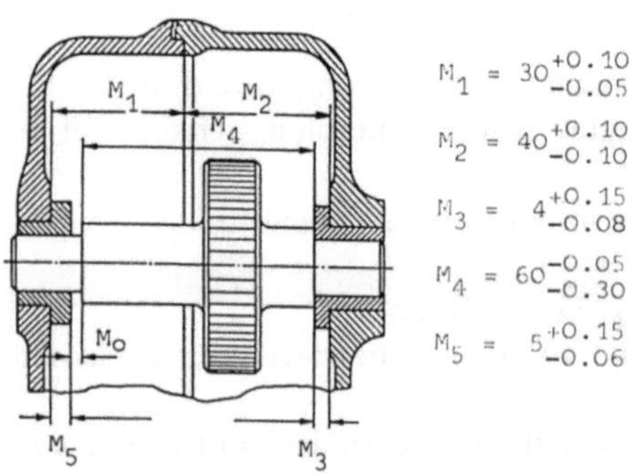

Fig. 8.4. Section of a gear unit with linear dimensional chain, from Smirnow

process: dimensional chain in figure 8.4									
date:		positive score direction			negative score direction			aritnm. calculation	
column	1	2	3	4	5	6	7	8	9
line	size No.	nominal size N_i	up. deviation A_o	low. deviation A_u	nominal size N_i	up. deviation A_o	low. deviation A_u	tolerance T_i	T_i^2
1	1	30	+0.10	-0.05				0.15	
2	2	40	+0.10	-0.10				0.20	
3	3				4	+0.15	+0.08	0.07	
4	4				60	-0.05	-0.30	0.25	
5	5				5	+0.15	+0.06	0.09	
6									
7									
8									
9									
10									
11									
12									
13									
14	Σ =	70	+0.20	-0.15	① 69	② +0.25	③ -0.16	0.76	
15		- ① -69	- ③ +0.16	- ② -0.25					
16	N_S =	1	+0.36	-0.40	← check of calculation				
17	C_S =	0.98	+0.38	-0.38					

Fig. 8.5. Form for the tolerance calculation for which the solution is given in Exercise 8.1

Provided all the entries in the form are made in pencil, eraser and pencil can be used for possible and repeated changes from design.

Advantages of arithmetical toleration.

—They can always be carried out, include the worst case and contain a high value for security regarding function fulfillment.

—They constitute the ideal starting point and a necessary pre-requisite for a statistical tolerance calculation if this is necessary.

Disadvantages of arithmetical tolerance. The larger the number of members of the dimensional chain,

—the larger the tolerance of the closing value, resulting from individual tolerances deemed necessary for technical production reasons;

—the smaller the individual tolerances, if the functional tolerances of the closing value are arithmetically divided to conform or not to conform to individual values. Individual tolerances can then no longer be produced with efficient methods.

8.6 Examples for Arithmetically Calculated Linear Dimensional Chains

In this section examples for arithmetically calculated linear dimensional chains with different assumptions of position, spread and shape of the individual size distributions and their effects on the closing value distributions will be discussed. It is recommended that examples are calculated and modified.

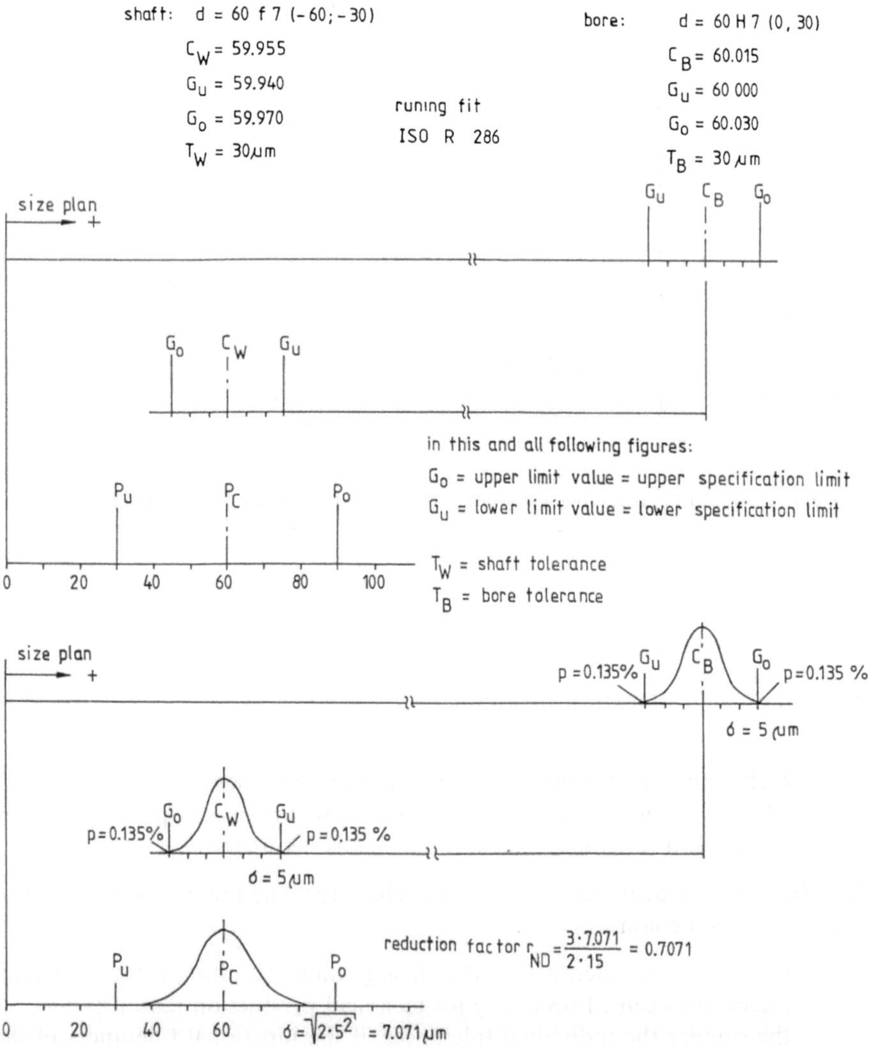

Fig. 8.6. Size plan for the bearing $d = 60\,H7/f7$. In the lower part of the figure it is assumed that the batches for bores and shafts are normally distributed and fill the tolerance zones. Then the normally distributed clearances do not fill the fit tolerance zone

In Fig. 8.6 there is a size plan for the bearing $d = 60\ H7/f7$, additionally with details of tolerance zones. The tolerances $T_B = 30\,\mu m$ and $T_W = 30\,\mu m$ give the fit tolerance $T_P = 60\,\mu m$. The limit fits are lower fits $P_u = 30\,\mu m$ and upper fits $P_0 = 90\,\mu m$. This size plan has no direct bearing on production, other than that during production it is possible to check (half way) the tolerances for bores and shafts.

In Fig. 8.6 below it is assumed that the batches for bores and shafts lie about the middle value and are normally distributed with the standard deviation $\sigma = 5\,\mu m$. Then the clearances about the middle clearance P_C are also normally distributed with the standard deviation $\sigma = 7.071\,\mu m$. This distribution does not fill the arithmetically calculated clearance tolerance zone.

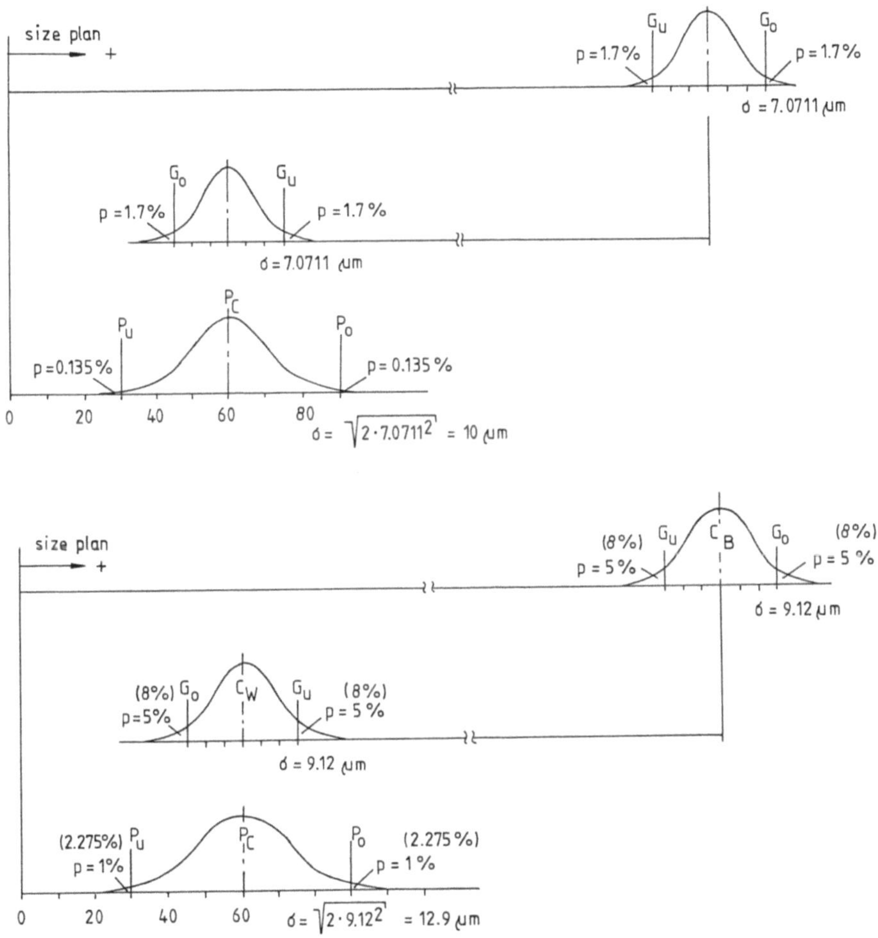

Fig. 8.7. Size plan for the bearing $d = 60H7/f7$ with different assumptions about the spread of the normally distributed batches for bores and shafts

In Fig. 8.7 too all distributions lie on the tolerance zone centre. If the clearance tolerance zone is filled from the 6σ range, then the standard deviation for bores and shafts must total $\sigma = 7.071$ respectively, so that $p = 3.4\%$ of the actual values must be non-conforming, half of which lie on both sides. If higher proportions are accepted in actual clearances, for example 1% (2.275%), then the non-conformity proportion in the component batches must total 5% (8%) both sides, Fig. 8.7 below.

Figure 8.8 above is a repeat example. If Fig. 8.8 below the case is shown where the distributions for the shafts and bores do not fill the tolerance zones

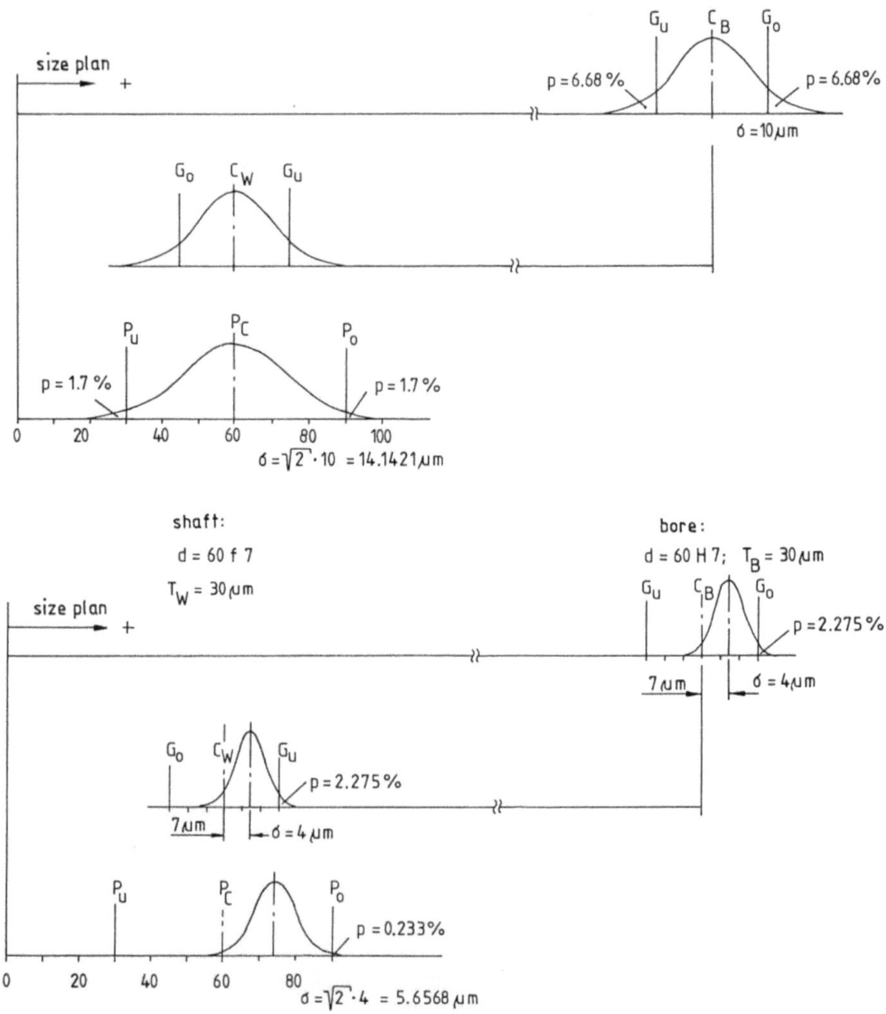

Fig. 8.8. Size plan for the bearing $d = 60$ H7/f7 with different assumptions about the position and spread of the normally distributed bores and shafts

($\sigma = 4\,\mu m$), but are shifted in the same direction to $T/4$; the non-conformity proportion in the clearances totals $p = 0.233\%$.

In Fig. 8.9 the distributions are shifted in the direction of a smaller clearance with the effects as calculated and given.

With statistical tolerance calculation the assumption of the existence of rectangular distributions which fill the tolerance zone plays a major role. Thus the mixed distribution is approached which develops from a trend (Fig. 6.2).

In Fig. 8.10 the RD for the components of the bearing $d = 60H7/f7$ are assumed, along with the resulting effects. If it is assumed that a two-sided error proportion of 2.78% means no reduction in function of the clearances, then the error proportion in the components should be about 16 (sixteen!)%.

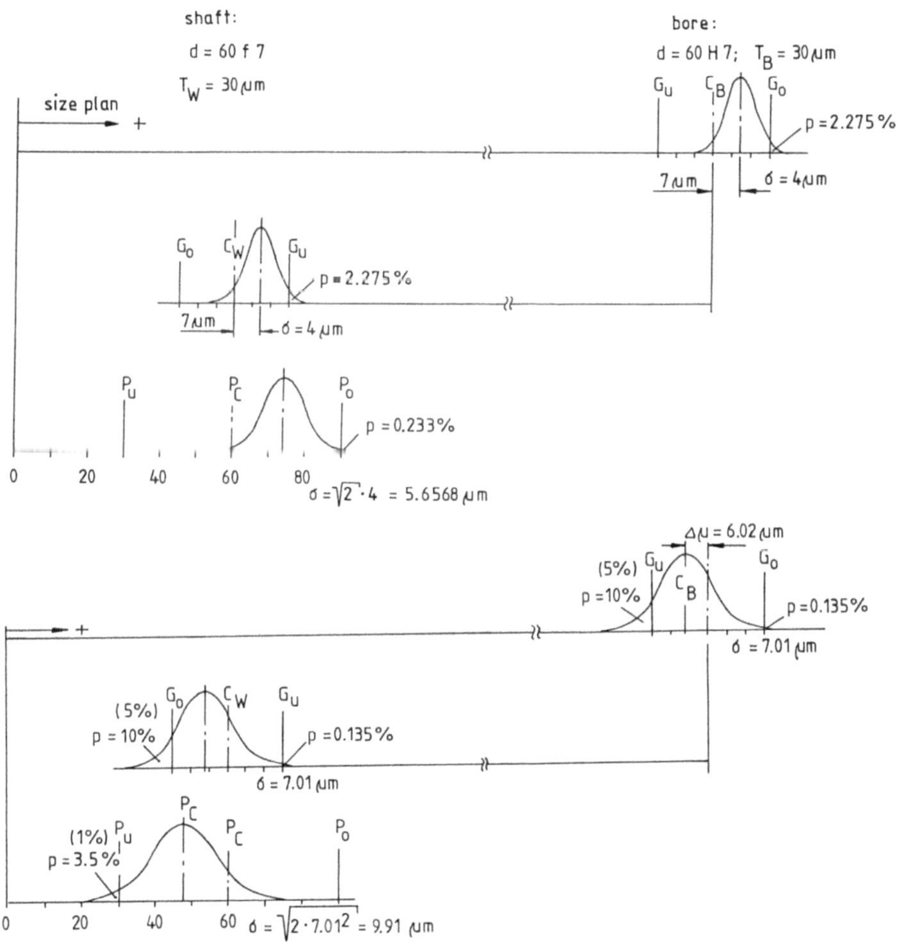

Fig. 8.9. Size plan for the bearing $d = 60H7/f7$ with different assumptions about the position and the spread of the normally distributed batches of bores and shafts

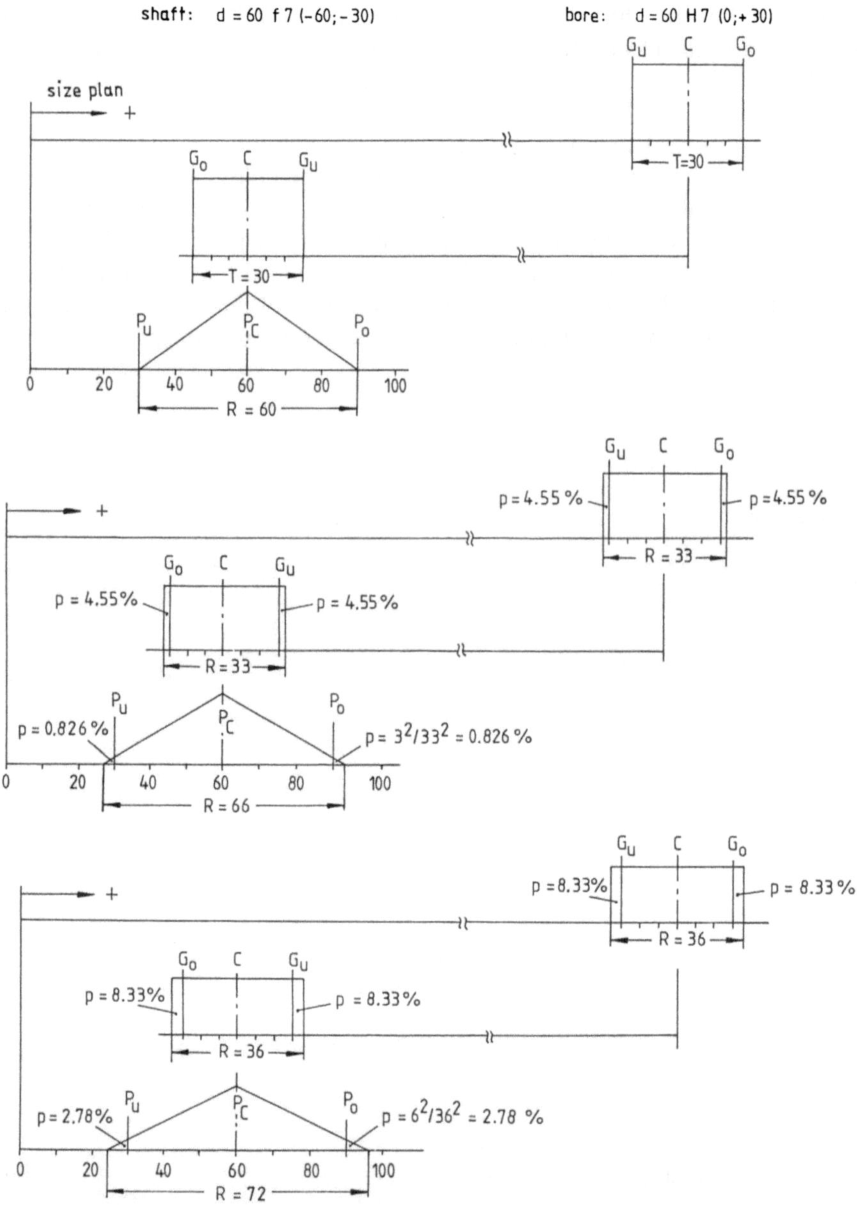

Fig. 8.10. Size plan for the bearing $d = 60H7/f7$ and the assumption that the batches of shafts and bores are similarly distributed (rectangular distributions) with different ranges R

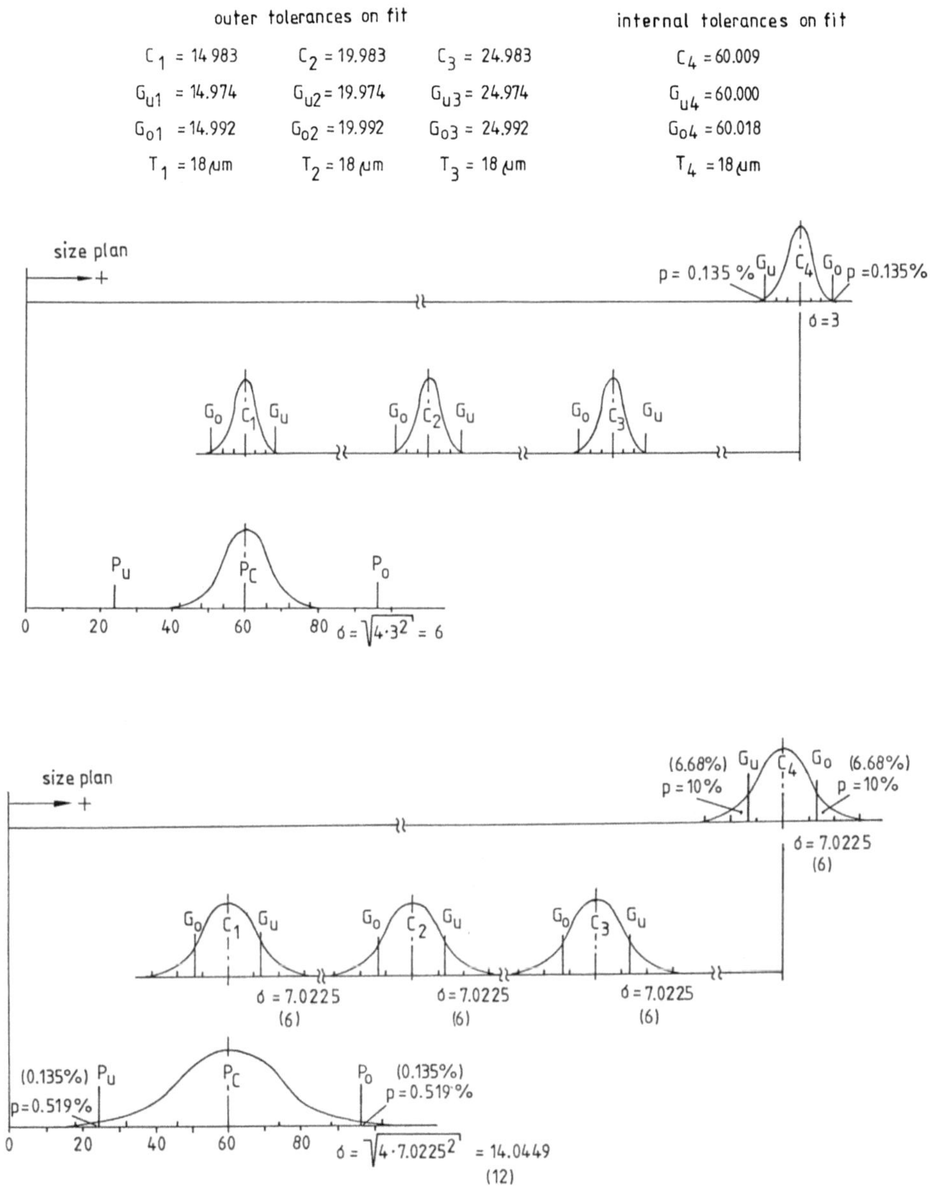

Fig. 8.11. Size plan for a four-member dimensional chain with different assumptions about the spread in the normally distributed individual batches

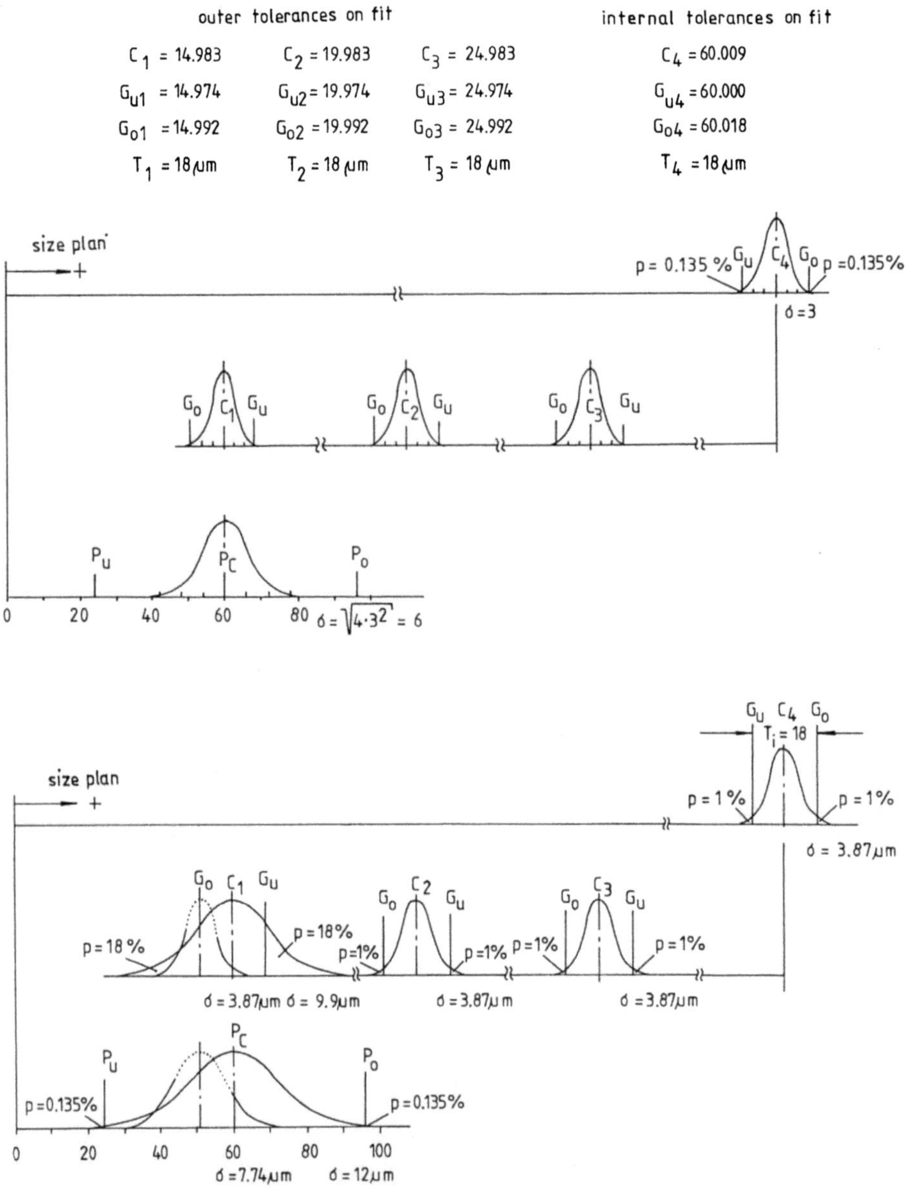

Fig. 8.12. Size plan for a four-member dimensional chain with different assumptions about the conforming or non-conforming spreads in the normally distributed individual batches

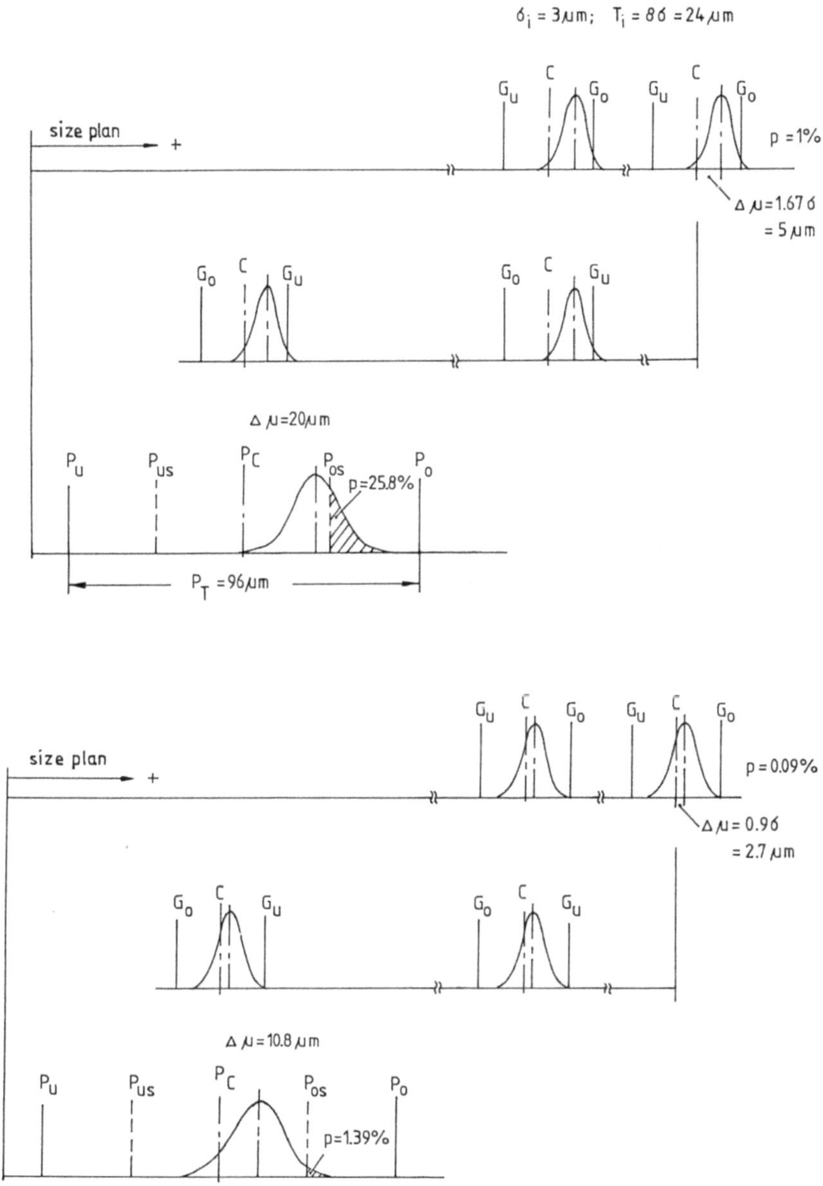

Fig. 8.13. Size plan for a four-member dimensional chain and assumptions that all batch distributions deviate in the same effective direction from the middle of their tolerance zones.

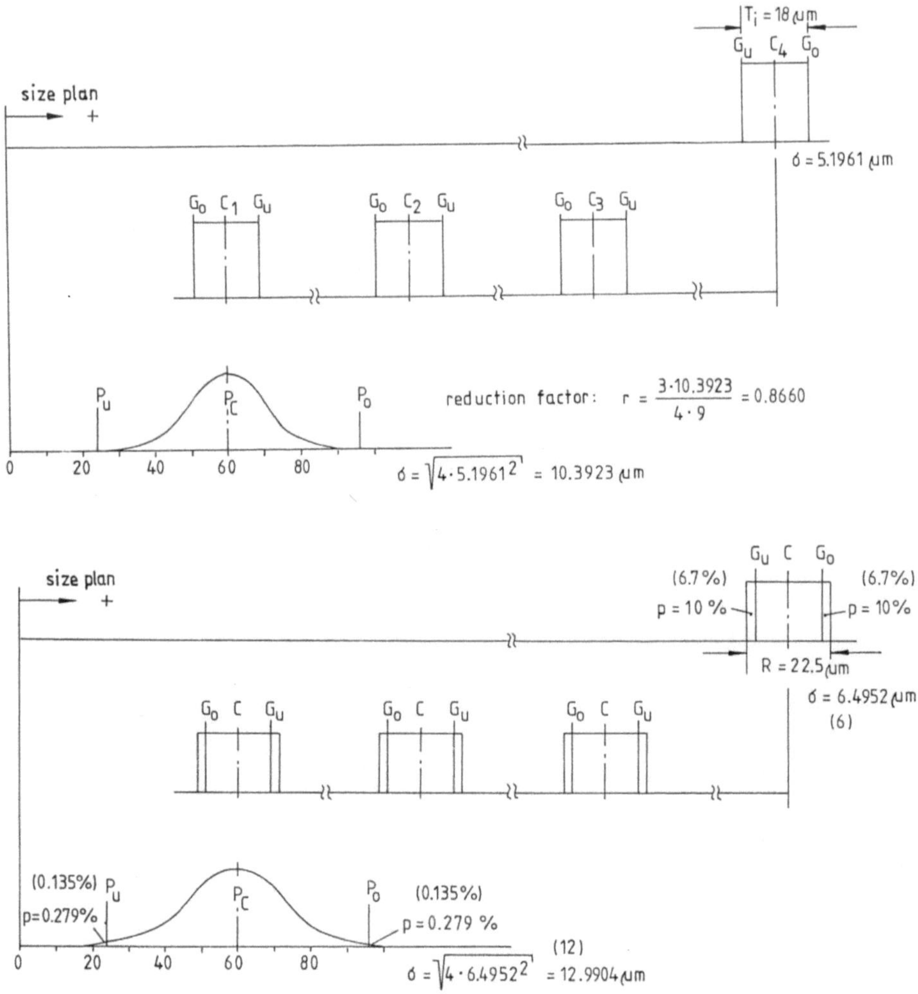

Fig. 8.14. Size plan for a four-member linear dimensional chain with different assumptions about the range of the rectangular distributed individual size batches

With multiple, arithmetically calculated dimensional chains, Figs. 8.11 to 8.14, considerable non-conformity proportions in the components can be accepted, provided that the limit clearances are not nominally exceeded. Two-sided AQL values of 10% endanger the quality standard. Even with $k=4$ members in a dimensional chain it should be considered whether to widen the individual tolerances and set up smaller AQL values.

Figure 8.12 below is especially interesting. With three of the four size members the tolerance is completely filled, so that both sides of the error proportion amount to $p = 1\%$, so that with the fourth member either

—both sides 18% or
—one side 50%

can be accepted, without the limit clearance being exceeded.

There are considerable objections to the acceptance of non-conformity proportions of 36 to 50%.

—If it concerns an assembly which will never be repaired (domestic machine motors, hydraulic systems for vehicle brakes), then these objections are unjustified.
—If exchangeability must be efficient, there is the possibility of
  —sorting both sides to 18% or one side to 50% and scrapping
  —sorting the required 2 to 5% necessary for the repair service from close to the middle size and stocking these in the replacement part store.

In Figs. 8.15 and 8.16 there are cases of six-member arithmetically calculated dimensional chains. The figures speak for themselves and for necessity, either

—the total tolerance is narrowed or
—the individual tolerances are widened or
—both—coordinated together—can be done.

8.7 The Sorting of Non-Conforming Parts

In the previous section it was shown that with fits not endangering functional fulfillment, non-conforming batches (batches with non-conforming items) can be accepted, particularly with multiple dimensional chains.

If non-conforming batches continue to appear, then it must be determined whether these batches are

—given freely to assembly or
—a sorting test must be applied.

In many businesses regular discussions are set up, in business jargon often called

—meetings or
—prayer meetings,

the latter because non-conforming batches should be "faith-healed". Here experience plays a significant role to be able to "balance non-conformities" for assembly.

In order to be sure a test assembly will often be assembled with say the first 20% of the batch to go—with luck—into the assembly. After assembly each set of components will be tested for function:

—"with clearance present" or
—with an interference and a spring element in the dimensional chain:

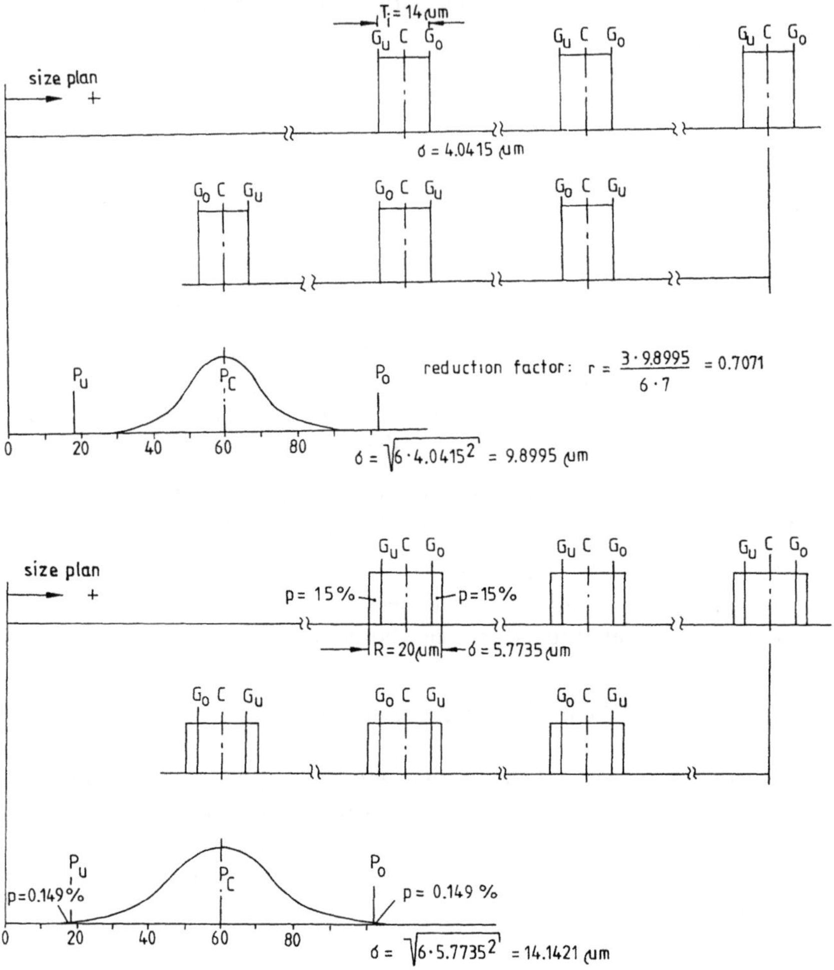

Fig. 8.15. Size plan for a six-member linear dimensional chain with different assumptions about the range of the rectangular distributed individual size batches

"spring under tension" on one side or "sufficient spring travel" on the other side.

Only if the test assembly shows no non-conforming products, will the remaining proportion of the batch be allowed into assembly.

In this way the analysis of non-conforming batches is not a matter of faith or chance but a matter for tangible testing and calculation. The common practice in many concerns is that every non-conforming = waste and thus non-conformities are basically not accepted, this is not an up to date view and is also costly.

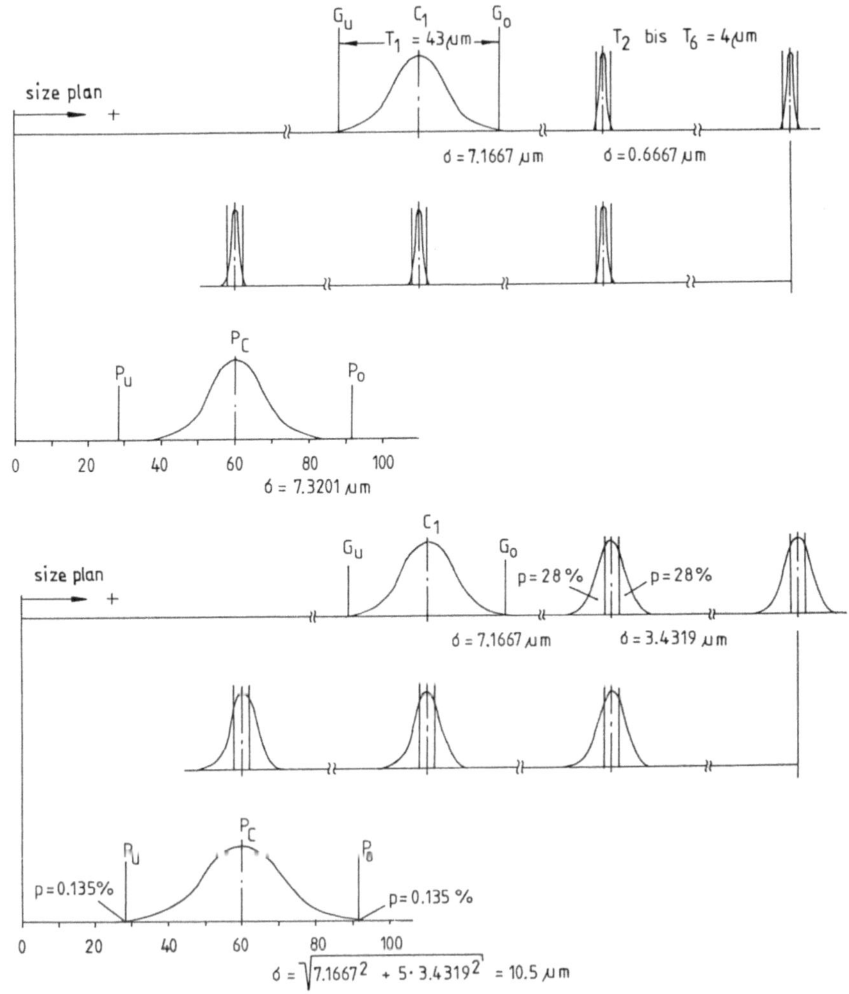

Fig. 8.16. Size plan for a six-member linear dimensional chain with representation of the effects of the fact that one tolerance zone is very large compared to the other tolerance zones

It is recommended that you work through the following examples. Fig. 8.23 can be consulted for abnormal distributions and the calculation of non-conformity proportions within them.

In Fig. 8.17 it is assumed that the drill-jig bush is in order but that the shafts show an non-conformity proportion of approx. 11%. After assembly of the parts however, all actual clearances lie in the clearance tolerance field. If—because of consequent fear of non-conformities—the $p = 10.7\%$ shafts had been sorted out and scrapped or—in the case of shafts lying as a reflected image

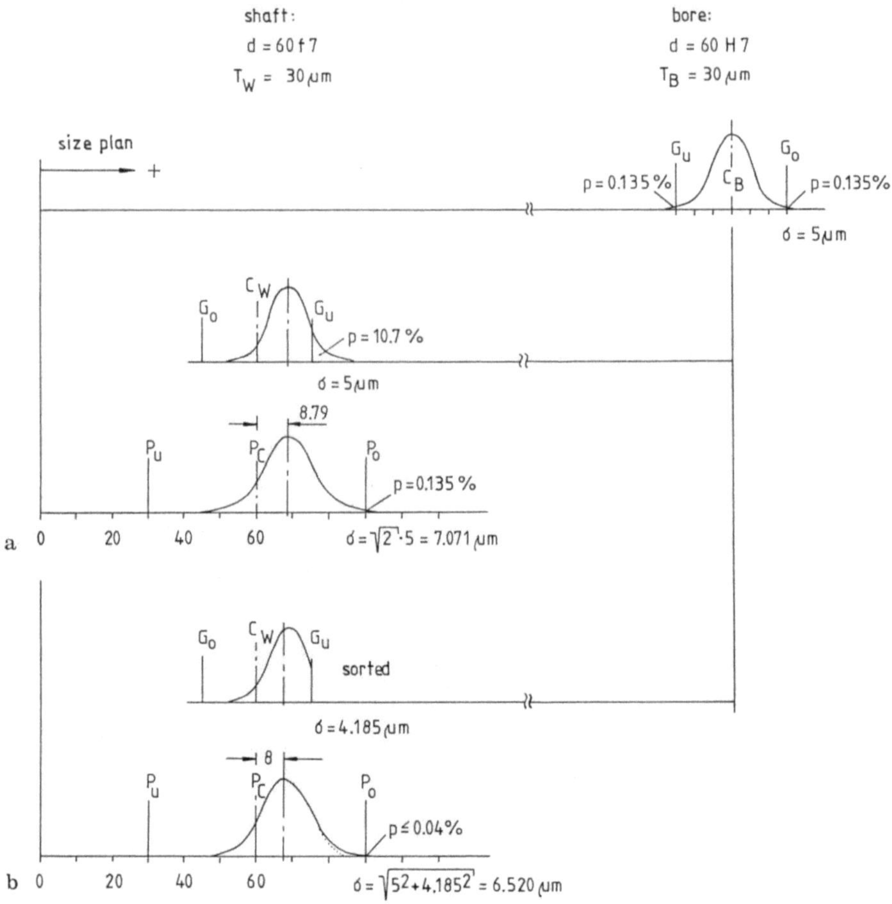

Fig. 8.17 a, b. Size plan for a two-member dimensional chain. The assembly follows **a** without sorting the non-conforming shafts; **b** after sorting the non-conforming shafts

on the repair side—then virtually nothing would be changed in the distribution of the fits (clearances), Fig. 8.17 below.

Figure 8.18 is a similar example with 16% non-conformity proportion in the bores with rather different assumptions.

With small and medium size runs, occasionally, especially with expensive parts, even with large runs the repair side is worked. In Fig. 8.19 above there are rectangular distributions with $R + T/2$ on the repair side about $T/4$ distant from the middle value; all clearances are in the clearance tolerance zone. In Fig. 8.19 below an increase of the range to $R = 20$ μm is assumed with otherwise similar conditions.

The non-conformity proportion in the components is 12.5%; in the assemblies however it is only 3.13%.

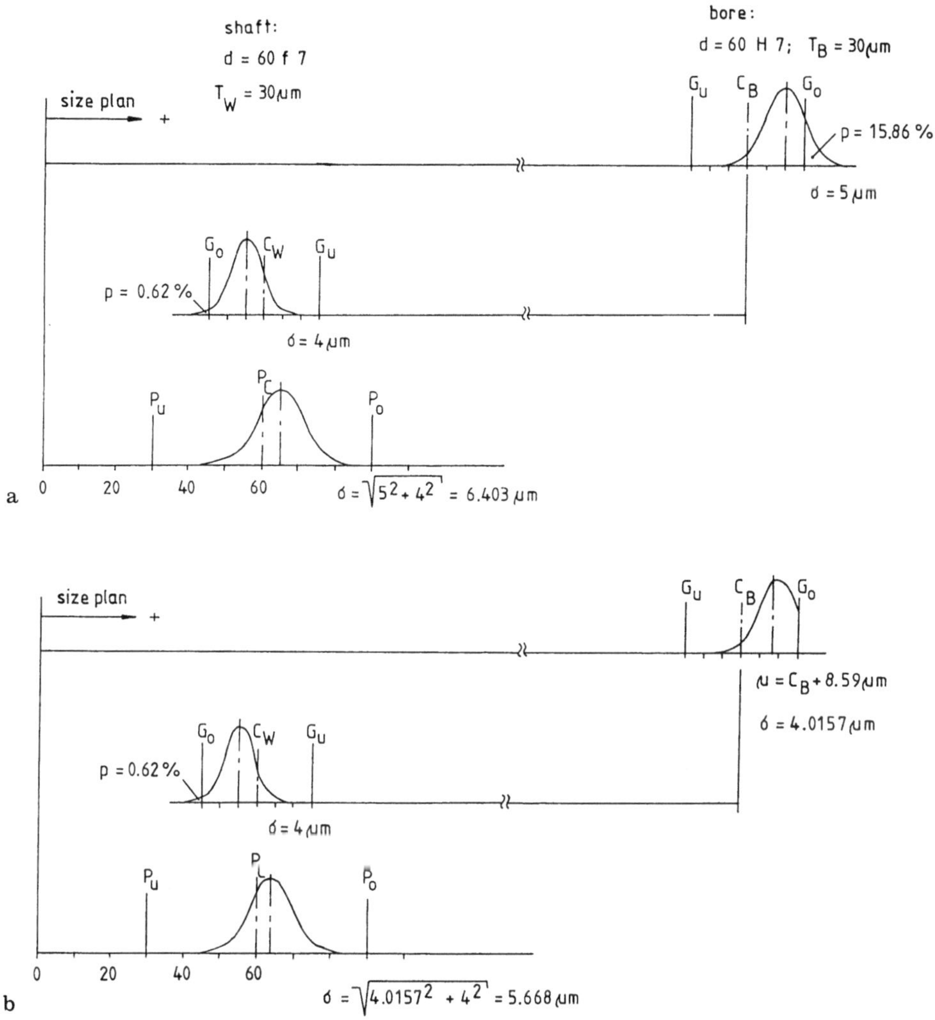

Fig. 8.18 a, b. Size plan for a bearing. Assembly follows **a** without sorting out the non-conforming bores and **b** after sorting out the non-conforming bores

In Fig. 8.20 the interval to the centre is larger compared with Fig. 8.19, so that the component batch includes 27.5% of non-conformities. Every practical person (and also theorist) would consider this a catastrophe and would find it essential to sort out the batch and repair the non-conforming components.

The calculation suggests that 15.13% of the clearances lie below the lowest clearance P_u, if the batch remains unsorted and is assembled. The smallest clearance (smallest actually presented) would then be about 20 µm.

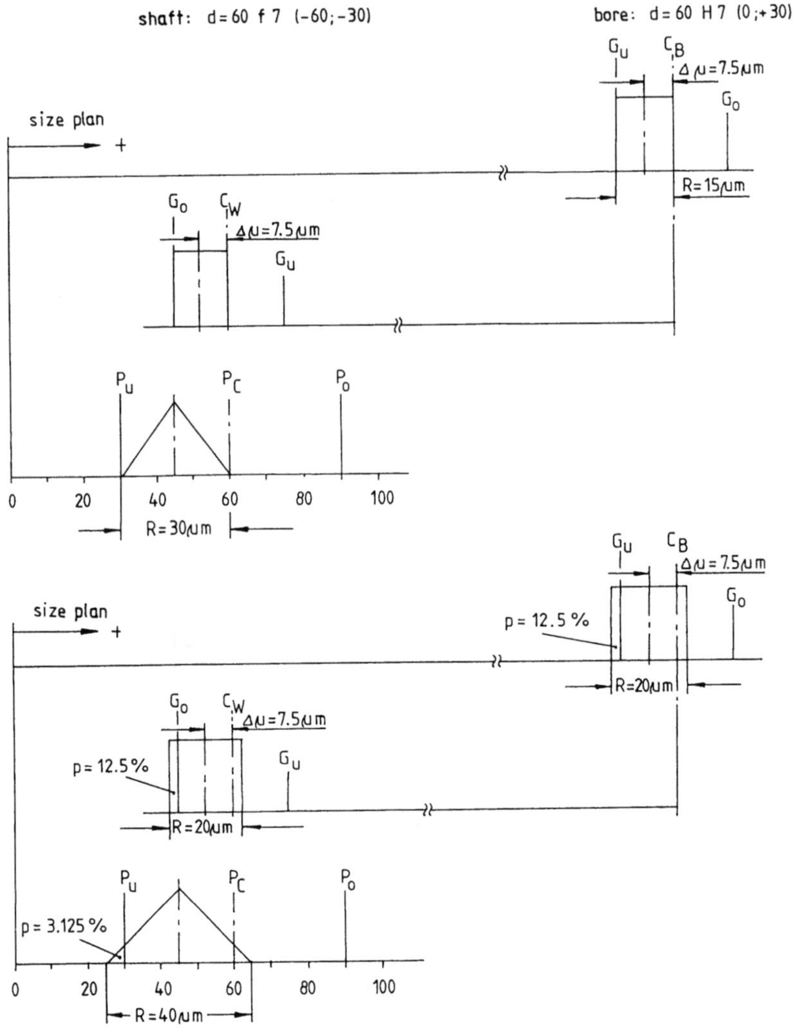

Fig. 8.19. Size plan for a bearing with different assumptions about the position and the range of the rectangular distributed batches of bores and shafts.

After inspection, the norm is established that the design intended $d = 60H7/f7$ (running fit), in fact production has produced "better quality" and $d = 60H7/g6$ (narrower running fit) is produced.

It is conceivable that the design department would consider the narrower running fit better; only from cost grounds it was not specified. Sorting the non-conforming items would be completely senseless.

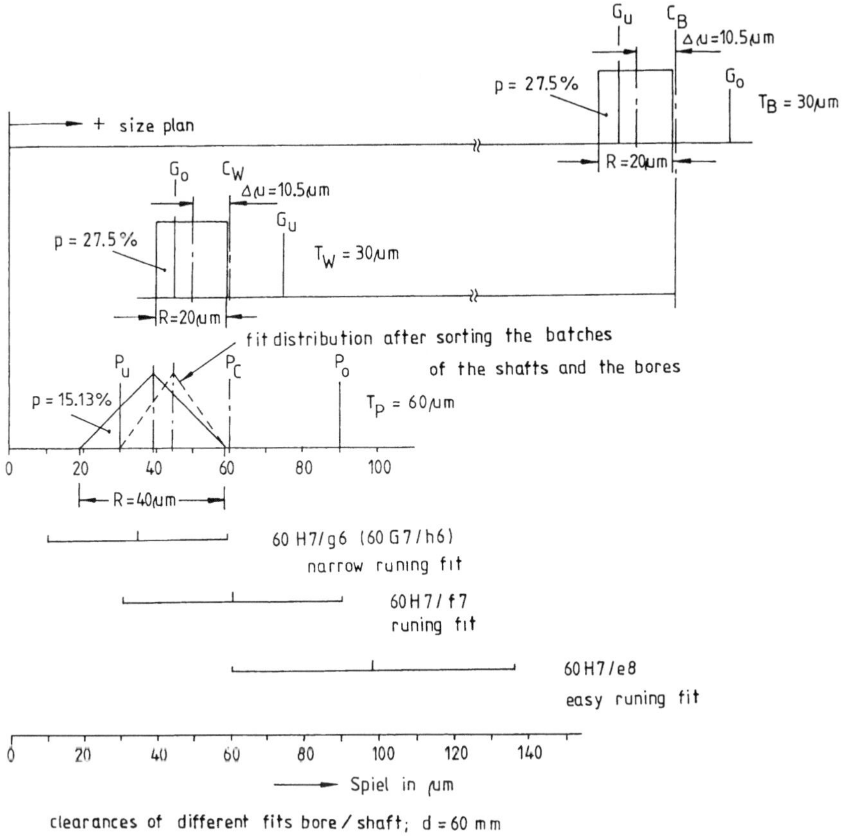

Fig. 8.20. Size plan for a bearing with representation of the difference between unsorted and sorted assemblies

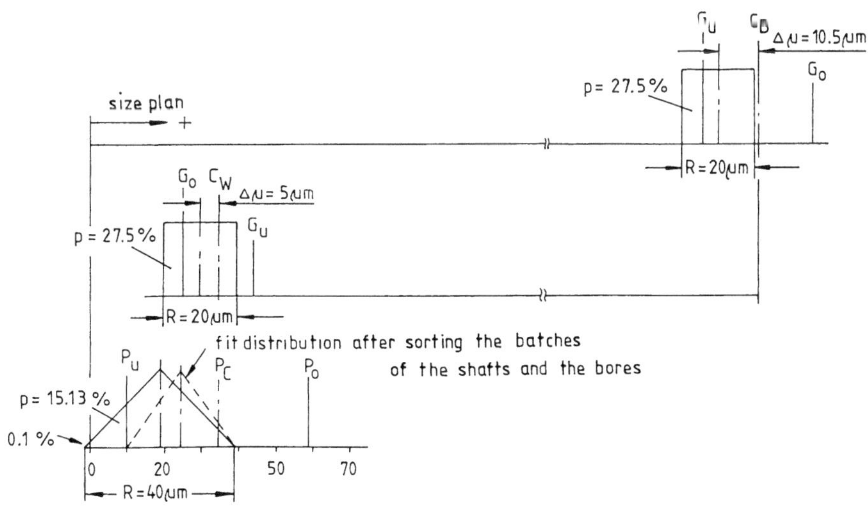

Fig. 8.21. Size plan for a bearing with representation of the tolerance on fit distributions after assembly of unsorted and sorted batches

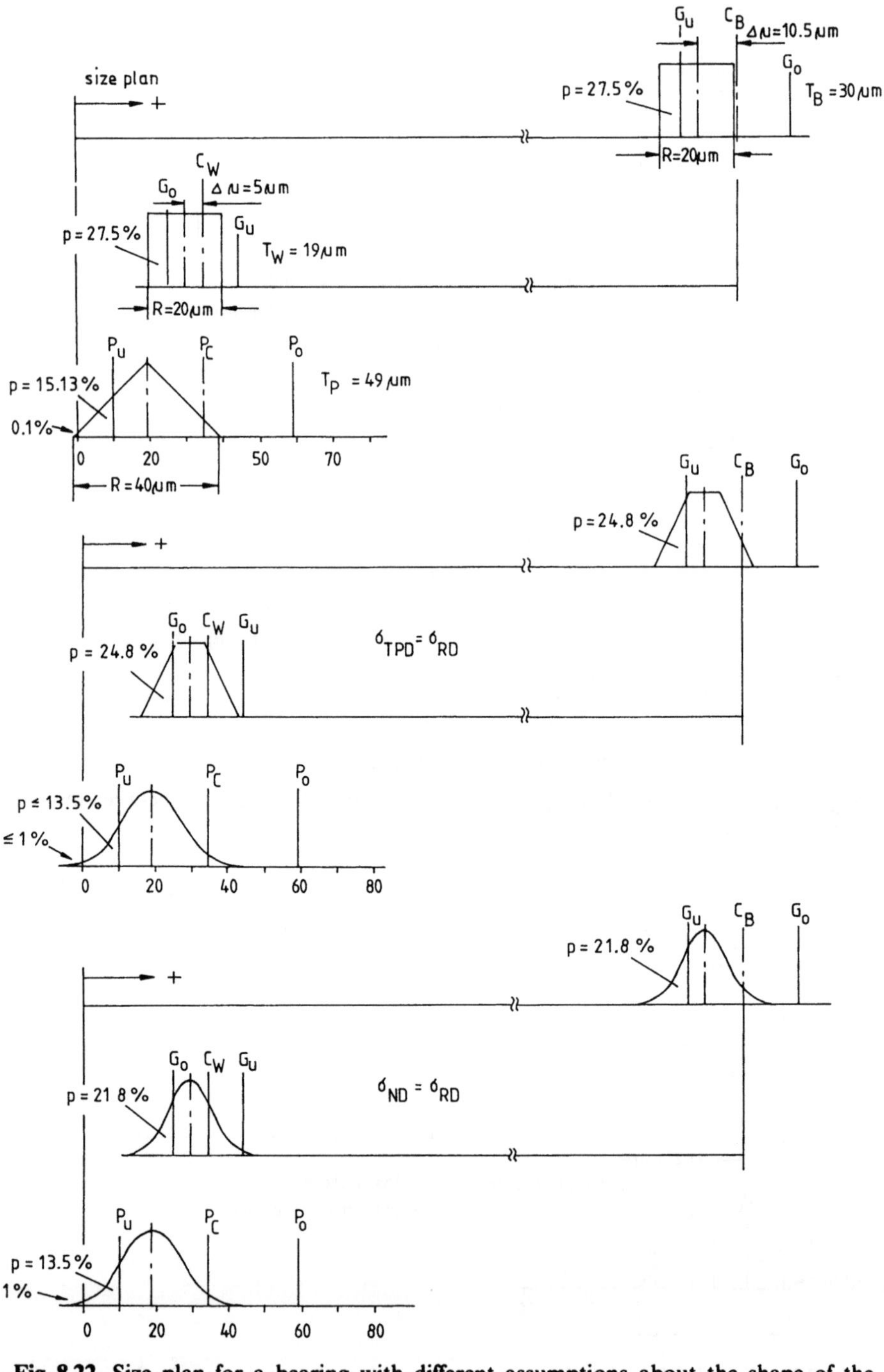

Fig. 8.22. Size plan for a bearing with different assumptions about the shape of the individual size distributions

1 distribution	2 range proportion R/R_{RD}	 R/R_{ND}	3 high h	4 proportion of non-conformities p		5 example for p with u = 1.5
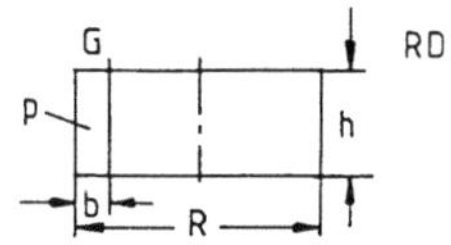 RD	1	0.5774	$h = \frac{1}{R}$	$p = \frac{b}{R}$	for $b \leq R$	$p = 6.7000\ \%$
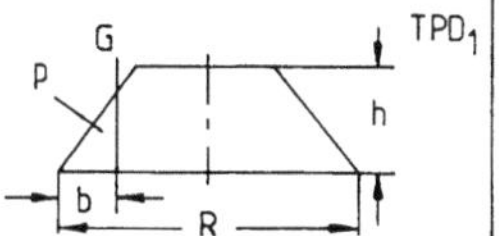 TPD_1	1.2649	0.7303	$h = \frac{4}{3 \cdot R}$	$p = \frac{8 \cdot b^2}{3 \cdot R^2}$	$b \leq \frac{R}{4}$	$p = 6.6300\ \%$
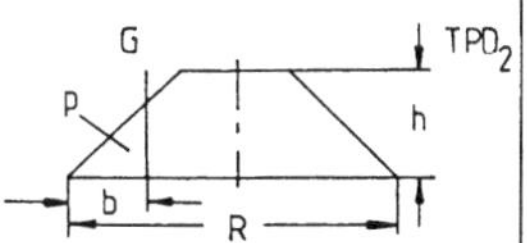 TPD_2	1.3416	0.7746	$h = \frac{3}{2 \cdot R}$	$p = \frac{9 \cdot b^2}{4 \cdot R^2}$	$b \leq \frac{R}{3}$	$p = 7.0691\ \%$
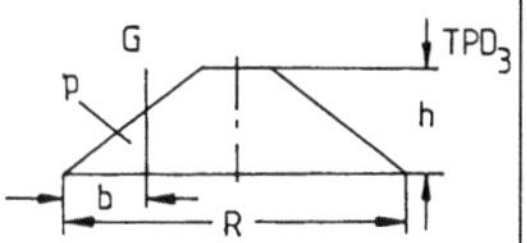 TPD_3	1.3868	0.8006	$h = \frac{5}{3 \cdot R}$	$p = \frac{25 \cdot b^2}{12 \cdot R^2}$	$b \leq \frac{2 \cdot R}{5}$	$p = 7.3435\ \%$
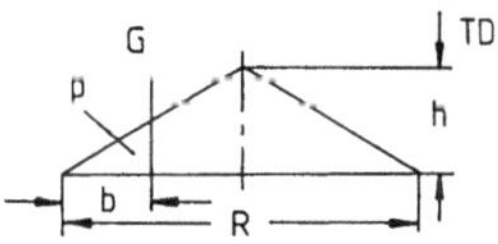 TD	1.4142	0.8165	$h = \frac{2}{R}$	$p = \frac{2 \cdot b^2}{R^2}$	$b \leq \frac{R}{2}$	$p = 7.5120\ \%$
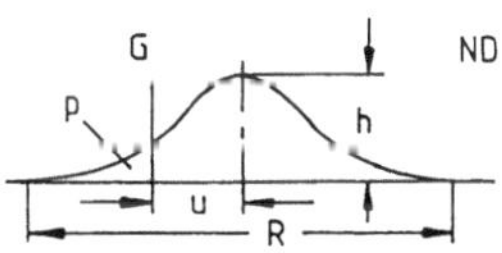 ND	1.7321	1 $(R_{ND} = \sqrt{6})$	$h = \frac{6}{R\sqrt{2\pi}}$	u – table		$p = 6.681\ \%$

Fig. 8.23. Comparison of different probability distributions from Fig. 4.9 and information about the error proportion p left of the limit value LV, which is about u distant from the middle value. The standard deviation is similar for all distributions with $\sigma = 1$; in all cases $b = \frac{R}{2} - u$

From these examples it should not be understood that it is always superfluous or inappropriate to sort out non-conforming batches. "Basically don't sort" would be as incorrect as "basically sort".

In Fig. 8.21 it is assumed that the design department have specified $d = 60H7/g6$ and as in Fig. 8.20 it is produced with an error proportion of 27.5% in the component batches on the repair side.

Now an non-conformity proportion of $p = 15.13\%$ lies under the lowest clearance $P_u = 10\,\mu\text{m}$. This is unacceptable. Shafts get warmer than the (cooled) bearings. Thus with a narrower running fit the lowest clearance can in no way be exceeded. They must be sorted and repaired.

The summary shown in Fig. 8.21 is as far as the non-conformity proportion is concerned independent in order of magnitude of the distribution shape; this comes from Fig. 8.22. The upper part is identical with Fig. 8.21. In the middle part it is assumed that the component batch shows a trapezoid distribution, in the lower part the component batch is normally distributed with otherwise similar conditions.

In order to calculate all examples the formulae for the calculation of the non-conformity proportion is in Fig. 8.23, column 4 under a limit value an interval of u distant from the mean of the distribution.

Figure 8.24 is also calculated with these formulae. The distributions are different in their width; so the differences between small non-conformity proportions are considerable. If an non-conformity proportion of $p = 7\%$ is given with a normal distribution, then it will be roughly the same for other distributions.

Exercise 8.2

Given: Slide bearing position with bore $d = 60H7$ (0; 30) and shaft $d = 60f7$ (-60; -30). The tolerances of internal dimensions on fit are always more difficult to produce than the tolerances of outer dimensions on fit (shafts).

Required: Sub-conformity proportion which can be accepted as maximal in the bearing batch, with the assumption that the bearings and the shafts are centred on the tolerance zone centre and the shafts fill their tolerance zone with their 6σ range in the extreme case.
It is also accepted that the clearances of the fit limits are allowed to exceed by 1%.

Solution: According to the outline the standard deviation of the clearance is

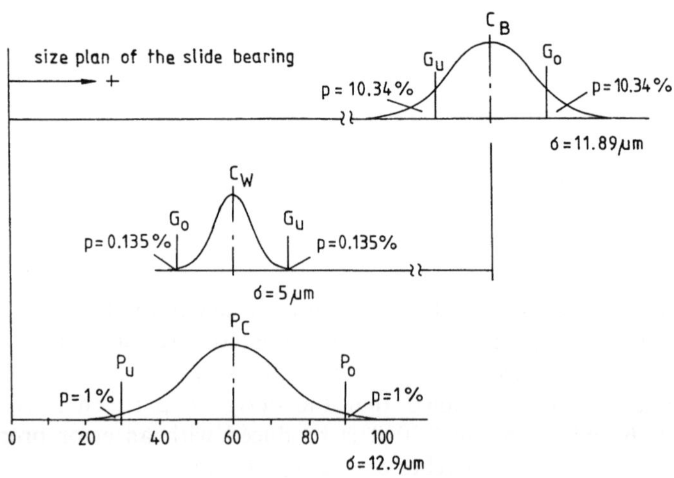

$$\sigma = \frac{T/2}{u_{99}} = \frac{30}{2.3263} = 12.9\ \mu\text{m}.$$

The shafts have a maximal standard deviation of

$$\sigma = \frac{T/2}{3} = \frac{15}{3} = 5\,\mu\text{m}.$$

Then the bores are allowed a standard deviation of

$$\sigma_B^2 = 12.9^2 - 5^2,$$

$$\sigma_B = 11.89\,\mu\text{m}.$$

Then the interval to the limit values is

$$u = \frac{G_0 - G_B}{\sigma_B} = \frac{15}{11.89} = 1.262$$

and thus $p = 10.34\%$ is two-sided.

Exercise 8.3

Given: Model of a normal distribution with the parameters $\mu = 7.5$ and $\sigma = 2$ as in Fig. 4.7.

Required: Change of parameters if about 10% of the batch is sorted out.

1. from the middle of the distribution,
2. from the end of the distribution.

Solution:

$ND_{unsorted}$		ND_{sorted}		
		1) from the middle	2) from the end	
x_j	N_j	N_j	N_j	
1	1	1	0	
2	5	5	0	
3	15	15	0	
4	40	40	0	
5	103	103	64	
6	150	150	150	
7	190	140	190	
8	190	140	190	
9	150	150	150	
10	103	103	103	
11	40	40	40	
12	15	15	15	
13	5	5	5	
14	1	1	1	
N	1008	908	908	calculated with
μ	7.5	7.5	7.8733	$\bar{x}/s$ automatic of
σ_N	2	2.1007	1.7162	a calculator
$\frac{\sigma_{sort.}}{\sigma_{unsort.}}$		1.0504	0.8581	
$\frac{\Delta\mu}{\sigma_{unsort.}}$			$\frac{0.3733}{2}$	= 0.1867

Note. The solution of this exercise needed for the solution of Exercises 8.4 to 8.6.

Exercise 8.4

Given: Bearing with bore $d_B = 60H7 = 60^{+30}_{-30}$ ($T_B = 30\,\mu m$) and shaft $d_W = 60f7 = 60^{-30}_{-60}$($T_W = 30\,\mu m$). The middle of the clearance is $P_C = 60.015 - 59.955 = 0.060$ mm. It is assumed that the batch with the bores in the middle of the tolerance zone would be produced with a standard deviation of $\sigma_B = 5\,\mu m$; the normally distributed bores fill the tolerance zone with their 6σ range. The produced, normally distributed shafts also have a standard deviation of $\sigma_W = 5\,\mu m$; the distribution is however around $\Delta\mu = 2\sigma_W$ shifted in the direction of the lowest value, so that the proportion of non-conforming shafts which are too small becomes $p = 15.866\%$, Fig. 8.25.

Required:
1. non-conformity proportion in the tolerance on fit where both batches are assembled unsorted.
2. non-conformity proportion in the tolerance on fit (clearances) where 10% of the shafts are sorted out of the middle of the normal distribution for the replacement part store before assembly.
3. non-conformity proportion in the clearances where 10% of the shafts are sorted from the middle of the tolerance zone (from the end of the ND) before assembly.

Solution:
1. After assembly the standard deviation of the clearance $\sigma_S = \sqrt{2} \cdot 5 = 7.0711\,\mu m$ and the middle clearance lies at $70\,\mu m$. Thus the interval to

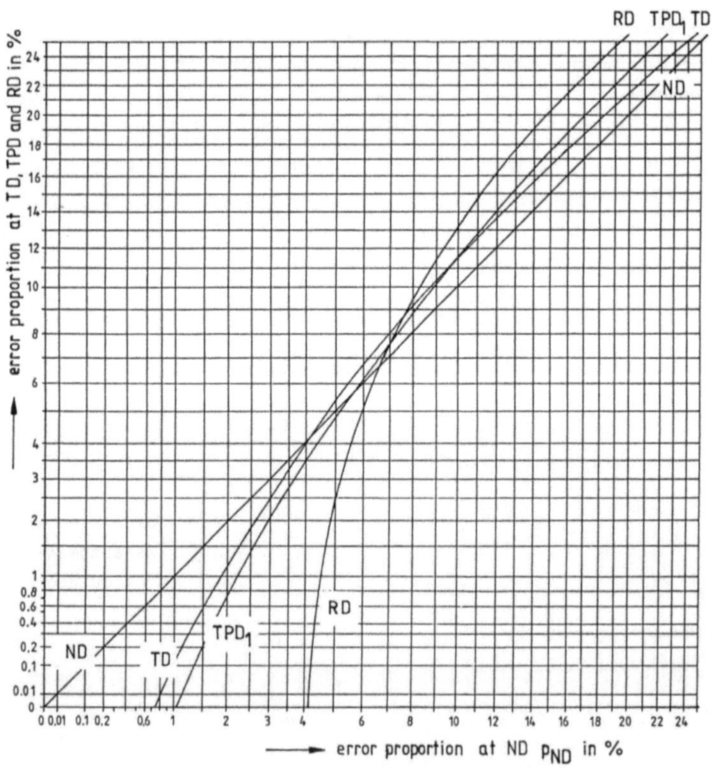

Fig. 8.24. Error proportion with different distributions over the normal distribution; all distributions have the same mean value and the same standard deviation

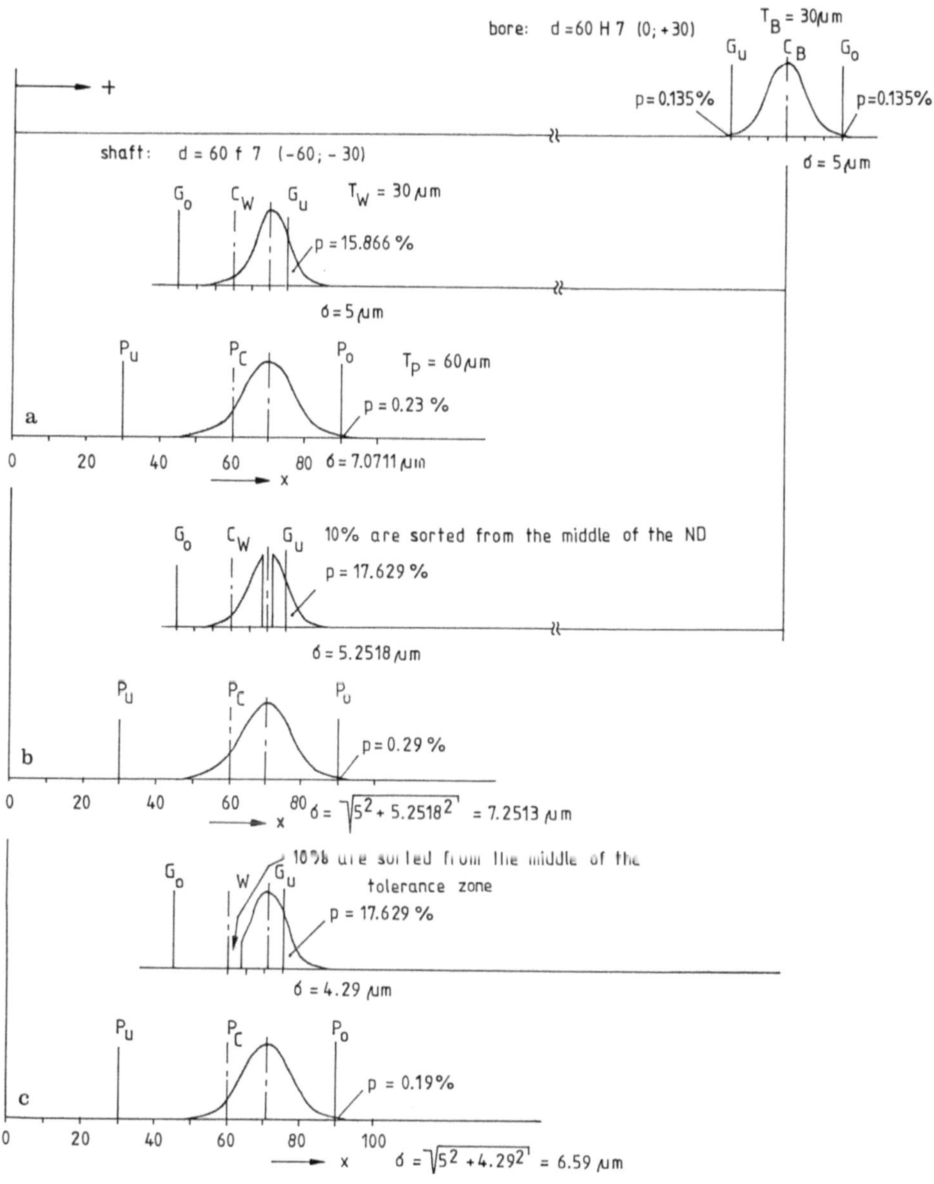

Fig. 8.25 a–c. Two-member dimensional chain (shaft/bore) with a produced error proportion of $p = 15.866\%$ in the shafts **a** assembled unsorted, **b** about 10% sorted out from the middle of the distribution, and **c** about 10% sorted out from the middle of the tolerance zone

the upper limit value is in σ units

$$u = \frac{90 - 70}{7.071\,1} = 2.828$$

and the proportion with greater clearance is $p = 0.23\%$.
Sorting and scrapping the shafts which are too small results in the mean value of the shafts becoming slightly larger and the standard deviation somewhat smaller, so that the non-conformity proportion in the clearances would sink to approximately nil.

2. If sorting and scrapping of approx. 16% of the shafts which were too small was avoided and instead—to protect exchangeability—a proportion of about 10% is sorted out from the centre of the distribution and taken to the replacement part store, then the standard deviation will increase, given an unchanged mean value.
If, as in exercise 8.3, 50 values are taken out from both middle classes of the ND model, the standard deviation is raised by a factor of $\sigma_{sorted}/\sigma_{unsorted} = 1.050\,4$; by sorting 10% of the shafts from the middle of the distribution the standard deviation $\sigma = 5 \cdot 1.0504 = 5.251\,8\,\mu m$ and the standard deviation in the clearances is raised to $\sigma = \sqrt{5^2 + 5.251\,8^2} = 7.251\,3\,\mu m$. Thus the non-conformity proportion in the clearances becomes $p = 0.29\%$ with $u = (30 - 10)/7.251\,3 = 2.76$. By sorting the non-conformity proportion of the shafts is increased to $p = 17.629\%$, Fig. 8.25b.
3. If 10% of the shafts from the upper end of the ND are sorted, then the interval of the mean value from the tolerance zone centre increases to $10 + 5 \cdot 0.186\,7 = 10.93$ and the standard deviation decreases to $\sigma = 5 \cdot 0.858\,1 = 4.29\,\mu m$. Thus the spread in the clearances becomes $\sigma = \sqrt{5^2 + 4.29^2} = 6.59\,\mu m$. Then the interval to the upper limit value is $u = (90 - 70.93)/6.59 = 2.89$; the non-conformity proportion to larger clearances is then $p = 0.19\%$, Fig. 8.25c.

Exercise 8.5

Given: Bearing with bore $d_B = 40H7 = 40^{+25}_{-0}$ ($T_B - 25\,\mu m$) and shaft $d_W = 40f7 = 40^{-25}_{-50}$ ($T_W = 25\,\mu m$). The middle of the tolerance on fit as $P_C = 40.012\,5 - 39.962\,5 = 0.05$ mm.
It is assumed that the batch with the bores is produced in the middle of the tolerance zone with a standard deviation of $\sigma_B = 3\,\mu m$.
The shafts have a standard deviation of $\sigma_W = 4\,\mu m$; if working on the repair side the normal distribution is shifted to $\Delta\mu = 2.5 \cdot \sigma = 10\,\mu m$ in the direction of the highest value, so that the proportion of non-conforming shafts which are too large becomes $p = 26.6\%$.

Required:
1. non-conformity proportion in the clearances where both batches are assembled unsorted.
2. non-conformity proportion in the clearances where 10% of the shafts are sorted out from the centre of the normal distribution for the replacement part store before assembly.
3. non-conformity proportion in the clearances where 10% of the shafts are shorted out from the end of the ND near the middle of the tolerance zone before assembly.

Solution:
1. After assembly the standard deviation of the clearances is $\sigma = \sqrt{3^2 + 4^2} = 5\,\mu m$. The middle clearance is $\mu = 40\,\mu m$. The interval to the lower limit value in σ units is $u = (25 - 40)/5 = -3$; thus the proportion of clearances which are too small is $p = 0.135\%$, Fig. 8.26a. Sorting and repairing of 26.6% of the larger shafts is not necessary.

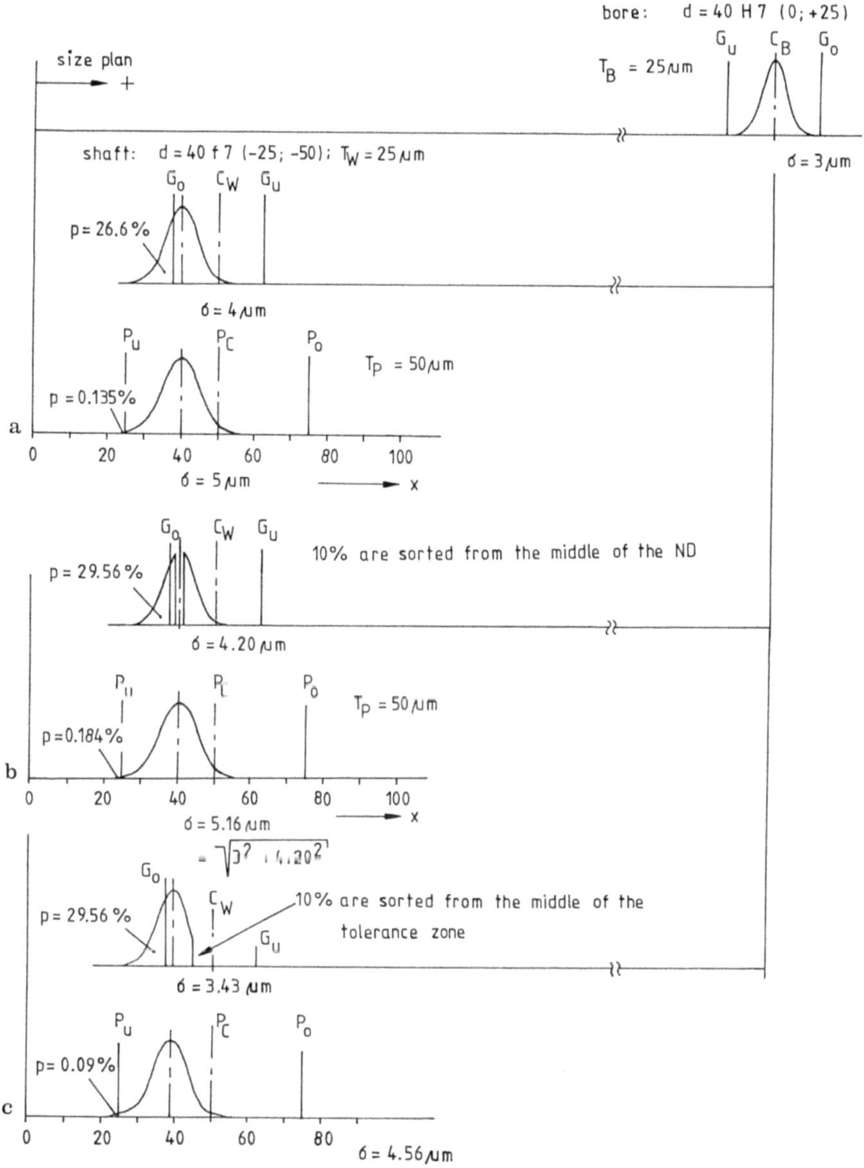

Fig. 8.26 a–c. Two-member dimensional chain (shaft/bore) with a produced error proportion of 26.6% in shafts **a** assembled unsorted, **b** about 10% sorted out from the middle of the distribution; **c** about 10% sorted out from the middle of the tolerance zone

2. In order to decide whether the non-conforming shafts should be used as replacement parts, 10% of the shafts are sorted from the middle of the normal distribution, where—with mean value remaining the same—the standard deviation is raised to $\sigma = 4 \cdot 1.0504 = 4.20\,\mu\text{m}$. Thus the standard deviation in the clearances becomes $\sigma = \sqrt{3^2 + 4.20^2} = 5.1627$; the

Folding of two sorted normal distributions;

from one side a proportion of 100/1008 = 0.0992 = 9.92 % is sorted.

N = 908; μ = 7.8733; σ = 1.7162

x	g(x)	5	6	7	8	9	10	11	12	13	14
		64	150	190	190	150	103	40	15	5	1 · 1/908
5	64	10 4096									
6	150	11 9600	12 22500								
7	190	12 12160	13 28500	14 36100							
8	190	13 12160	14 28500	15 36100	16 36100		continuation symmetrical				
9	150	14 9600	15 22500	16 28500	17 28500	18 22500					
10	103	15 6592	16 15450	17 19570	18 19570	19 15450	20 10609				
11	40	16 2560	17 6000	18 7600	19 7600	20 6000	21 4120	22 1600			
12	15	17 960	18 2250	19 2850	20 2850	21 2250	22 1545	23 600	24 225		
13	5	18 320	19 750	20 950	21 950	22 750	23 515	24 200	25 75	26 25	
14	1 ·1/908	19 64	20 150	21 190	22 190	23 150	24 103	25 40	26 15	27 5	28 1

fold product:

x_j	N_j	$\sum N_j$	G_j in %
10	4096	4096	0.4968
11	19200	23296	2.8256
12	46820	70116	9.5044
13	81320	151436	18.3678
14	112300	263736	31.9888
15	130384	394120	47.8032
16	129120	523240	63.4643
17	110060	633300	76.8135
18	81980	715280	86.7570
19	53428	768708	93.2373
20	30509	799217	96.9378
21	15020	814237	98.7595
22	6570	820807	99.5564
23	2530	823337	99.8633
24	831	824168	99.9641
25	· 230	824398	99.9920
26	55	824453	99.9987
27	10	824463	99.9999
28	1	824464	100.0000

Parameters of the fold product:

$N = 824464 = 908^2$

$\mu = 15.7467 = 2 \cdot 7.8733$

$\sigma = 2.4271 = \sqrt{2} \cdot 1.7162$

Fig. 8.27. Fold process and fold product of two normal distributions, from whose ends about 10% are sorted from one side

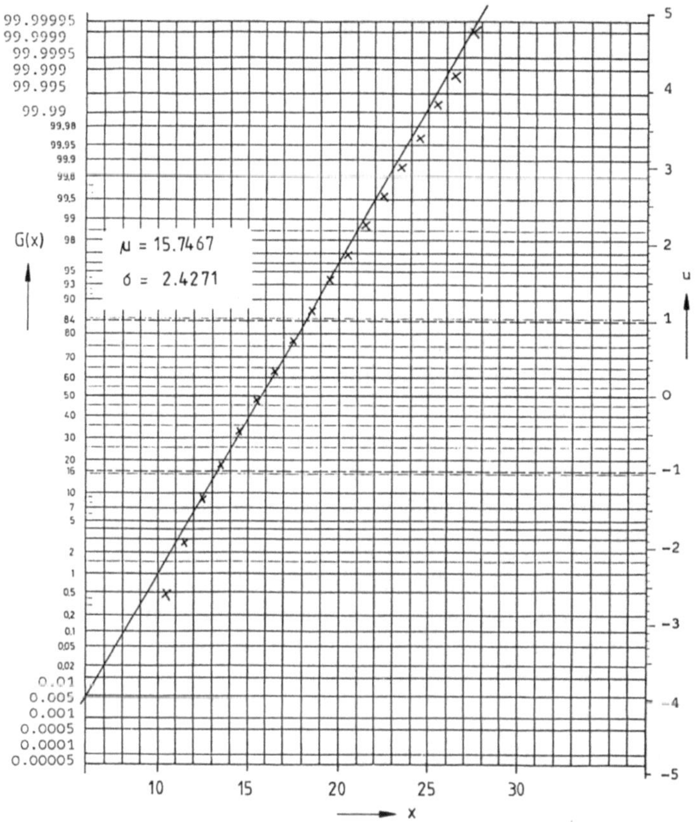

Fig. 8.27. (Continued)

interval to the lower limit value in σ units is then $u = -15/5.1626 = -2.91$ which includes an non-conformity proportion of $p = 0.184\%$, Fig. 8.26b.

3. Thus the mean value becomes larger and the standard deviation smaller. The interval from the mean value to the middle of the tolerance zone increases to $10 + 4 \cdot 0.1867 = 10.75\,\mu m$; thus the middle clearance lies at $\mu = 39.25\,\mu m$. The spread of the clearances narrows to $\sigma = \sqrt{3^2 + (4 \cdot 0.8581)^2} = 4.5587\,\mu m$. With $u = (25 - 39.25)/4.5587 = -3.1256$ the error proportion is $p = 0.09\%$, Fig. 8.26c.

Note to Exercise 8.4 and 8.5: With the decision to sort the replacement part requirements from the middle of the distribution or from its end, it should be considered that by sorting from one end (almost) the whole batch must be measured; the parts from the middle are much quicker to find.

Exercise 8.6

Given: Normal distributions in which at one side a proportion of 9.92% is sorted out; $N = 908$; $\mu = 7.8733$; $\sigma = 1.7162$ (see Exercise 8.3).

Required: Fold product of two of this kind of distributions.

Solution: Figure 8.27, the representation of the distribution function in the PG shows that the fold product is very close to a normal distribution.

9 Tolerance Calculation Using Square Law

If a three-member or multiple dimensional chain is calculated arithmetically and because of the required tolerance for the overall value, the individual tolerance is so narrow that it cannot be produced, then a widening of the individual tolerances by statistical tolerance calculation can be undertaken.

As long as the production processes with which the members of the dimensional chain are produced are sufficiently well known and forecasts are based on the facts that

—a normal distribution or a possible distribution will be shown,
—with a mean value which lies near the respective mean value, and
—a total standard deviation which is so small that the 6σ range is not greater than the respective tolerance zone produced from quadratic tolerance calculation,

then quadratic tolerance calculation may be used.

The quadratic closing value tolerance is

$$T_q = \sqrt{\sum_{i=1}^{i=k} T_{si}^2}. \tag{9.1}$$

If the statistical individual tolerances are all the same size, it is

$$T_q = \sqrt{k}T_{si}. \tag{9.2}$$

or

$$T_{si} = \frac{T_q}{\sqrt{k}}. \tag{9.3}$$

By arithmetical calculation it was

$$T_a = kT_i \tag{9.4}$$

or

$$T_i = \frac{T_a}{k}. \tag{9.5}$$

If the arithmetically calculated overall tolerance T_a is retained for the quadratic calculation then the proportion of the new individual tolerances compared with those before is

$$\frac{T_{si}}{T_i} = \sqrt{k}. \tag{9.6}$$

Formula (9.6) is shown graphically in Fig. 9.1. If the individual tolerances are not the same size, the tolerance extensions to the individual sizes are smaller. The explanation is shown in Fig. 9.2 with numerical examples. If all tolerances are the same size, line 1, then they can be doubled, where there are $k = 4$ members of a dimensional chain, line 2; T_q is then as large as T_a according to the arithmetical calculation.

If the tolerances are not as large as each other but come to the same total, line 3 and line 6, then after doubling, the individual tolerances T_q are greater than T_m, lines 4 and 7. Only when the large tolerances are less than doubled will $T_q = T_a$ again, lines 5 and 8 in Fig. 9.2.

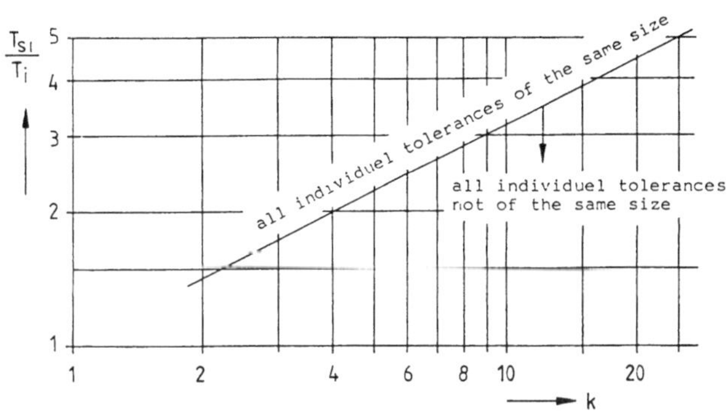

Fig. 9.1 Behaviour of the individual tolerances T_{si}/T_i depending on the number k of members of the dimensional chain, if all individual tolerances are the same size

line	arith-metical	quadra-tical	T_1	T_2	T_3	T_4	$T_a = \sum T_i$	$T_q = \sqrt{\sum T_{si}^2}$
1	x		10	10	10	10	40	
2		x	20	20	20	20		$\sqrt{1600} = 40$
3	x		5	5	15	15	40	
4		x	10	10	30	30		$\sqrt{2000} = 44.72$
5		x	10	10	26.45	26.45		$\sqrt{1600} = 40$
6	x		3	3	17	17	40	
7		x	6	6	34	34		$\sqrt{2384} = 48.83$
8		x	6	6	27.64	27.64		$\sqrt{1600} = 40$

Fig. 9.2. Numerical explanation that with quadratic tolerance calculation and $k = 4$ members of a linear dimensional chain, a doubling of the individual tolerances occurs, if these are the same size

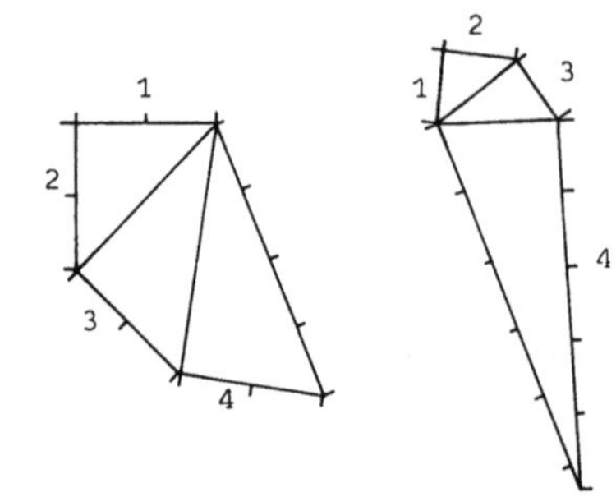

i	T_i	T_i^2		T_i	T_i^2
1	2	4		1	1
2	2	4		1	1
3	2	4		1	1
4	2	4		5	25
T_a	= 8	16	T_a	= 8	28
	T_q	= 4		T_q	= 5.29

Fig. 9.3. Pythagoras of tolerances

A plausible explanation for this is offered to tolerances by Pythagoras, i.e. that dimensional chains with the same value totals have higher quadratic totals (and roots) if the individual values are unequal, Fig. 9.3. Both arithmetical closing tolerances are equal with $T_a = 8$. However the quadratic tolerance is greater if the individual tolerances are not equal to each other.

In geometric terms: right angle triangles with the same totals for the short sides have hypotenuses which differ in length; the hypotenuse is smallest when the short sides of the triangle are equally long.

Quadratic tolerance calculation involves the following steps:

1. First the arithmetical tolerance calculation as described in Sect. 8.5 is carried out with the appropriate form, if necessary with constructive corrections.

2. It should be checked whether the arithmetical overall tolerance T_a should be retained for the quadratic calculation as a representative statistical tolerance T_{sv}, or whether T_{sv} can be made smaller.

3. After insertion of T_{sv} in line 18 of the form, its square T_{sv}^2 is inserted into line 19. T_{sv}^2 is the required total which must not be exceeded by the $\sum T_{si}^2$ widening of the individual tolerances.

4. Before the individual T_{si}^2 is formed in column 10, the square of the previous individual tolerances is formed and inserted into column 9.

With equal or similar sizes for individual tolerances the values for column 10 can be chosen to be approx. k times as large as in column 9, so that the required total at the end of column 10 is not exceeded. When choosing the values for column 10 care should be taken as to which tolerances are particularly narrow; also the square roots of the values in column 10 are necessary for column 11. So that this can be an iterative process, it is better to use pencil and eraser.

5. Since we are talking about values which belong to a statistical dimensional chain, it is especially important that the nominal value conforms with the mean value, column 12, and is steered towards the centre of the tolerance zone.

Exercise 9.1

Given: Exercise 8.1 with all values and the arithmetical tolerance calculation (old forms not necessary).

Required: Quadratic tolerance calculation.

Solution: The solution is given in Fig. 9.4. The solution from Fig. 8.5 would be changed so that both bearings have the same collar height. Also the clearance would be narrowed. To balance this the value for the shaft section would be changed. With $k=5$ members of a dimensional chain and with the use of a quadratic tolerance calculation, the individual tolerances are more than doubled. Since this is not held to be necessary, the new tolerance for clearance on $T_{sv}=0.6$ is established, line 18. The solution in Fig. 9.4 is one of many possibilities.

After carrying out a quadratic tolerance calculation and an appropriate widening of the first arithmetically calculated individual tolerances, from now on special demands must be fulfilled by the distributions of the individual values. In no case are the individual values permitted to lie "possibly" in the (widened) tolerance zone.

In Figs. 9.5 and 9.6 different examples are shown which fulfill or do not fulfill these demands. These examples are explained as follows:

Figure 9.5

Line 1: Ideal case for a quadratic calculation. All individual values show normal distributions, which lie in the middle of the tolerance zone and fill the respective tolerance zones with their 6σ ranges. Obviously the requirement is also fulfilled if the standard deviation are all or partly smaller than $\sigma=0.5$.
Line 2: The spread of individual values is nil with all $k=4$ individual values (pointed distributions). Then these individual value distributions must in all cases deviate around $T/4$ from the middle of the tolerance zone, assuming equal effective directions. If this extreme case occurs or if the standard deviation is only $\sigma<0.25$, then the tolerance widening is not necessary. The widened individual tolerances are again narrowed according to the arithmetical calculation.
Line 3: All distributions have $\sigma=0.5=T/6$ but are however shifted around $T/4$ in the same effective direction with respect to the tolerance zone centre. The effect is catastrophic. 50% of all closing values lie outside the overall value tolerance zone. This would also be exactly the case if the standard deviations in the individual values were all similarly larger or smaller.
Line 4: All individual value distributions have $\sigma=0.5=T/6$, but are however shifted around $T/4$ from the tolerance zone centre, with only 3 in the same

error proportion p %	10	5	4.55	2.5	2	1	0.27	0.1	0.01
conforming prop. 1-p %	90	95	95.45	97.5	98	99	99.73	99.9	99.99
variable of ND u_{1-p}	1.6449	1.9600	2	2.2414	2.3263	2.5758	3	3.2905	3.8906
conversion factor	0.9497	1.1316	1.1547	1.2941	1.3431	1.4871	1.7321	1.8998	2.2462
order or choice									

process: dimensional chain in figure 8.4													
date:		positive score direction			negative score direction			arithm. calculation		new values, statistical calculated			
column	1	2	3	4	5	6	7	8	9	10	11	12	13
line	size No.	nominal size N_i	up. deviation A_o	low. deviation A_u	nominal size N_i	up. deviation A_o	low. deviation A_u	tolerance T_i	T_i^2	T_{si}^2	T_{si}	centre size C_i	deviations $A_{o,u}$
1	1	30	+0.10	-0.05				0.15	0.0225	0.09	0.30	30	±0.15
2	2	40	+0.10	-0.10				0.20	0.0400	0.09	0.30	40	±0.15
3	3				5	+0.15	+0.08	0.07	0.0049	0.04	0.20	5	±0.10
4	4				59.5	-0.05	-0.30	0.25	0.0625	0.09	0.30	59.5	±0.15
5	5				5	+0.15	+0.06	0.09	0.0081	0.04	0.20	5	±0.10
6													
7													
8													
9													
10													
11													
12													
13													
14	Σ =	70	+0.20	-0.15	① 69.5	② +0.25	③ -0.16	0.76	0.1380	0.35			
15		- ① -69.5	- ③ +0.16	- ② -0.25						0.36	= riquired total		
16	N_S =	0.5	+0.36	-0.40	← check of calculation								
17	C_S =												

new clearance: $0.5\,^{+0.30}_{-0.30}$

provisional calculation:

18	T_a =	T_{sv} = 0.60
19	ND: T_{sv}^2 = 0.36	$= \sum T_{si}^2$
20	RD: $(T_{sv}/r_u)^2$ =	$= \sum T_{si}^2$

Fig. 9.4. Form with solution of Exercise 9.1

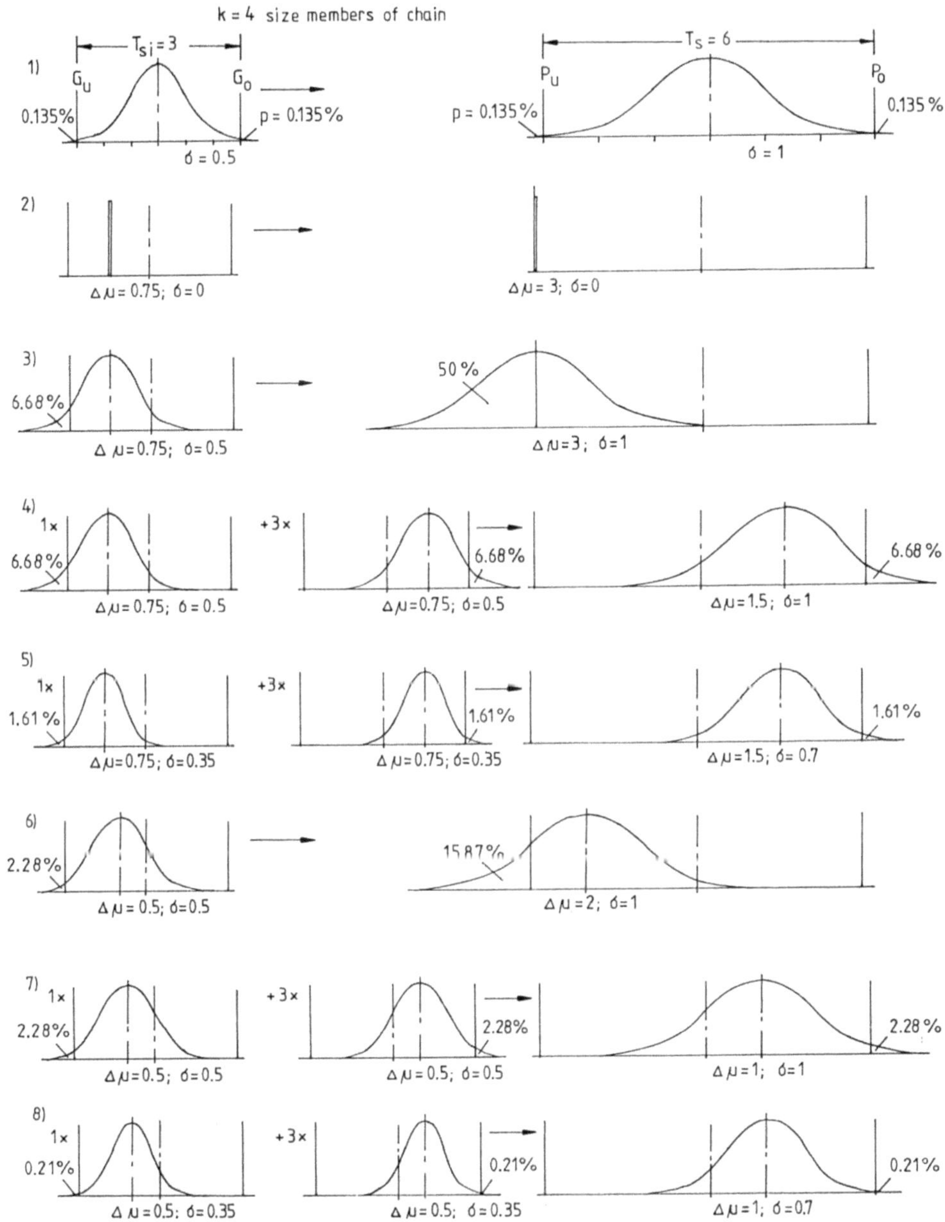

Fig. 9.5. By quadratic tolerance calculation, linear dimensional chains with $k = 4$ members and examples showing the parameters of the actual distributions deviating from the ideal case (line 1)

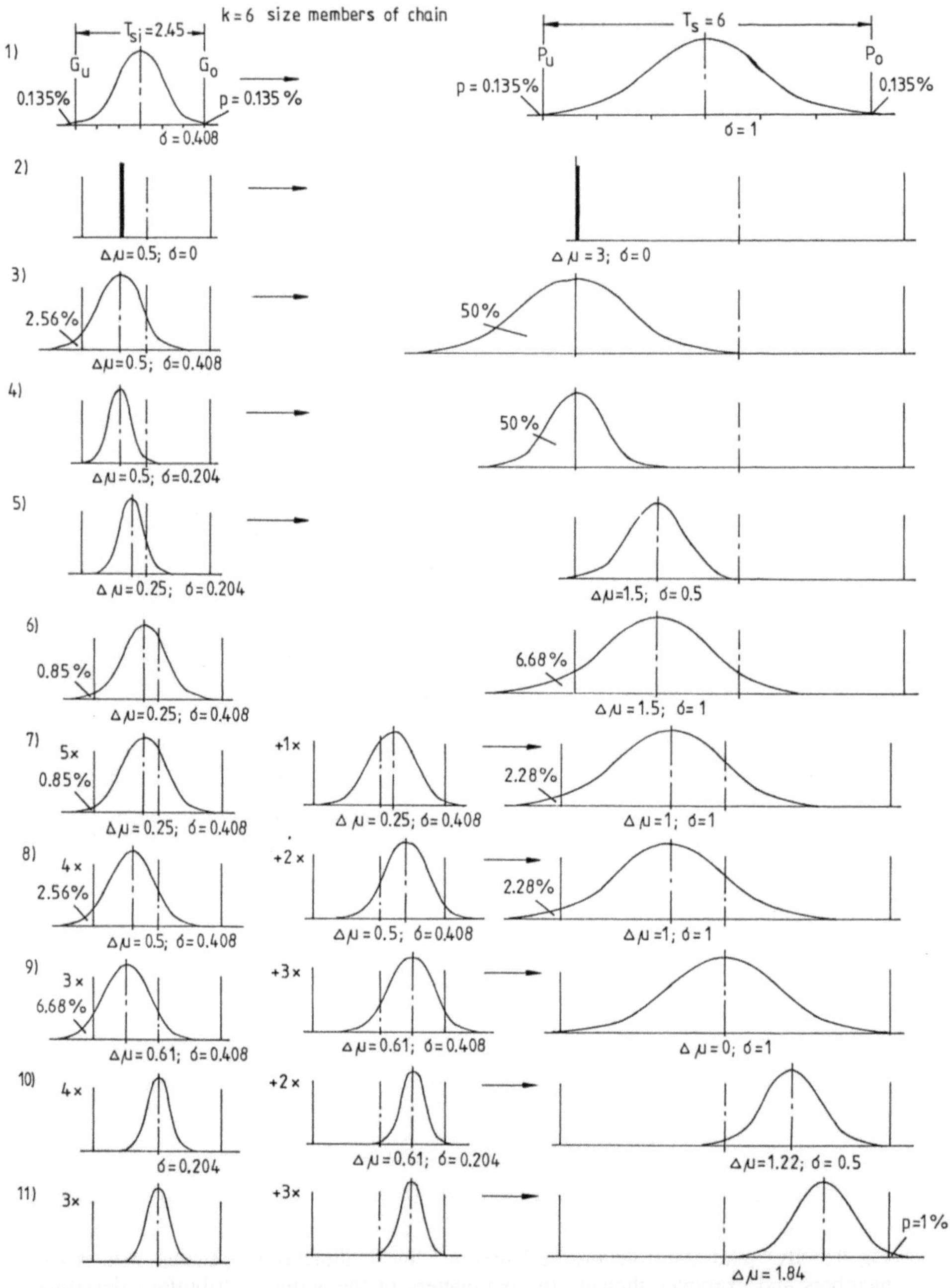

Fig. 9.6. By quadratic tolerance calculation, linear dimensional chains with $k = 6$ members and examples showing the parameters of the distributions deviating from the ideal case in line 1.

effective direction. In spite of this the error proportion in the overall values at $p = 6.68\%$ is unacceptably high.
Line 5: As in line 4, however the spread at $\sigma = 0.35$ is smaller than in line 4. That result is acceptable.
Line 6: As in line 3, however the deviation with respect to the tolerance zone centre is only $T/6$ in the same effective direction. The error proportion in the overall values is at $p = 15.87\%$ not acceptable.
Line 7: As line 6, however only three deviations lie in the same effective direction. This is acceptable.
Line 8: As line 7, however, in all cases the standard deviation with $\sigma = 0.35$ is smaller. This is definitely acceptable.

Figure 9.6

Line 1: Ideal case, for the quadratic calculation of the dimensional chain with $k = 6$ members. All individual value distributions are normal, lie within the tolerance zone centre and fill the tolerance zones with their 6σ ranges. Obviously it would be permissible if the standard deviations were all or partly smaller than $\sigma = 0.408$.
Line 2: All overall values lie by the lower tolerance on fit P_u; for this case the spread-less individual values lie only about 0.5 from the tolerance zone centre. Here quadratic tolerance calculation would not be necessary and it should be referred back to arithmetically calculated tolerances.
Line 3: The distributions of all $k = 6$ members deviate about 0.5 from the tolerance zone centre in the same effecive direction with catastrophic results, 50% of the overall values lie outside the tolerance zone for closing values.
Line 4: Although the individual values are only half as strongly spread, the effects are the same as in column 3.
Line 5: As line 4, however the deviation with respect to the tolerance zone centre is only half as great. This is acceptable.
Line 6: As line 5, however the spread is as large as in line 4, This is not acceptable, although not nearly as bad as in line 4.
Line 7: As line 6, however has only five deviations from respective tolerance zone centre in the same direction. The effect on the distribution of the overall values is acceptable.
Line 8: As line 3, however only four deviations have the same effective direction. This is (almost) acceptable.
Line 9: The standard deviation is in all individual distributions as in line 1; the deviations are equal however up to half are in the other direction. This is acceptable, however here exchangeability is strongly in question. Chapter 8 goes into the problems of exchangeability more extensively.
Line 10: Four individual distributions lie in the middle; then the other distributions can deviate considerably from the middle, without the overall limit values being exceeded.
Line 11: Analysis as in line 10.

The examples represented in Figs. 9.5 and 9.6 are possible resumed, in particular on the assumption that the standard deviations are different within the individual value distributions.

Advantage of quadratic tolerance calculations. By quadratic tolerance calculation—compared with arithmetical calculation—a maximal widening of the individual tolerances is possible.

Disadvantage of quadratic tolerance calculations. The maximal widening of the individual tolerances is based on the assumption that from now on the individual values in production distributions lie on the mean value with a 6σ range which is smaller or at the most the same as the width of the tolerance zone.

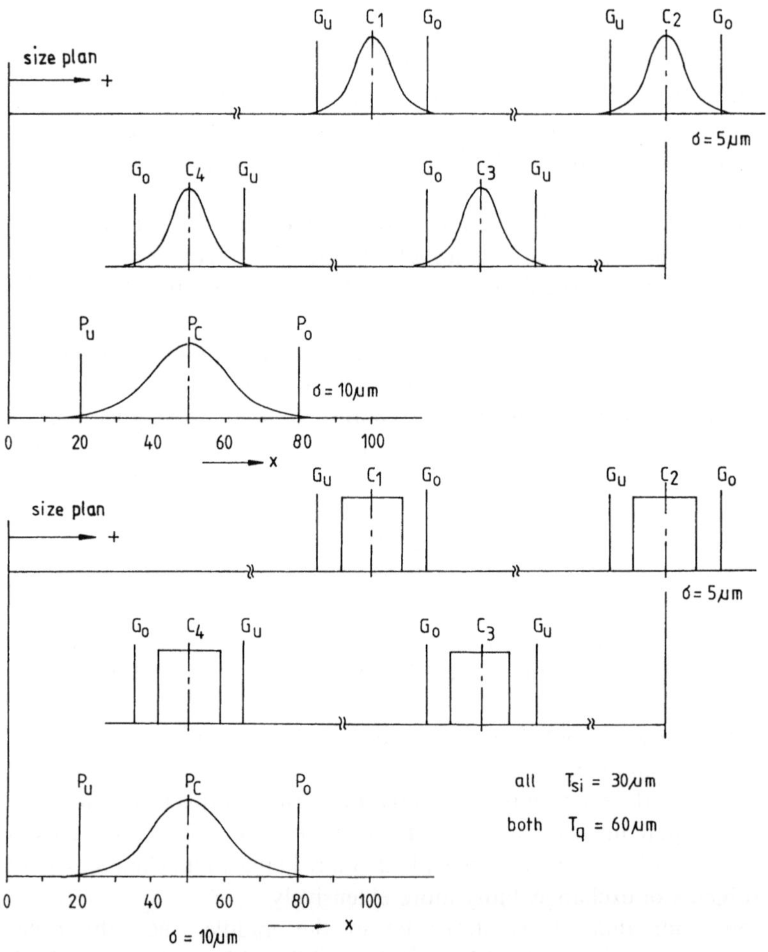

Fig. 9.7. By quadratic tolerance calculation, linear dimensional chain with $k = 4$ members and with distributions of permissible position and width in the individual values, above: with normal distributions and below: with rectangular distributions

This assumption is fulfilled if all individual value distributions are normally distributed so that the tolerance zone is filled, Fig. 9.7.

However if there are mixed distributions in individual values which are narrower than normal distributions with the same standard deviation, then these distributions should be allowed to fill only the middle 70% of the tolerance zone.

This case applies if, for example, the production distributions are at one instant normal with a constant trend, so that the total distribution in the batches tends towards a rectangular distributions, Fig. 9.7. Then this assumption can better be fulfilled so that the calculation of the previous rectangular distributed individual batches are set out and allowed to fill the respective tolerance zones, Chap. 10.

10 Statistical Calculation of Tolerances of Individual Values with Rectangular Distributions

10.1 General

From the disadvantages of quadratic tolerance calculations, Chap. 9, comes the basis for statistical tolerance calculation on the assumption of the presence of rectangular distributions in the individual values.

The rectangular distribution is the narrowest of all single-pointed distributions. Since actual mixed distributions can only be wider, not narrower, the calculation on the assumption of the presence of a rectangular distribution represents the most unfavourable case.

The statistical tolerance calculation on the assumption of the presence of rectangular distributions is covered by an industry standard simplified tolerance calculation. The calculation is not easier; however, production control and testing is simplified, since a test of distribution shape is not necessary.

Frequently the rectangular distribution is called—with respect to the normal distribution—"unfavourable". This is not the case. Unfavourable is merely the assumption that a tolerance zone of a rectangular distribution is filled. That does not happen with the shape of this distribution, but with its greater standard deviation with respect to the normal distribution which with its 6σ range fills the same tolerance zone.

10.2 Deviation of Reduction and Expansion Factors with Equal Individual Tolerances

The derivation follows from the assumption that—even with tolerance expansion—the tolerance zones are filled by all individual values of a dimensional chain in rectangular distributions.

If all T_i are equal, then the arithmetically calculated total tolerance is

$$T_a = kT_i. \tag{10.1}$$

With a rectangular distribution as in Fig. 4.9

$$R = \sqrt{12} \cdot \sigma$$

with $R = T_i$ is

$$T_i = 2\sqrt{3} \cdot \sigma_i \tag{10.2}$$

or

$$\sigma_i = \frac{T_i}{2\sqrt{3}}. \tag{10.3}$$

With k members it is independent of the distribution shape

$$\sigma_k = \sqrt{k}\sigma_i \tag{10.4}$$

and with (10.3)

$$\sigma_k = \frac{\sqrt{k}T_i}{2\sqrt{3}}. \tag{10.5}$$

The statistical total tolerance

$$T_s = 2u_{1-p}\sigma_k \tag{10.6}$$

gives with (10.5)

$$T_s = \frac{2u_{1-p}\sqrt{k}T_i}{2\sqrt{3}}. \tag{10.7}$$

Thus the reduction factor used to narrow the arithmetically calculated total tolerance can be

$$r = \frac{T_s}{T_a} = \frac{u_{1-p}\sqrt{k}T_i}{\sqrt{3}kT_i}$$

and shortened

$$r = \frac{u_{1-p}}{\sqrt{3}\sqrt{k}}. \tag{10.8}$$

The reciprocal value of r is the expansion factor, to widen the individual tolerances if the total tolerance is retained

$$e = \frac{\sqrt{3}\sqrt{k}}{u_{1-p}}. \tag{10.9}$$

The increase (decrease) of the expansion factor (reduction factor) over the number of values k follows from Fig. 10.1.

Also inserted are the trends of factors for the TPD, and the TD; more of this in Chap. 11.

Also the trends of factors for the normal distribution are inserted, which differ from the rectangular distribution in that their range R is greater by a factor of $\sqrt{3}$. So

$$r_{ND} = \frac{u_{1-p}}{3\sqrt{k}}.$$

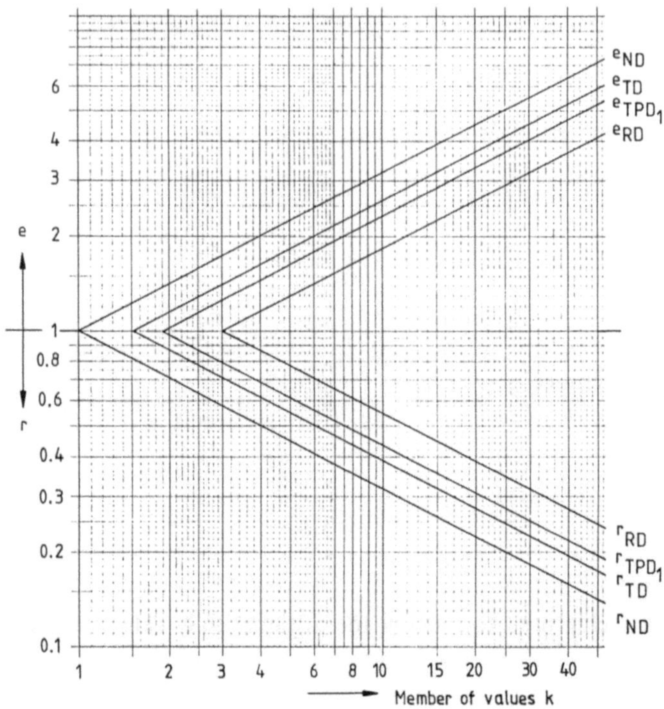

Fig. 10.1. Reduction factors and expansion factors dependent on the number of values k for the case that all individual tolerances T_{si} are equal and all distributions fill their respective tolerance zones; $p \leqq 27\%$ in the closing values

In case—as in individual values—in the closing value an excess proportion (error proportion) of $p = 0.27\%$ is accepted, then $u_{1-p} = 3$ and

$$r_{ND} = \frac{1}{\sqrt{k}} \tag{10.10}$$

and the expansion factor

$$e_{ND} = \sqrt{k}. \tag{10.11}$$

Thus the e_{ND} line in Fig. 10.1 agrees with the trend of T_{si}/T_i in Fig. 9.1.

The expansion and reduction factors in Fig. 10.1 are valid for the case where an error proportion of $p = 0.27\%$ is accepted in the closing values.

In Fig. 10.2 the trends of expansion and reduction factors are given for other different error proportions p in the closing values. For higher error proportions the straight lines shift nearer the lines for the normal distribution in Fig. 10.1. Higher error proportions in the closing values can always then be accepted although these rarely occur in practice.

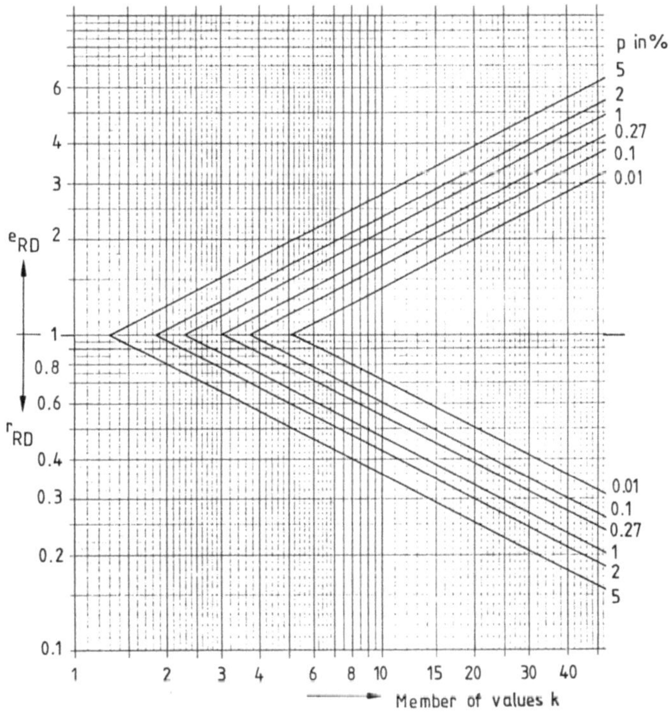

Fig. 10.2. Reduction factors and expansion factors dependent on the number of values k for equal individual tolerances T_{si} and for different error proportions in the overall values; all tolerance zones are filled by rectangular distributions

This is the case when either

—the tolerance zones of the individual values are not filled or
—the individual distributions of a rectangular distribution deviate in the sense that they fall sloping down the sides.

In Fig. 10.3 the reduction factor is plausibly explained. A linear dimensional chain with $k = 6$ members is represented. The individual tolerances are all equal with $T_i = 18\ \mu m$. The arithmetically calculated tolerance of the overall value is

$$T_a = kT_i = 6 \cdot 18 = 108\ \mu m.$$

In the middle of the figure a statistical tolerance premiss is taken under the assumption that all individual values have rectangular distribution, which fill the previous tolerance zones. Then the standard deviations in the individual values are

$$\sigma_i = \frac{R}{\sqrt{12}} = \frac{T_{si}}{\sqrt{12}} = \frac{18}{\sqrt{12}} = 5.196\ \mu m.$$

The overall value is then a normal distribution with the standard deviation

$$\sigma_k = \sqrt{k}\sigma_i = \sqrt{6}\cdot 5.196 = 12.726\,\mu\text{m}.$$

If an error proportion of $p = 1\%$ is accepted in the overall values, then the required tolerance is

$$\begin{aligned} T_s &= 2\cdot u_{1-0.01}\cdot\sigma_k \\ &= 2\cdot 2.5758\cdot 12.726 = 65.569\,\mu\text{m}. \end{aligned}$$

The tolerance reduction is therefore

$$r = \frac{T_s}{T_a} = \frac{65.569}{108} = 0.6071$$

in agreement with the reduction factor (10.8)

$$r = \frac{u_{1-p}}{\sqrt{3}\sqrt{k}} = \frac{2.5758}{\sqrt{3}\sqrt{6}} = 0.6071.$$

In the lower part of Fig. 10.3 it is shown that the arithmetically calculated tolerance of the overall value is retained. A tolerance expansion is undertaken assuming that the new tolerance zones are filled by rectangular distributions.

The standard deviation of the normal distribution in the overall value is then

$$\sigma_k = \frac{T_s}{2\cdot u_{1-p}} = \frac{108}{2\cdot 2.5758} = 20.964\,\mu\text{m}.$$

Therefore the standard deviation of the rectangular distributions in the individual values is

$$\sigma_i = \frac{\sigma_k}{\sqrt{k}} = \frac{20.964}{\sqrt{6}} = 8.559\,\mu\text{m}.$$

The range and there the individual tolerance is then

$$R_i = T_{si} = \sqrt{12}\cdot\sigma_i = \sqrt{12}\cdot 8.559 = 29.648\,\mu\text{m}.$$

The individual tolerances can therefore be expanded by the factor

$$e = \frac{29.648}{18} = 1.647$$

in agreement with the factor in (10.9)

$$e = \frac{\sqrt{3}\cdot\sqrt{k}}{u_{1-p}} = \frac{\sqrt{3}\cdot\sqrt{6}}{2.5758} = 1.647.$$

At this stage in the individual values is a proportion of up to

$$p = \frac{29.648 - 18}{29.648} = 39.29\%$$

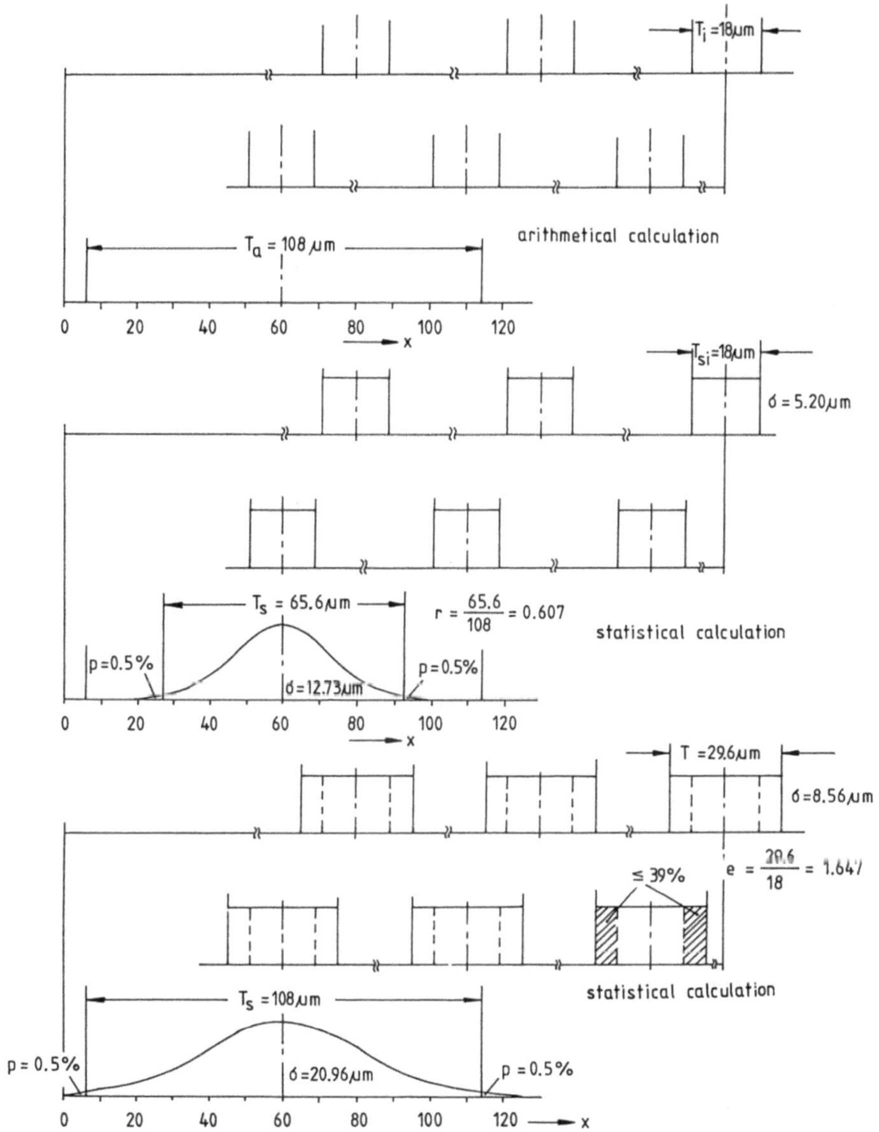

Fig. 10.3. Dimensional chains with $k = 6$ members for illustrations of the reduction and expansion factors

admissible, which before—with arithmetical tolerance calculation—was inadmissible (non-conforming).

For frequent calculations the reduction factor

$$r = \frac{u_{1-p}}{\sqrt{3}\sqrt{k}}$$

could be shown in a table. If however the $\sqrt{k}$ is left in the denominator, then

$$r_u = \frac{u_{1-p}}{\sqrt{3}} \tag{10.12}$$

is an conversion factor which is independent of k and thus even easier to show, see form for calculation of tolerances, Fig. 10.6.

The conversion factor r_u when multiplied by $1/\sqrt{k}$ gives the reduction factor r.

10.3 Calculation of Dimensional Chains with Unequal Individual Tolerances and Rectangular Individual Distributions

If the individual tolerances are unequal then their quadratic mean is formed

$$T_m = \sqrt{\frac{\sum T_i^2}{k}}. \tag{10.13}$$

The quadratic mean is illustrated in Figs. 10.4 and 10.5.

If T_m instead of T_i in (10.7) is inserted, then

$$T_s = \frac{u_{1-p}\sqrt{k}\sqrt{\sum T_i^2}}{\sqrt{3}\sqrt{k}} = r_u\sqrt{\sum T_i^2}. \tag{10.14}$$

So that

$$\sum T_i^2 = (T_s/r_u)^2 \tag{10.15}$$

is the target size for the new division of the tolerances, line 20 in the form for the calculation of tolerances, Fig. 10.6.

Exercise 10.1

Given: Exercise 9.1 (gear unit, old forms not necessary).

Required: Statistical tolerance calculation on the assumption that the individual distributions are rectangular and will fill the tolerance zones.

Solution: The solution in Fig. 10.6 is one of many possibilities.

A peculiarity of conversion factors r_u which are unimportant in terms of calculation but very important for the understanding of the somewhat complicated connections, is thus (10.14).

The factor $\sqrt{\sum T_i^2}$ is the quadratically calculated tolerance T_q of the overall value and r_u is the factor which must be expanded around T_q if the individual tolerance zones cannot be filled by normal distributions but by rectangular distributions.

line	T_1	T_2	T_3	T_4	T_5	$\sum T_i$	$\sum T_i^2$	$T_m = \sqrt{\frac{\sum T_i^2}{k}}$
1	10	10	10	10	10	50	500	10
2	20	15	5	5	5	50	700	11.832
3	30	5	5	5	5	50	1000	14.142
4	40	4	2	2	2	50	1628	18.044
5	46	1	1	1	1	50	2120	20.591
6	50	0	0	0	0	50	2500	22.361

Fig. 10.4. Quadratic mean T_m of $k = 5$ members equal and unequal individual tolerances; all tolerance chains have the same total

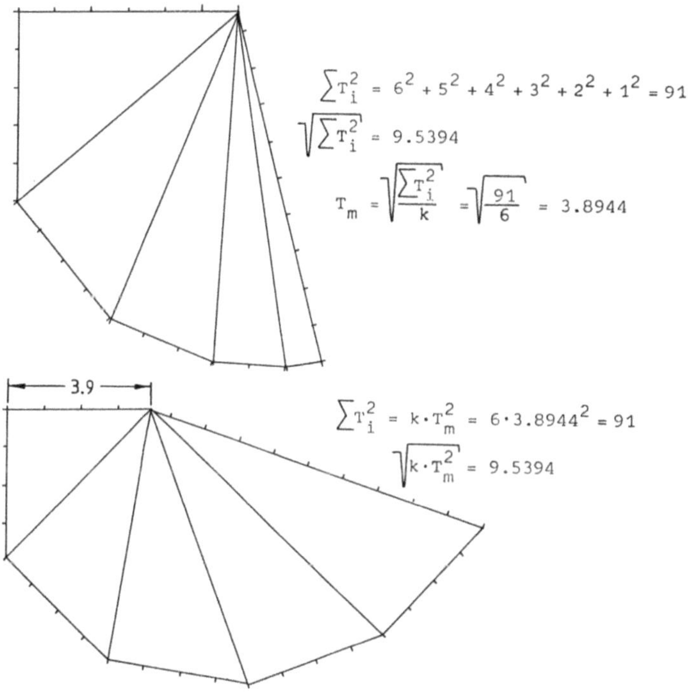

Fig. 10.5. Pythagoras for tolerances, below are all tolerance as large as the quadratic mean of the unequal tolerances in the upper tolerance chain

For the example in the middle of Fig. 10.3 it would be (9.2)

$$T_q = \sqrt{k}\, T_i = \sqrt{6} \cdot 18 = 44.091\ \mu\text{m}.$$

If this T_q is multiplied by r_u (according to the form), then the result is

$$T_s = r_u T_q = 1.4871 \cdot 44.091 = 65.567\ \mu\text{m}$$

as in Sect. 10.2 calculated in another way.

error proportion p %	10	5	4.55	2.5	2	1	0.27	0.1	0.01
conforming prop. 1-p %	90	95	95.45	97.5	98	99	99.73	99.9	99.99
variable of ND u_{1-p}	1.6449	1.9600	2	2.2414	2.3263	2.5758	3	3.2905	3.8906
conversion factor	0.9497	1.1316	1.1547	1.2941	1.3431	1.4871	1.7321	1.8998	2.2462
order or choice						×			

process: dimensional chain in figure 8.4

date:		positive score direction			negative score direction			arithm. calculation		new values, statistical calculated			
column	1	2	3	4	5	6	7	8	9	10	11	12	13
line	size No.	nominal size N_i	up. deviation A_o	low. deviation A_u	nominal size N_i	up. deviation A_o	low. deviation A_u	tolerance T_i	T_i^2	T_{si}^2	T_{si}	centre size C_i	deviations $A_{o,u}$
1	1	30	+0.10	−0.05				0.15	0.0225	0.0576	0.24	30	±0.12
2	2	40	+0.10	−0.10				0.20	0.0400	0.0576	0.24	40	±0.12
3	3				5	+0.15	+0.08	0.07	0.0049	0.01	0.10	5	±0.05
4	4				59.5	−0.05	−0.30	0.25	0.0625	0.09	0.30	59.5	±0.15
5	5				5	+0.15	+0.06	0.09	0.0081	0.01	0.10	5	±0.05
6													
7													
8													
9													
10													
11													
12													
13													
14	Σ =	70	+0.20	−0.15	① 69.5	② +0.25	③ −0.16	0.76	0.1380	0.2252			
15		− ① −69.5	− ③ +0.16	− ② −0.25						0.2216 = riquired total			
16	N_S =	0.5	+0.36	−0.40	← check of calculation								
17	C_S =												

provisional calculation:

18	T_a =	T_{sv} = 0.70
19	ND: T_{sv}^2 =	$= \sum T_{si}^2$
20	RD: $(T_{sv}/r_u)^2$ =	$= \sum T_{si}^2$ = $(0.70/1.4871)^2$

new clearance:

$0.5^{+0.35}_{-0.35}$

Fig. 10.6. Form with solution to Exercise 10.1

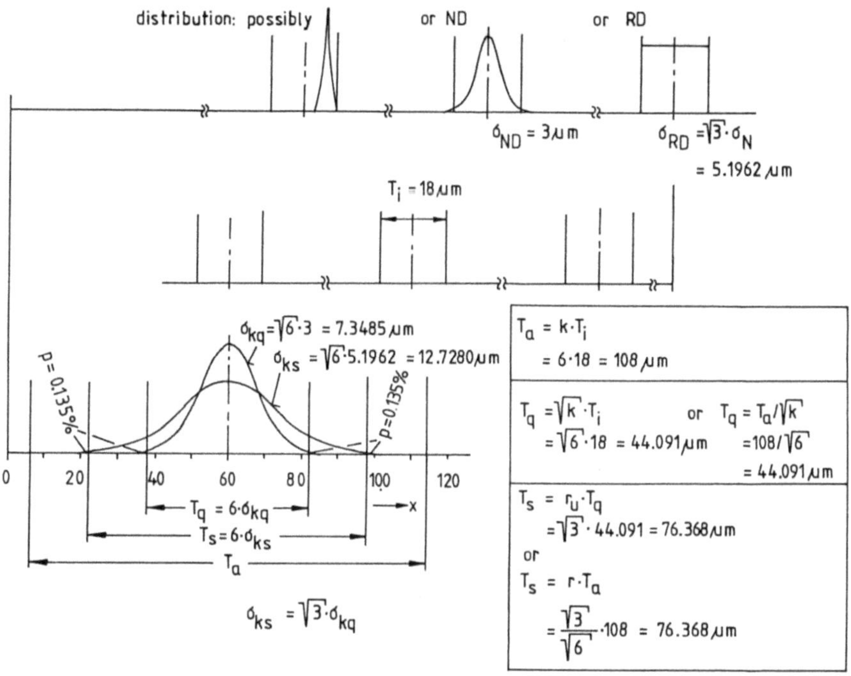

Fig. 10.7. Comparison of the closing tolerances T_a, T_q and T_s in a linear dimensional chain with $k = 6$ members and equal individual tolerances T_i

The simplest derivation of the reduction factor is possible as follows:

$$T_s = r_u T_q = \frac{u_{1-p}}{\sqrt{3}} T_q. \tag{10.16}$$

With equal individual tolerances it is

$$T_q = \frac{T_a}{\sqrt{k}} \tag{10.17}$$

and inserted in (10.16)

$$T_s = \underbrace{\frac{u_{1-p}}{\sqrt{3}\sqrt{k}}}_{\text{Reduction Factor}} T_a$$

The closing value tolerances T_a, T_q and T_s are compared with one another in Fig. 10.7.

10.4 Advantages of the Assumption of the Presence of Rectangular Distributions

The statistical tolerance calculation on the assumption of the presence of rectangular distributions in the individual values which fill the respective tolerance zones has the following advantages:

- —The result lies on the "safe side"; actual mixed distributions—arising from normal distributions—have sloping sides (bell-shaped).
- —No test processes for distribution shape are required; so the costs are kept down. In addition the test processes which exist for this are less effective

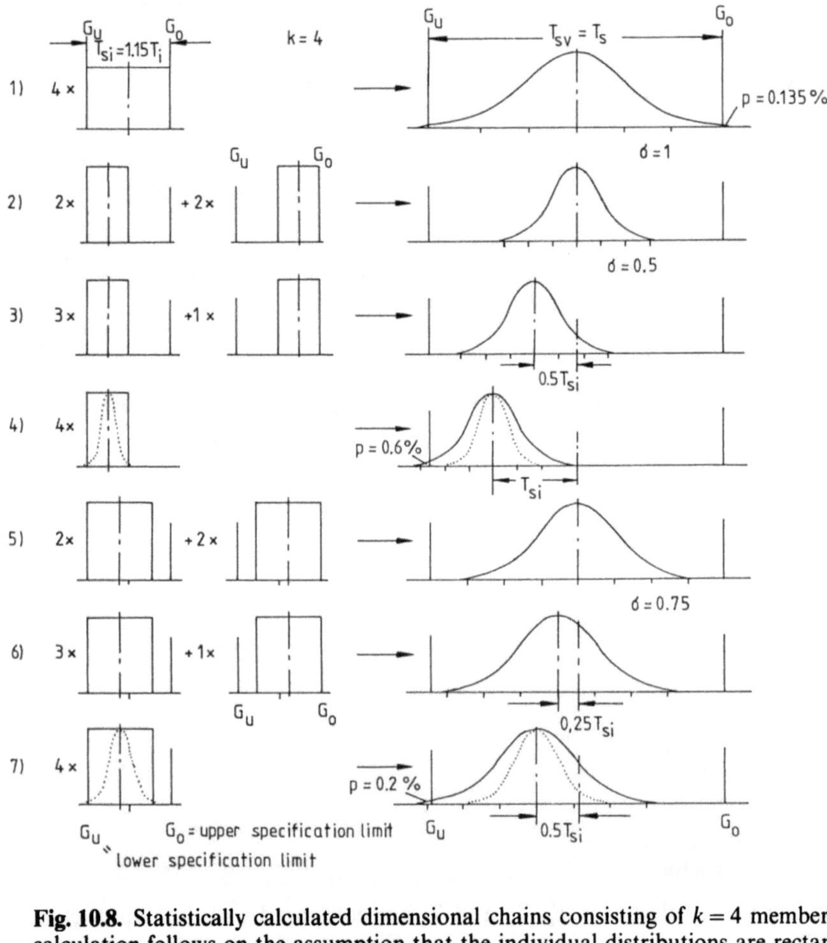

Fig. 10.8. Statistically calculated dimensional chains consisting of $k = 4$ members. The calculation follows on the assumption that the individual distributions are rectangular, which fill the respective (expanded) tolerance zones. Lines 2) to 7) are examples in which that is not the case

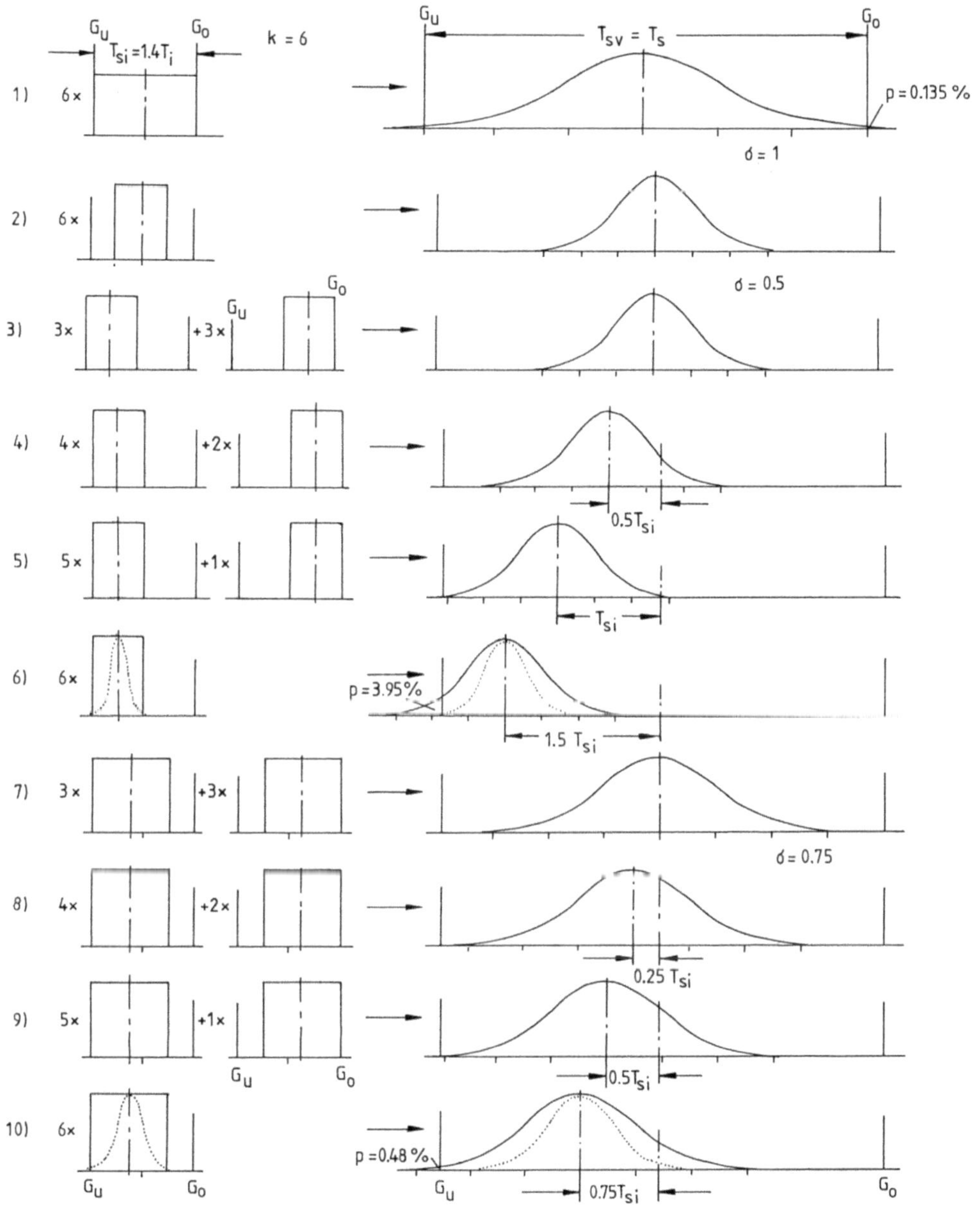

Fig. 10.9. Statistically calculated dimensional chains with $k = 6$ members and examples where the assumptions are not valid.

—If the distributions fill the tolerance zones, they lie on the mean value too. If the actual distributions are slightly narrower, then deviations from the mean value are not critical, because then the deviations are compensated by a smaller spread.
In Fig. 10.8 this is shown for a four-member dimensional chain. If the tolerance of the closing value from arithmetical calculation can be retained, the individual tolerances can be expanded about the factor

$$e_{RD} = \frac{\sqrt{3}\sqrt{k}}{u_{1-p}} = \frac{\sqrt{4}}{\sqrt{3}} = 1.1547.$$

In the lines 2) to 7) there are different assumptions for the position and width of the actual distributions met. In all cases the tolerance of the closing value is not exceeded.

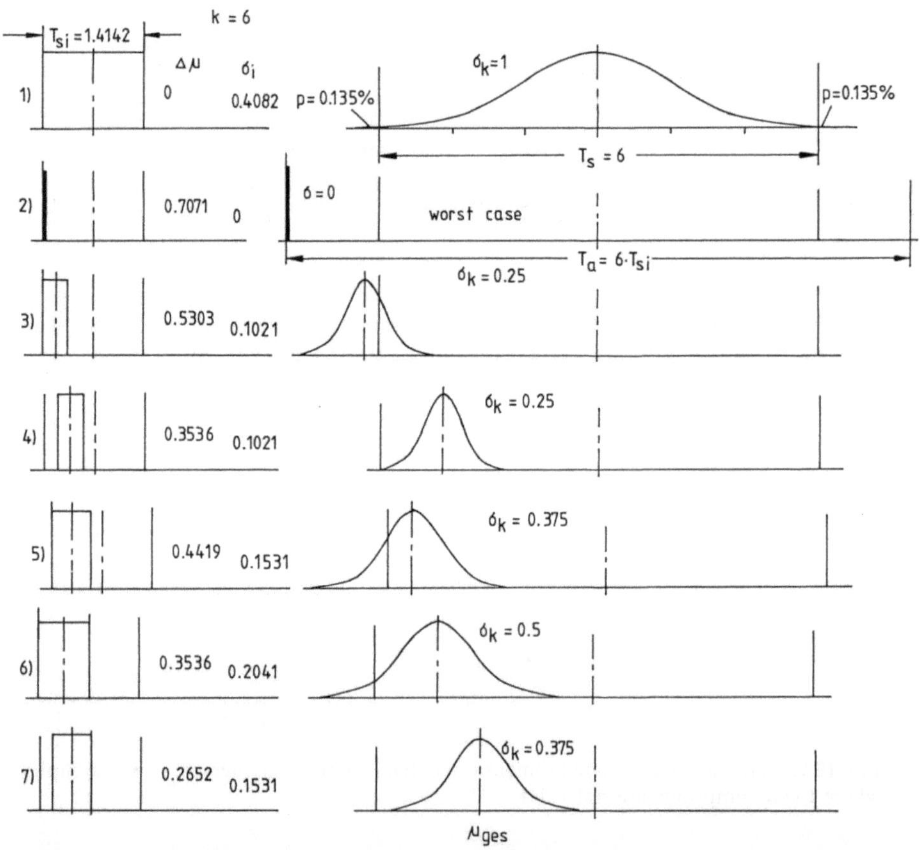

Fig. 10.10. Dimensional chains with $k = 6$ equal individual tolerances. The statistical tolerance calculation follows with the assumption that the individual tolerances are filled by rectangular distributions. The lines 2) to 7) contain the results when this assumption is not met

The same is shown in Fig. 10.9 for a six-member dimensional chain, with the individual tolerances expanded about the factor

$$e_{RD} = \frac{\sqrt{6}}{\sqrt{3}} = 1.4142.$$

—If the distributions are considerably narrower than the tolerance zones, then the assumption must be fulfilled, that the mean values lie close to the middle size, in no case must the distributions lie by the limit values, Fig. 10.10. This case must however rarely occur, for if the distribution is considerably narrower than the tolerance zone, then the tolerance expansion through statistical calculation would not be necessary and a return could be made to arithmetically calculated tolerances. Possibly a comprehensive narrowing of the tolerances would be sensible.
The case in which the random spread is considerably narrower than the tolerance zone is given with the so-called hard tools (cutting tools, die cast metal housings). Here the distribution wanders in the course of the service life of the tool from one limit value to another through the tolerance zone. Thus all values which are produced on hard tools must not be used in a statistical tolerance calculation.
—The improvements for production achieved under the assumption of the presence of rectangular distributions are usually sufficient.

Exercise 10.2

Given: Figure 10.10.
Required: Further execution of the calculations; in addition for the lines 3) to 7) the error proportions in the overall values are to be determined with the aid of the u-table A1.

Solution:

line	$u = \dfrac{3 - 6\Delta\mu}{\sqrt{6}\sigma;}$	p in %
3	−0.7269	76.64
4	3.5122	0.02
5	0.9296	17.62
6	1.7568	3.95
7	3.7568	0.00

11 Statistical Tolerance Calculation of Individual Values with Trapezoid Distribution or with Triangular Distribution

11.1 Method of Calculation

In the cases in which the expansion of individual tolerances on the assumption of the presence of rectangular distributions does not apply, assumptions can be made that the individual actual distributions cannot fill the tolerance zones but can be approximated by trapezoid or triangular distributions.

In the case of the assumption of rectangular distribution not being appropriate, more sloping distributions can be deduced in order to estimate the reliability of the rectangular distribution.

In Fig. 11.1 the parameters and factors for different distributions are summarised. For type TPD_2 these factors are calculated here by way of example;

The variance is

$$\sigma^2 = \frac{5R^2}{108}$$

as in Fig. 4.9. If the distribution fills the tolerance zone, it is

$$T_i = R = \sqrt{\frac{108}{5}} \cdot \sigma = 2\sqrt{\frac{27}{5}} \cdot \sigma = 2 \cdot 2.323\,8 \cdot \sigma.$$

Thus the reduction factor is

$$r_{TPD_2} = \frac{1.732\,1}{2.323\,8} \cdot r = 0.745\,4 \cdot r$$

and the conversion factor

$$r_{uTPD_2} = 0.745\,4 \cdot r_u.$$

Then the required total is

$$\sum T_i^2 = \left(\frac{T_{sv}}{r_{uTPD_2}}\right)^2 = \frac{T_{sv}^2}{0.745\,4^2 \cdot r_u^2} = 1.8\left(\frac{T_{sv}}{r_u}\right)^2.$$

The required total is greater by a factor of 1.8 than with the assumption of a rectangular distribution; thus the individual tolerances are expanded by a factor $\sqrt{1.8} = 1.342$ or the closing value tolerance could be smaller by a factor of $1/1.342 = 0.745$.

Distribution	Variance	Tolerance filled with R	Reduction factor	Conversion factor	Required total of tolerance calculation	Statistical total tolerance for $u_{1-p}=3$
RD 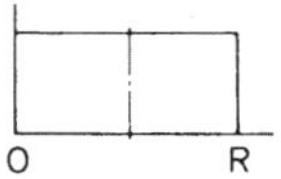	$\sigma^2 = \frac{R^2}{12}$	$T_i = 2\sqrt{3}\cdot\sigma_i = 2\cdot 1.7321\cdot\sigma_i$	$r = \frac{u_{1-p}}{\sqrt{3}\sqrt{k}}$	$r_u = \frac{u_{1-p}}{\sqrt{3}}$	$\Sigma T_i^2 = \left(\frac{T_{sv}}{r_u}\right)^2$	$T_s = 1.7321\sqrt{\Sigma T_{si}^2}$
TPD_1 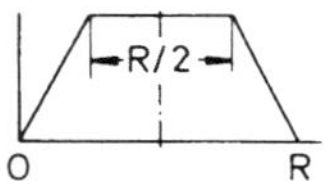	$\sigma^2 = \frac{10\cdot R^2}{192}$	$T_i = 2\sqrt{\frac{48}{10}}\cdot\sigma_i = 2\cdot 2.1909\cdot\sigma_i$	$r_{TPD_1} = 0.7906r$	$r_{uTPD_1} = 0.7906r_u$	$\Sigma T_i^2 = 1.6\left(\frac{T_{sv}}{r_u}\right)^2$	$T_s = 1.3694\sqrt{\Sigma T_{si}^2}$
TPD_2 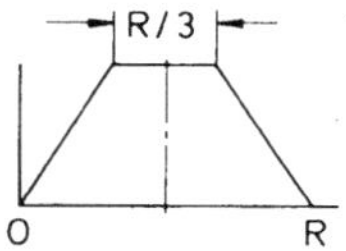	$\sigma^2 = \frac{5\cdot R^2}{108}$	$T_i = 2\sqrt{\frac{27}{5}}\cdot\sigma_i = 2\cdot 2.3238\cdot\sigma_i$	$r_{TPD_2} = 0.7454r$	$r_{uTPD_2} = 0.7454r_u$	$\Sigma T_i^2 = 1.8\left(\frac{T_{sv}}{r_u}\right)^2$	$T_s = 1.2910\sqrt{\Sigma T_{si}^2}$
TPD_3 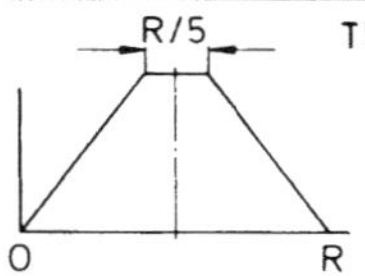	$\sigma^2 = \frac{13\cdot R^2}{300}$	$T_i = 2\sqrt{\frac{75}{13}}\cdot\sigma_i = 2\cdot 2.4019\cdot\sigma_i$	$r_{TPD_3} = 0.7211r$	$r_{uTPD_3} = 0.7211r_u$	$\Sigma T_i^2 = 1.923\left(\frac{T_{sv}}{r_u}\right)^2$	$T_s = 1.2490\sqrt{\Sigma T_{si}^2}$
TD 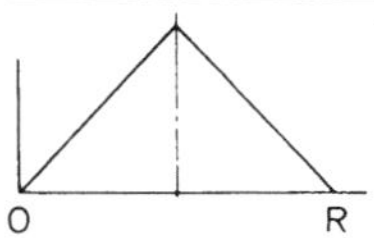	$\sigma^2 = \frac{R^2}{24}$	$T_i = 2\sqrt{6}\cdot\sigma_i = 2\cdot 2.4495\cdot\sigma_i$	$r_{TD} = 0.7071r$	$r_{uTD} = 0.7071r_u$	$\Sigma T_i^2 = 2\left(\frac{T_{sv}}{r_u}\right)^2$	$T_s = 1.2247\sqrt{\Sigma T_{si}^2}$
ND 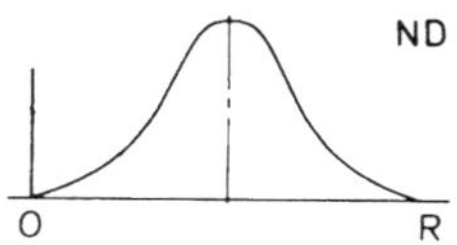	$\sigma^2 = \frac{R^2}{36}$	$T_i = 2\cdot 3\cdot\sigma_i$ proportion outside limit values $p = 0.27\%$	$r_{ND} = 0.5774r$	$r_{uND} = 0.5774r_u$	$\Sigma T_i^2 = 3\left(\frac{T_{sv}}{r_u}\right)^2$	$T_s = 1.0000\sqrt{\Sigma T_{si}^2}$

Fig. 11.1. Summary of parameters for different distributions

The same result is presented if with an RD calculation an error proportion of $p = 2.5\%$ in closing values is accepted. Then the behaviour of the conversion factors according to the form would be

$$\frac{r_{u,1.5\%}}{r_{u,0.27\%}} = \frac{1.294\,1}{1.732\,1} = 0.747\,1$$

and the target size for the total of the squared individual tolerances would also be greater by the factor

$$\frac{1}{0.747\,1^2} = 1.791\,6 \approx 1.8.$$

It is worth noting that the different TPD types and the TD can hardly be distinguished in operation. This is clearest in Fig. 10.1 in which the straight lines for the reduction factors and for the expansion factors of the distributions TPD, and TD are inserted. The straight lines for TPD_2 and TPD_3 would run between if they are drawn in.

Exercise 11.1

Given: Solution of the Exercise 10.1 (RD calculation) with

C_i	T_i
30	0.24
40	0.24
5	0.10
59.5	0.30
5	0.10

Required: Security for the case when the actual production distributions fill the tolerance zones but have approximated the shape of a trapezoid distribution of the type TPD_2.

Solution: The target total is now

$$\sum T_i^2 = 1.8 \cdot \left(\frac{T_{sv}}{r_u}\right).$$

The calculated total is

$$T_i^2 = 0.225\,2.$$

So

$$T_{sv}^2 = \frac{\sum T_i^2 r_u^2}{1.8},$$

$$T_{sv} = \frac{\sqrt{0.225\,2 \cdot 1.487\,1}}{\sqrt{1.8}} = 0.526\,\text{mm}.$$

In words: according to the considerable security given by the RD calculation, the tolerance of the clearance $T_{sv} = 0.70$ mm, which is exceeded by the actual clearance by 1%. If the actual distributions should have the shape TPD_2,

1% of the actual clearances lie outside the tolerance zone with the width $T_{TD2} = 0.526$ mm.
The clearance is then instead of $0.5^{+0.35}_{-0.35}$, in truth $0.5^{+0.26}_{-0.26}$.

Exercise 11.2

Given: Hub fitting on a axle-end pivot; dimensional chain with $k = 4$ members as in Fig. 11.2.
A suggestion for the size measurement with arithmetical tolerance calculation is shown in Fig. 11.3.

Required:
1. Analysis of the arithmetically calculated tolerance chain.
2. Quadratic tolerance calculation
3. Calculation under different assumptions for distribution shapes in the individual actual values and for acceptable error proportions in the overall values.
4. Summarised analysis of this example.

Solution:
1. The lowest clearance is achieved with $P_u = 0.2$ mm; the highest clearance is measured at $P_0 \cdots 0.66$ mm, but should on no account be greater, because otherwise the axial power is likely to produce "knocking". The individual tolerances are somewhat tight and should be widened.

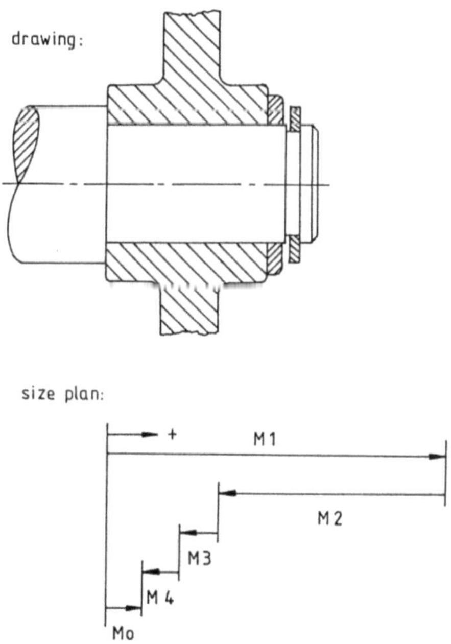

size table:

axle length	$M1 = 51.3^{0}_{-0.1}$
width of hub	$M2 = 45.0^{0}_{-0.1}$
lollar	$M3 = 4.0^{0}_{-0.1}$
securing ring	$M4 = 2.0^{0}_{-016}$

Fig. 11.2. Drawing, Size plan and Size table for the linear dimensional chain of Exercise 11.2

error proportion p %	10	5	4.55	2.5	2	1	0.27	0.1	0.01
conforming prop. 1−p %	90	95	95.45	97.5	98	99	99.73	99.9	99.99
variable of ND u_{1-p}	1.6449	1.9600	2	2.2414	2.3263	2.5758	3	3.2905	3.8906
conversion factor	0.9497	1.1316	1.1547	1.2941	1.3431	1.4871	1.7321	1.8998	2.2462
order or choice									

process: hub fitting

date:		positive score direction			negative score direction			arithm. calculation		new values, statistical calculated			
column	1	2	3	4	5	6	7	8	9	10	11	12	13
line	size No.	nominal size N_i	up. deviation A_o	low. deviation A_u	nominal size N_i	up. deviation A_o	low. deviation A_u	tolerance T_i	T_i^2	T_{si}^2	T_{si}	centre size C_i	deviations $A_{o,u}$
1	1	51.3	0	−0.1				0.1	0.01	0.04	0.2	51.3	±0.1
2	2				45.0	0	−0.1	0.1	0.01	0.04	0.2	45	±0.1
3	3				4.0	0	−0.1	0.1	0.01	0.04	0.2	4	±0.1
4	4				2.0	0	−0.16	0.16	0.0256	0.0256	0.16	1.92	±0.08
5													
6													
7													
8													
9													
10													
11													
12													
13													
14	$\sum$ =	51.3	0	−0.1	① 51.0	② 0	③ −0.36	0.46	0.0556	0.1456			
15		− ① −51.0	− ③ +0.36	− ② 0						0.1600 = riquired total			
16	N_S =	0.3	+0.36	−0.1	← check of calculation								
17	C_S =	0.43	+0.23	−0.23									

provisional calculation:

18	T_a = 0.46 T_{sv} = 0.40
19	ND: T_{sv}^2 = 0.16 $= \sum T_{si}^2$
20	RD: $(T_{sv}/r_u)^2$ = $= \sum T_{si}^2$

Fig. 11.3. Form with solution of Exercise 11.2

2. The quadratic tolerance calculation is shown in Fig. 11.3. $T_{sv} = 0.4$ mm gives $T_{sv}^2 = 0.16$ as target total in column 10. The actual total in column 10 lies with 0.1456 below; however the tolerances for the values 1 to 3 can be doubled. The tolerance for the value 4 would be retained, because it is a standard part.
   The mean values in individual column 12 give, in association with the number direction of the middle clearance $C_s = 0.38$ mm. Lowest clearance and highest clearance lie symmetrically about this.
3. The analysis for the quadratic calculation is in line no. 1 of the Fig. 11.4; further assumptions and analyses are contained in lines 2 to 5. It may at

1	assumption		4	5
line No	2 induviduel tolerances filled by	3 error proportion outside closing tolerance p	calculation of T_S	expected clearance
1	ND	0.27 %	$T_S^2 = \sum T_{si}^2$ $T_S = \sqrt{0.1456} = 0.3816$	$C_S = 0.38 \begin{smallmatrix}+0.19\\-0.19\end{smallmatrix}$
2	RD	5 %	$T_S^2 = r_u^2 \cdot \sum T_{si}^2$ $T_S = 1.1316 \cdot 0.3816$ $T_S = 0.4318$	$C_S = 0.38 \begin{smallmatrix}+0.22\\-0.22\end{smallmatrix}$
3	TPD_2	5 %	$T_S^2 = 0.7454^2 \cdot r_u^2 \cdot \sum T_{si}^2$ $T_S = 0.7454 \cdot 1.1316 \cdot 0.3816$ $T_S = 0.3218$	$C_S = 0.38 \begin{smallmatrix}+0.16\\-0.16\end{smallmatrix}$
4	ND	5 %	$T_S^2 = 0.5774^2 \cdot r_u^2 \cdot \sum T_{si}^2$ $T_S = 0.5774 \cdot 1.1316 \cdot 0.3816$ $T_S = 0.2493$ (1)	$C_S = 0.38 \begin{smallmatrix}+0.125\\-0.125\end{smallmatrix}$
5	RD	0.27 %	$T_S = 1.7312 \cdot 0.3816$ $T_S = 0.6609$ (2)	$C_S = 0.38 \begin{smallmatrix}+0.33\\-0.33\end{smallmatrix}$

(1) or $T_S = 0.3816 \frac{1.96}{3} = 0.2493$ (2) or $T_S = 0.4318 \frac{3}{1.96}$

formulae in figure 11.1

Fig. 11.4. Analysis of the statistically tolerated linear dimensional chain of Exercise 11.2 with reference to the actual value distributions in the individual tolerance zones and the error proportions in the closing values

first strike you that the tolerance of the closing value in Line 3 (TPD_2) is smaller than in the quadratic calculation in Line 1; this is based on the fact that in this line there is an error proportion of 5% in the closing values which is held to be acceptable, while in the quadratic calculation—as in the individual values- an error of $p = 0.27\%$ is accepted in the closing values.

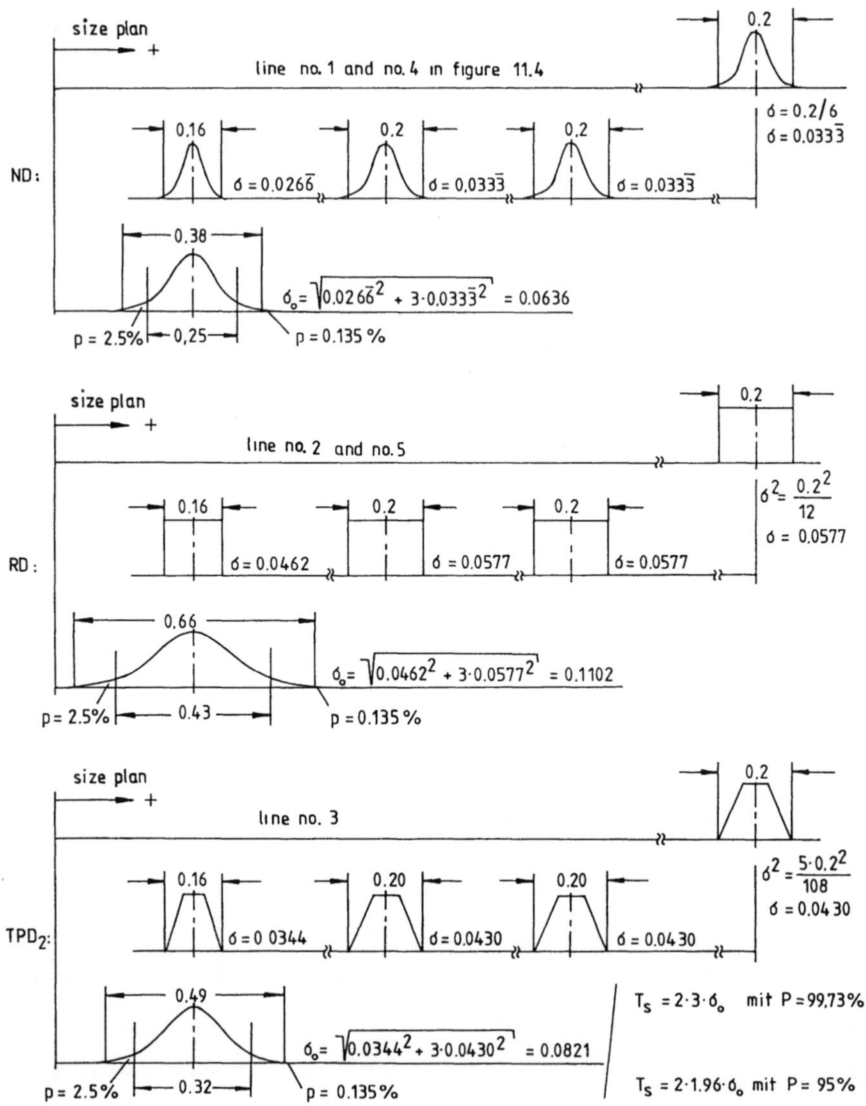

Fig. 11.5. Representation of the assumptions of Exercise 11.2 in Fig. 11.4 and their effects

A repetition of the quadratic calculation with $p = 5\%$ is contained in line 4. The following formula says the same:
The limit values in line 4 for the overall value are calculated so that 95% of the assemblies can be expected within them, as long as all tolerance zones are filled by individual values of normal distributions.
Where the production distributions are rectangular, which fill the (widened) tolerance zones, then the lowest clearance is always sufficient, Fig. 11.4, line 5.

4. The example is very clear because of its simplicity. In Fig. 11.5 are given the assumptions and calculations produced in Fig. 11.4 and their effects. In addition the standard deviations are given through which a further analysis of all calculations is easily and clearer possible.

drawing.

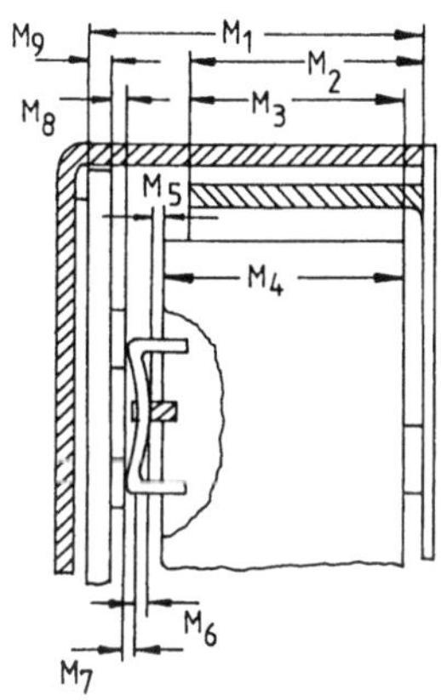

size table:

size no.	characteristic	N	A_o	A_u
M_1	housing inside size	12.5	0.1	-0.1
M_2	frontage outside size	10.5	0	-0.1
M_3	frontage inside size	8.6	0.1	0
M_4	switch width	8.0	0	-0.1
M_5	spring nipple	1.3	0.05	-0.05
M_6	wire diameter	0.9	0.11	-0.01
M_7	spring deviation	0.1	0.1	-0.05
M_8	contact strip (on)	0.2	0.02	-0.02
M_9	contact plate	1.0	0.1	-0.1

size plan:

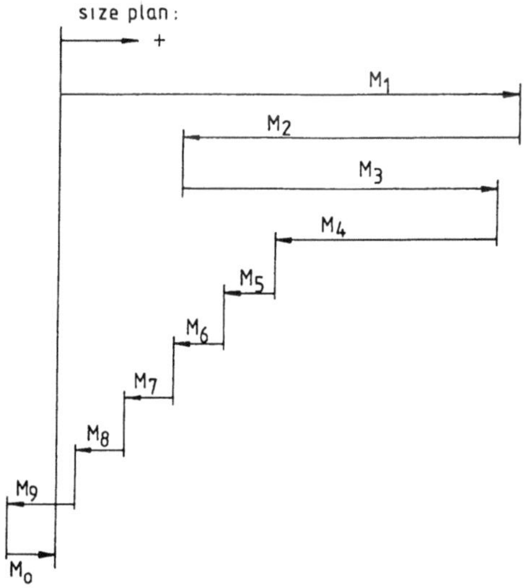

Fig. 11.6. Drawing, size table and size plan for the contact line in Exercise 11.3

Exercise 11.3

Given: Contact line switch/housing of a clock with $k=9$ size members (from Braun AG, Kronberg im Taunus). Summary drawing, size table and size plan are given in Fig. 11.6.
From this, the clearance is negative. This exceeded size will be compared to nil by the spring nipple on the switch to which the contact wire is attached.

Required: Calculation of the middle closing value and its variation under different assumptions.
The aim of the calculation is not the widening of the individual tolerances. Rather the function of the overall value should be tested; the function is given if on one side the spring is deflected and thus a lowest contact power is provided and on the other side the spring deflection suffices at $s = 1.2$ mm.

Solution: 1. Arithmetical calculation; from Fig. 11.7 the arithmetical calculation gives the overall value

$$M_o = P_C = -0.825^{+0.555}_{-0.555}$$

or

$$P_u = -1.38 \text{ resp. } P_o = -0.27 \text{ mm.}$$

2. Statistical calculation on the assumption that all individual values are normally distributed and with their 6σ range fill the respective tolerance zones.
—Limit values for the overall value for an excess proportion of $p = 0.27\%$.

$$T_{sv}^2 = \sum T_{si}^2 = 0.1585$$

(Fig. 11.7 line 14 in column 10)

$$T_{sv} = 0.398\,1$$

so

$$P_C = -0.825^{+0.199}_{-0.199}$$

or

$$P_u = -1.024 \text{ resp. } P_o = -0.626.$$

—Limit values for the overall value for an excess proportion of $p = 1\%$

$$T_{sv} = \frac{0.398\,1 \cdot 2 \cdot 2.575\,8}{6} = 0.341\,8$$

or

$$P_u = -0.996 \text{ resp. } P_o = -0.654.$$

3. Statistical calculation on the assumption that all individual values are rectangularly distributed and the right angles fill the respective tolerance zones.
—Limit values for the closing value for an excess proportion of $p = 0.27\%$

$$\sum T_{si}^2 = 0.158\,5 = (T_{sv}/r_u)^2$$

$$T_{sv}^2 = 0.158\,5 \cdot r_u^2 = 0.158\,5 \cdot 1.732\,1^2$$

$$T_{sv} = 0.689\,6$$

conforming prop. 1-p %	90	95	95.45	97.5	98	99	99.73	99.9	99.99
variable of ND u_{1-p}	1.6449	1.9600	2	2.2414	2.3263	2.5758	3	3.2905	3.8906
conversion factor	0.9497	1.1316	1.1547	1.2941	1.3431	1.4871	1.7321	1.8998	2.2462
order or choice									

process: contact line switch/housing of a clock

date:		positive score direction			negative score direction			arithm. calculation		new values, statistical calculated			
column	1	2	3	4	5	6	7	8	9	10	11	12	13
line	size No.	nominal size N_i	up. deviation A_o	low. deviation A_u	nominal size N_i	up. deviation A_o	low. deviation A_u	tolerance T_i	T_i^2	T_{si}^2	T_{si}	centre size C_i	deviations $A_{o,u}$
1	1	12.5	+0.10	-0.10				0.20	0.0400	0.0400			
2	2				10.5	0	-0.10	0.10	0.0100	0.0100			
3	3	8.6	+0.10	0				0.10	0.0100	0.0100			
4	4				8.0	0	-0.10	0.10	0.0100	0.0100			
5	5				1.3	+0.05	-0.05	0.10	0.0100	0.0100			
6	6				0.9	+0.11	-0.01	0.12	0.0144	0.0144			
7	7				0.1	+0.10	-0.05	0.15	0.0225	0.0225			
8	8				0.2	+0.02	-0.02	0.04	0.0016	0.0016			
9	9				1.0	+0.10	-0.10	0.20	0.0400	0.0400			
10													
11													
12													
13													
14	$\sum$ =	21.1	+0.20	-0.10	① 22.0	② +0.38	③ -0.43	1.11		0.1585			
15		-① -22.0	-③ +0.43	-② -0.38						0.1585 = riquired total			
16	N_S =	-0.9	+0.63	-0.48	← check of calculation c.k.								
17	C_S =	-0.825	+0.555	-0.555									

provisional calculation:

18	T_a = T_{sv} =	
19	ND: T_{sv}^2 =	$= \sum T_{si}^2$
20	RD: $(T_{sv}/r_u)^2$ =	$= \sum T_{si}^2$

Fig. 11.7. Calculation for Exercise 11.3

so:

$P_C = 0.825^{+0.3448}_{-0.3448}$

or

$P_u = -1.17$ resp. $P_o = -0.480$.

—Limit values for the overall value for an excess proportion of $p = 1\%$

$T_{sv}^2 = 0.158\,5 \cdot r_u^2 = 0.158\,5 \cdot 1.487\,1^2$

$T_{sv} = 0.592$

or

$P_u = -1.121$ resp. $P_o = -0.529$.

Summary of the calculations is shown in Fig. 11.8.

Analysis: If the unlikely worst case occurs, cases of jamming may be expected; the clock can not then be switched.
Even with the unfavourable assumption of the presence of rectangular distributions in the individual values, this does not happen.
As long as all individual tolerances are retained, it is to be expected that the actual distribution of the overall value lies between the RD and the ND calculation.

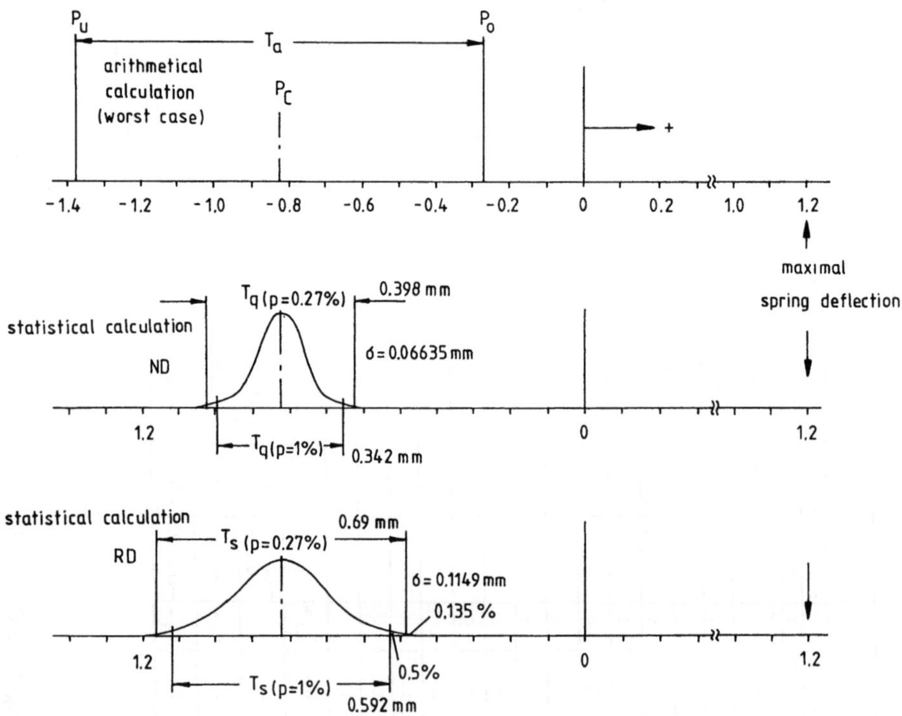

Fig. 11.8. Representation of the solution to Exercise 11.3

Exercise 11.4

Given: Three member dimensional chain with drawing and size table; the appropriate clearance should be 0.05 to 0.55 mm

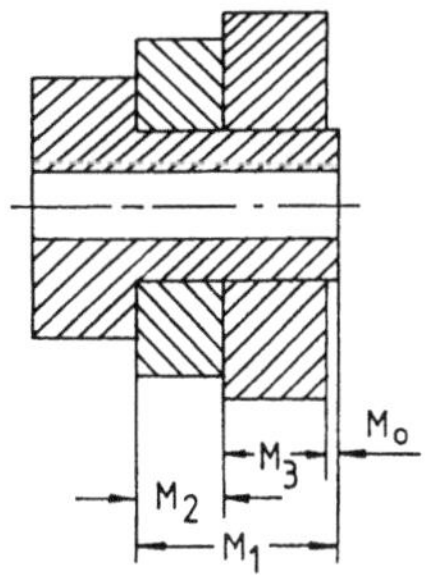

Bush $M_1 = x^{+y}_{-z}$,

Collar $M_2 = 6^{+0.10}_{-0.10}$,

Lever $M_3 = 7^{+0.10}_{-0.10}$.

Required:

1. Mean value x for the box assembly and the arithmetical calculation of the relevant deviations y and z.
2. Error proportion which can be accepted in the bush batch on the assumption that the normal distributed clearances and values M_1 and M_2 fill the respective tolerance zones with their 6σ ranges.
3. Quadratic tolerance calculation with the object of widening the tolerance for the bush assembly.

Solution:

1. With the form in connection with the calculation of tolerances, from lines 2 and 3 to lines 15 and 16, line 1 calculated to

$$x^{+y}_{-z} = 13.3^{+0.05}_{-0.05}.$$

The calculation control in column 8 states the correctness of the direction.

process: three member dimensional chain (bush with collar and lever)

date:		positive score direction			negative score direction			arithm. calculation		new
column	1	2	3	4	5	6	7	8	9	
line	size No.	nominal size N_i	up. deviation A_o	low. deviation A_u	nominal size N_i	up. deviation A_o	low. deviation A_u	tolerance T_i	T_i^2	
1	1	13.3	+0.05	-0.05				0.10		
2	2				6	+0.10	-0.10	0.20		
3	3	↑	↑	↑	7	+0.10	-0.10	0.20		
4										
5										
11										
12										
13										
14	Σ =				① 13	② +0.20	③ -0.20	0.50		
15		- ① -13	- ③ +0.20	- ② -0.20						
16	N_S =	+0.3	+0.25	-0.25	← check of calculation					
17	C_S =	+0.3	+0.25	-0.25						

2. So that the normally distributed clearances fill the tolerance on fit, their standard deviation must be

$$\sigma = \frac{T}{6} = \frac{0.5}{6} = 0.083\,3\,\text{mm}$$

So that the values M_2 and M_3 fill the tolerance zones, their standard deviation must be

$$\sigma = \frac{T}{6} = \frac{0.2}{6} = 0.033\,3\,\text{mm}$$

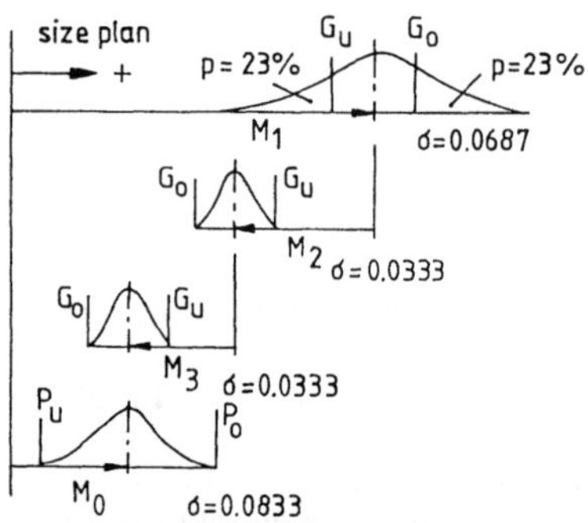

According to the deviation transmission principle

$$\sigma_1^2 = \sigma_{\text{ges}}^2 - \sigma_2^2 - \sigma_3^2$$
$$= 0.083\,3^2 - 0.033\,3^2 - 0.033\,3^2$$

$\sigma_1 = 0.0687\,\text{mm}.$

Thus the interval from mean value to the limit values in σ units is

$$\frac{G_o - C}{\sigma} = \frac{0.05}{0.068\,7} = 0.728 = u_{G_o},$$

according to the ND table the acceptable error proportion is two-sided

$p = 23.33\%$.

3. The quadratic calculation undertaken under 2, shows that the variations for the value M_1, must be

   $A_1 = 3 \cdot \sigma_1 = 3 \cdot 0.068\,7 = 0.206 \approx 0.20\,\text{mm}.$

   This is valid only on the assumption that in future all three values M_1 to M_3 are
   —normally distributed,
   —are centred on the middle of the tolerance zone, and
   —in all cases fill the respective tolerance zones with their 6σ range.

Since on one hand the ND assumption is disturbing and on the other hand for the value M_1 a tolerance of $T_1 = 0.4\,\text{mm}$ is certainly not necessary, a further statistical tolerance calculation is undertaken under the assumption that all three distributions show a TPD_2 shape and thus must fill the respective tolerance zones

The standard deviation of the overall value is

$$\sigma_s = 0.083\,3\,\text{mm}.$$

For the values M_2 and M_3 the random spread in Fig. 4.9 is

$$\sigma_{2.3} = \sqrt{\frac{5 \cdot R^2}{108}} = \sqrt{\frac{5 \cdot 0.2^2}{108}} = 0.043\,0\,\text{mm}.$$

Then the standard deviation of the value M_1 must be

$$\sigma_1 = \sqrt{0.083\,3^2 - 2 \cdot 0.043\,0^2} = 0.0568\,8\,\text{mm}.$$

Thus the tolerance becomes

$$T_1 = R_1 = 2\sqrt{\frac{27}{5}} \cdot \sigma = 2\sqrt{\frac{27}{5}} \cdot 0.056\,88 = 0.26\,\text{mm}.$$

Thus the value M_1 can be established as follows:

$$M_1 = 13.3^{+0.13}_{-0.13}.$$

Note: An RD analysis with $p \approx 1\%$ in the overall values leads to $T_1 = 0.20\,\text{mm}$.

Exercise 11.5

Given: Small E-drive according to combined drawing with size table (from Braun AG, Kronberg/Taunus).

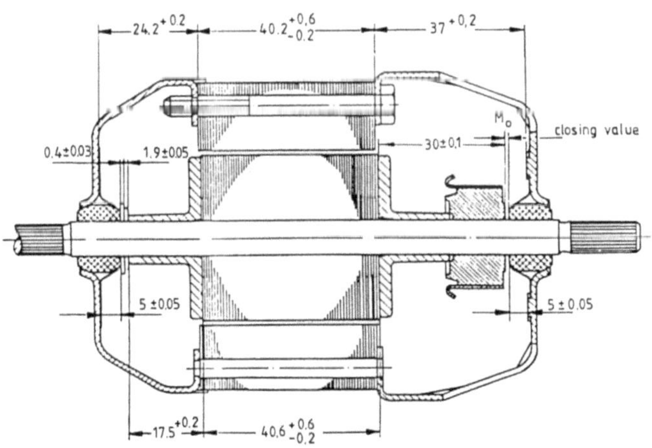

Size No.	direction	characteristic	N	A_o	A_u
1	p	bearing part left	24.2	+0.2	0
2	p	stator	40.2	+0.6	−0.2
3	p	bearing part right	37.0	+0.2	0
4	n	bearing	5.0	+0.05	−0.05
5	n	bearing	5.0	+0.05	−0.05
6	n	disc collar	0.4	+0.03	−0.03
7	n	disc	1.9	+0.05	−0.05
8	n	isolation	17.5	+0.2	0
9	n	rotor	40.6	+0.6	−0.2
10	n	isolation	30.0	+0.1	−0.1

Required: 1. Analysis of the closing value (clearance) under the assumption of the presence of a ND or RD or TPD_2 in the individual values; first an arithmetical calculation should be undertaken.
2. Diagrammatic representation of the analysis results.

Solution: 1. The arithmetical analysis of the dimensional chain using the form leads to $C_s = 1.1 \pm 1.38$; in the worst case there is an exceeded size = non-function to be expected.
—Quadratic calculation

$$\sum T_{si}^2 = 1.4736 \quad \text{according to column 10 of the form}$$

$$T_{sv} = 1.2139 \approx 1.22, \quad M_{oq} = 1.1^{+0.61}_{-0.61}.$$

—Statistical calculation, assuming RD, $p = 0.27\%$

$$\sum T_{si}^2 = 1.4736$$

$$T_{sv} = r_u \sqrt{T_{si}^2}$$

$$= 1.7321\sqrt{1.4736} = 2.1026,$$

$$M_{osRD} = 1.1^{+1.05}_{-1.05}.$$

—statistical calculation, assuming TPD_2, $p = 0.27\%$

$$T_{si}^2 = 1.4736$$

$$T_{sv} = r_u \sqrt{\frac{\sum T_{sv}^2}{1.8}}$$

$$= 1.7321 \sqrt{\frac{1.4736}{1.8}} = 1.5672,$$

$$M_{osTPD_2} = 1.1^{+0.78}_{-0.78}.$$

Analysis: With the RD calculation in almost all cases a positive clearance can be expected; the decision to make the middle clearance somewhat larger (1.1) depends on the effect of the largest clearance.
The above analysis appears so simple because the measurement in this example is concluded. In design practice, however, the setting up of the whole dimensional chain must be repeated a number of times. Even with later size or tolerance changes the following can be calculated in a simple way with the aid of the form.

error proportion p %	10	5	4.55	2.5	2	1	0.27	0.1	0.01
conforming prop. 1-p %	90	95	95.45	97.5	98	99	99.73	99.9	99.99
variable of ND u_{1-p}	1.6449	1.9600	2	2.2414	2.3263	2.5758	3	3.2905	3.8906
conversion factor	0.9497	1.1316	1.1547	1.2941	1.3431	1.4871	1.7321	1.8998	2.2462
order or choice									

process: small E-drive

date:		positive score direction			negative score direction			arithm. calculation		new values, statistical calculated			
column	1	2	3	4	5	6	7	8	9	10	11	12	13
line	size No.	nominal size N_i	up. deviation A_o	low. deviation A_u	nominal size N_i	up. deviation A_o	low. deviation A_u	tolerance T_i	T_i^2	T_{si}^2	T_{si}	centre size C_i	deviations $A_{o,u}$
1	1	24.2	+0.2	0				0.20	0.0400	0.0400			
2	2	40.2	+0.6	-0.2				0.80	0.6400	0.6400			
3	3	37	+0.2	0				0.20	0.0400	0.0400			
4	4				5	+0.05	-0.05	0.10	0.0100	0.0100			
5	5				5	+0.05	-0.05	0.10	0.0100	0.0100			
6	6				0.4	+0.03	-0.03	0.06	0.0036	0.0036			
7	7				1.9	+0.05	-0.05	0.10	0.0100	0.0100			
8	8				17.5	+0.2	0	0.20	0.0400	0.0400			
9	9				40.6	+0.6	-0.2	0.80	0.6400	0.6400			
10	10				30.0	+0.1	-0.1	0.20	0.0400	0.0400			
11													
12													
13													
14	$\sum$ =	101.4	1.0	-0.2	① 100.4	② +1.08	③ -0.48	2.76		1.4736			
15		- ① -100.4	- ③ +0.48	- ② -1.08						1.4736 = riquired total			
16	N_S =	1.0	+1.48	-1.28	← check of calculation								
17	C_S =	1.1	+1.38	-1.38									

provisional calculation:

18	T_a = T_{sv} =	
19	N: T_{sv}^2 =	$= \sum T_{si}^2$
20	R: $(T_{sv}/r_u)^2$ =	$= \sum T_{si}^2$

2. Diagrammatic representation of the analysis results:

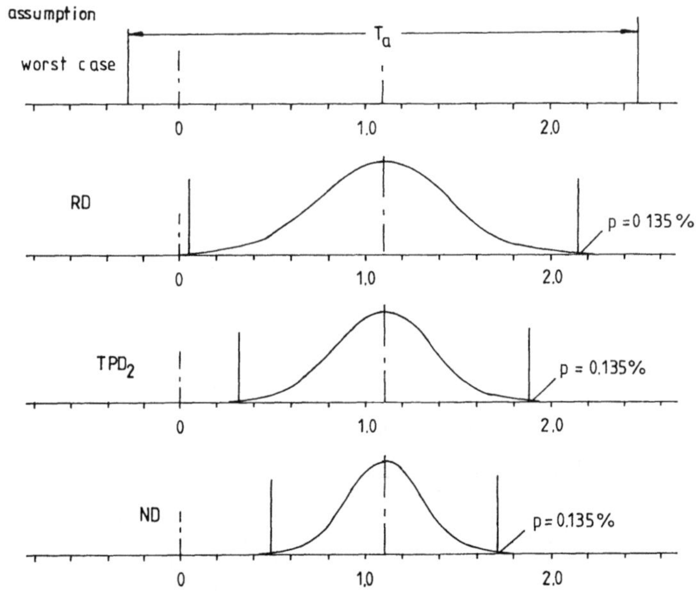

11.2 Simple Formula to Check Fixed Linear Dimensional Chains

At the end of the solutions to the Exercises 11.1 to 11.5 an analysis to check the closing value tolerance under differing assumptions for the distribution shape of individual values is undertaken.

From this the following simple formula (recipe) for the analysis of a running or final linear dimensional chain can be drawn up. This formula is divided into the following six steps:

1. Step: A linear dimensional chain has resulted from the realisation of a constructive exercise. From the functionally necessary tolerances of the overall value the individual tolerances can be calculated by arithmetical division. These individual tolerances can be already set up appropriately for the analysis as of differing size according to the technical production requirements. For the individual values of standard parts the tolerances are usually established in the Standard.

2. Step: If the individual tolerances established in the first step are all or only partly too small with the result that they cannot be achieved without error in technical yet economically efficient production, a statistical tolerance calculation is necessary.

This analysis can be undertaken as described in the form. Also there is the possibility—after long-term practice and experience—to widen all k individual tolerances or some of them by a factor which must be smaller than $\sqrt{k}$, taking into account technical production requirements. Any new tolerance setting must also take into account that all nominal sizes are at the same time mean values.

3. Step: According to the size and position of the closing value tolerance (lowest clearance and highest clearance) a limit-exceeding proportion, an error proportion, can be chosen for the overall value. Usually $p = 0.27\%$ is chosen.

4. Step: With the calculator, all quadratic individual tolerances can be added. This gives $\sqrt{\sum T_{si}^2}$.

5. Step: This value will be multiplied with the value in the last column of Fig. 11.1. This product is the statistical overall value tolerance.

6. Step: Comparison of the overall value tolerance in step 5 with the overall value tolerance established on technical production grounds in step 1. If necessary the analysis processes can be iterated after the moderation of one or the other individual tolerances, until the overall value tolerance in step 5 conforms with that in step 1 (approximately).

This schematic analysis in six steps with the aid of a calculator is just as fast and safer than with the aid of a computer program.

1	2	3		4	5
assumption for the distributions of the individuel sizes	closing tolerance p = 0.27 %	fit min. P_u	max. P_c	closing tolerance 1) p = 1.0 %	note
RD	$T_s = 1.7321 \cdot 0.4746$ $= 0.8220$	0.089	0.911	$T_s = 0.7057$	in accordance with exercise 10.1
TPD$_1$	$T_s = 1.3694 \cdot 0.4746$ $T_s = 0.6499$	0.175	0.825	$T_s = 0.5580$	
TPD$_2$	$T_s = 1.2910 \cdot 0.4746$ $T_s = 0.6126$	0.194	0.807	$T_s = 0.5260$	in accordance with exercise 11.1
TPD$_3$	$T_s = 1.2490 \cdot 0.4746$ $T_s = 0.5972$	0.204	0.796	$T_s = 0.5089$	
TD	$T_s = 1.2247 \cdot 0.4746$ $T_s = 0.5812$	0.209	0.791	$T_s = 0.4990$	
ND	$T_q = 0.4746$	0.263	0.737	$T_q = 0.4075$	

1) T_s-values of column 2 multiplied by the factor $\frac{1.4871}{1.7321}$

Fig. 11.9. Analysis of the solutions to Exercise 11.6

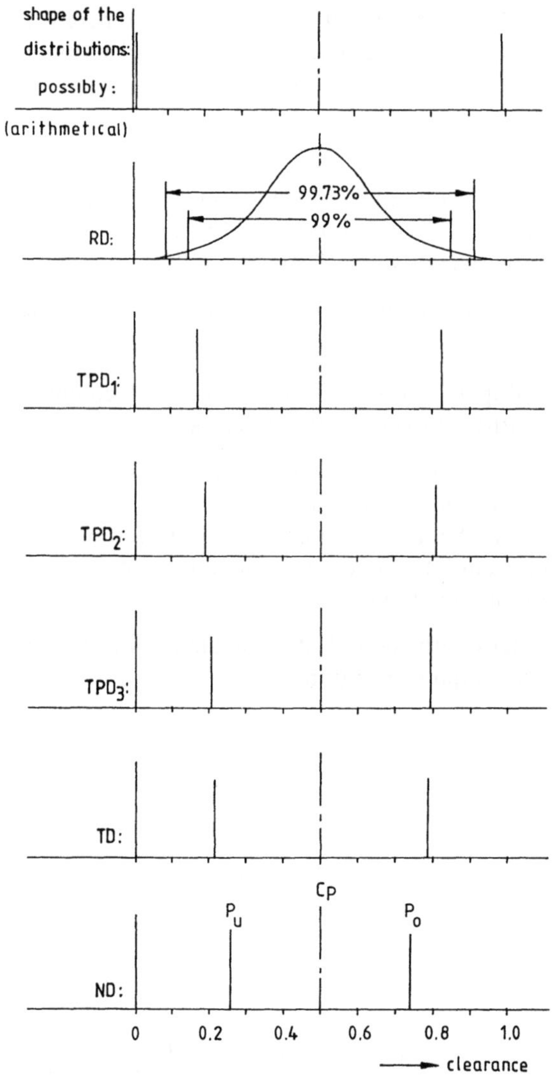

Fig. 11.10. Representation of the solutions to Exercise 11.6; the overall values have an error proportion of $p = 0.27\%$; for the case of an RD in the individual values an additional 99% range of overall values is inserted

Exercise 11.6

Given: Gear unit section as in Exercises 8.1, 9.1, 10.1, and 11.1 with the following final data:

C_i	T_{si}	T_{si}^2
30	0.24	
40	0.24	
5	0.10	
59.5	0.30	
5	0.10	
$C_P = 0.5$	0.98	0.2252

Required: New analysis with the aid of the formulae in the last column of Fig. 11.1 and the calculation for the case of an error proportion on $p = 1\%$ in the clearances.

Solution: The solutions are analysed in Fig. 11.9. For the error proportion of $p = 0.27\%$ the overall value tolerances are represented in Fig. 11.10, in which for the case where rectangular distributions exist in the individual values, the normal distribution of the overall value is drawn with the addition of the 99% range. From this it can be seen that with the choice of a larger error proportion in the overall values, only an "appearance" of reduction in the overall value tolerance is effected.

12 The Tolerance Model

The tolerance model, Fig. 12.1 serves to illustrate

—the central limit value principle,
—the deviation progression principle, and thus the
—statistical tolerance calculation.

A number of steel rods are fastened onto a wooden base. All the tubes, which are available in three batches of $N = 25$ in each in white ($L = 20$–40 mm), silver ($L = 40$–60 mm) and brown ($L = 60$–80 mm), pass onto these rods. The batches are in five classes divided between the quantities 1, 5, 13, 5 and 1. They are roughly normally distributed with an equal standard deviation in all cases of $\sigma_i = 4.2426$ mm.

A transparent tube belongs to the model which passes over the tubes of the three batches.

To begin with a systematic search is made for a tube with the middle length and this is stacked on the first rod from the right. They give the middle height of $\mu_{ges} = 150$ mm. This and the other achievable or possible heights are marked on the transparent tube by a circular mark.

Then the largest and the smallest tubes are stacked on to each other; this is the worst possible case from the three batches $x_{max} = 180$ mm and $x_{min} = 120$ mm with the difference B. For both worst cases the probability is $P = (1/25)^3 = 6.4 \cdot 10^{-5}$.

In Fig. 12.1 the worst case is shown which results when the batch is unlimited in number and width with otherwise similar parameters. The range A shows the arithmetical calculation.

Finally the range C is stacked with $C = 6\sqrt{3} \cdot \sigma = 44.09$ mm; this shows the quadratic calculation.

If now the systematically removed tubes are put back into their batches and randomly (blind) taken from one batch and stacked back together, then their sum lies in (according to the u-table) 99.73% of all cases in the range C. Since the three batches are not exactly normally distributed and are not continuous but discrete, in fact 99.7952% lie in the range C. To calculate this so exactly, the three batches were folded, Fig. 12.2. From the table for the fold product, it is shown that the total lies in range C.

$$P = 100 - 2 \cdot 0.1024 = 99.7952\%.$$

The tolerance model consists in 3 batches which are roughly normally distributed				
quantity	sizes white	silver	brown	sum
1	20	40	60	120 = sum of the smallest sizes
5	25	45	65	
13	30	50	70	150 = sum of the middle sizes
5	35	55	75	
1	40	60	80	180 = sum of the largest sizes
n	25	25	25	
μ	30	50	70	
σ	4.2426	4.2426	4.2426	$\sigma_{total} = \sqrt{3} \cdot \sigma = 7.3485$

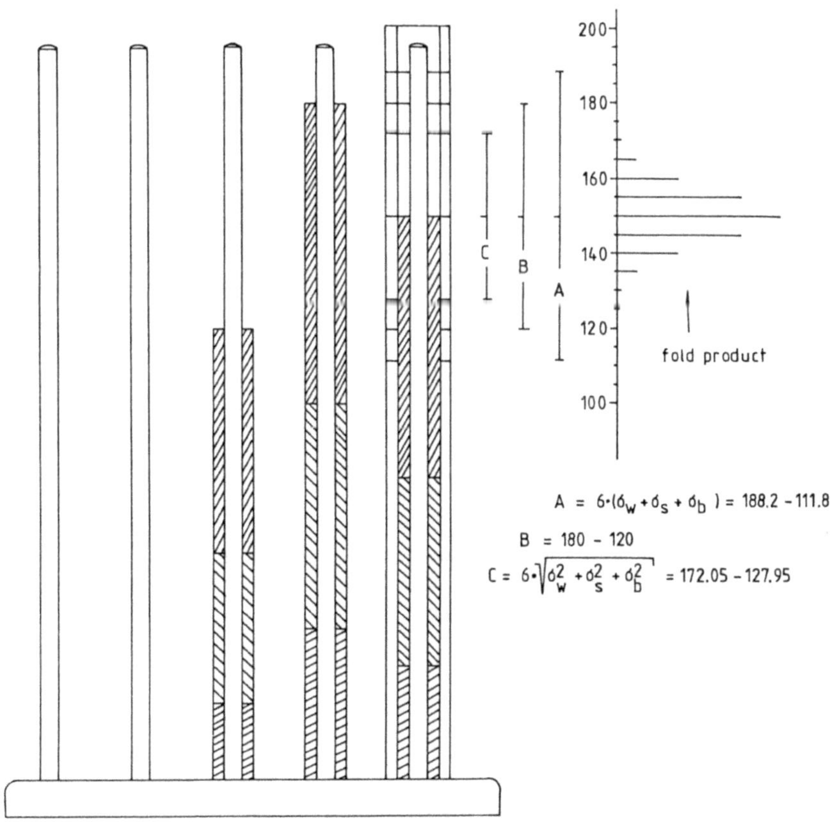

Fig. 12.1. Tolerance model

The line for the total probability for the fold product is shown in Fig. 12.3; the division of the fold product is very close to a normal distribution.

A random sample of $n = 100$ values from each batch gives 100 sum values, Fig. 12.4. The line for the sum frequency deviates as expected in a random way from a straight line.

x / g(x)	20 1	25 5	30 13	35 5	40 1 • 1/25
40 1	60 1	65 5	70 13	75 5	80 1 • 1/625
45 5	65 5	70 25	75 65	80 25	85 5
50 13	70 13	75 65	80 169	85 65	90 13
55 5	75 5	80 25	85 65	90 25	95 5
60 1	80 1	85 5	90 13	95 5	100 1
• 1/25					

x / g(x)	60 1	65 5	70 13	75 5	80 1 • 1/25
60 1	120 1	125 5	130 13	135 5	140 1 • 1/15625
65 10	125 10	130 50	135 130	140 50	145 10
70 51	130 51	135 255	140 663	145 255	150 51
75 140	135 140	140 700	145 1820	150 700	155 140
80 221	140 221	145 1105	150 2873	155 1105	160 221
85 140	145 140	150 700	155 1820	160 700	165 140
90 51	150 51	155 255	160 663	165 255	170 51
95 10	155 10	160 50	165 130	170 50	175 10
100 1	160 1	165 5	170 13	175 5	180 1
• 1/625					

fold product of the three batches

x_j	n_j	$\sum n_j$	G(x) in %
120	1	1	0.0064
125	15	16	0.1024
130	114	130	0.832
135	530	660	4.224
140	1635	2295	14.688
145	3330	5625	36.00
150	4375	10000	64.00
155	3330	13330	85.312
160	1635	14965	95.776
165	530	15495	99.168
170	114	15609	99.8976
175	15	15624	99.9936
180	1	15625	100.00

Parameters of the fold product:

$n = 15625$

$\mu = 150$

$\sigma = 7.3485$

Fig. 12.2. Folds of the three batches of the tolerance model with the parameters
$N_w = 25$ $N_s = 25$ $N_b = 25$
$\mu_w = 30\,\text{mm}$ $\mu_s = 50\,\text{mm}$ $\mu_b = 70\,\text{mm}$
$\sigma_w = \sigma_s = \sigma_b = 4.2426\,\text{mm}$

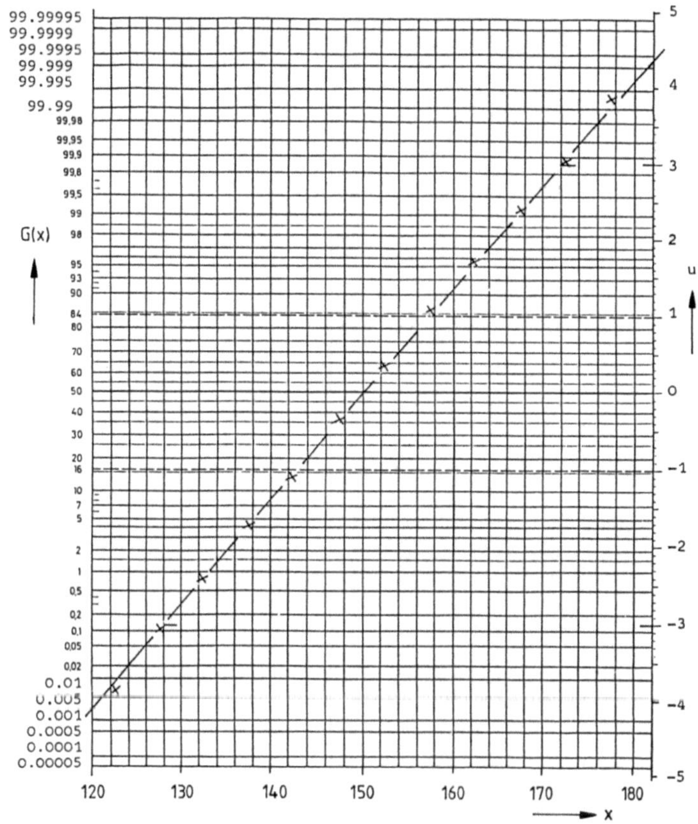

Fig. 12.3. Sum function of the fold product from the three batches of the tolerance model

Nr.	w	s	b	Σ	Nr.	w	s	s	Σ
1	30	50	70	150	51	25	55	60	140
2	30	50	70	150	52	20	55	70	145
3	30	55	70	155	53	35	50	70	155
4	30	45	70	145	54	35	50	65	150
5	30	50	70	150	55	30	50	70	150
6	25	50	70	145	56	30	55	60	145
7	35	60	70	165	57	25	55	65	145
8	25	45	75	145	58	30	50	70	150
9	35	50	70	155	59	30	50	70	150
10	25	50	65	140	60	30	45	65	140
11	30	55	80	165	61	35	50	70	155
12	25	50	75	150	62	30	45	70	145
13	30	40	75	145	63	35	50	75	160
14	30	55	65	150	64	20	50	70	140
15	30	45	70	145	65	35	55	70	160
16	30	50	70	150	66	30	50	70	150
17	35	50	65	150	67	30	50	65	145
18	25	45	75	145	68	35	50	70	155
19	30	45	75	150	69	35	40	75	150
20	30	60	70	160	70	30	60	70	160
21	30	45	65	140	71	30	55	80	165
22	30	45	70	145	72	30	45	75	150
23	25	50	65	140	73	35	40	70	145
24	40	50	70	160	74	35	50	75	160
25	35	55	70	160	75	30	50	70	150
26	25	40	80	145	76	30	55	65	150
27	30	50	70	150	77	30	55	70	155
28	20	50	70	140	78	30	50	70	150
29	30	45	75	150	79	25	50	70	145
30	30	45	65	140	80	25	55	60	140
31	25	50	75	150	81	30	45	70	145
32	30	55	70	155	82	35	50	60	145
33	25	45	65	135	83	30	50	70	150
34	30	55	75	160	84	35	50	70	155
35	20	50	65	135	85	30	55	70	155
36	35	55	70	160	86	30	50	65	145
37	35	45	75	155	87	30	45	60	135
38	40	60	70	170	88	30	50	65	145
39	30	45	75	150	89	30	50	65	145
40	25	45	75	145	90	40	55	70	165
41	25	50	70	145	91	35	50	70	155
42	35	55	70	160	92	25	50	75	150
43	25	55	70	150	93	25	40	80	145
44	30	50	70	150	94	40	50	70	160
45	30	50	70	150	95	25	50	75	150
46	30	50	70	150	96	30	50	65	145
47	30	50	75	155	97	30	50	65	145
48	30	45	70	145	98	25	50	75	150
49	35	55	65	155	99	30	50	75	155
50	30	55	75	160	100	30	50	70	150

n = 100 random samples out of the three batches and their sums

(w = white, s = silver, b = brown)

Fig. 12.4 a. Random samples from the tolerance model

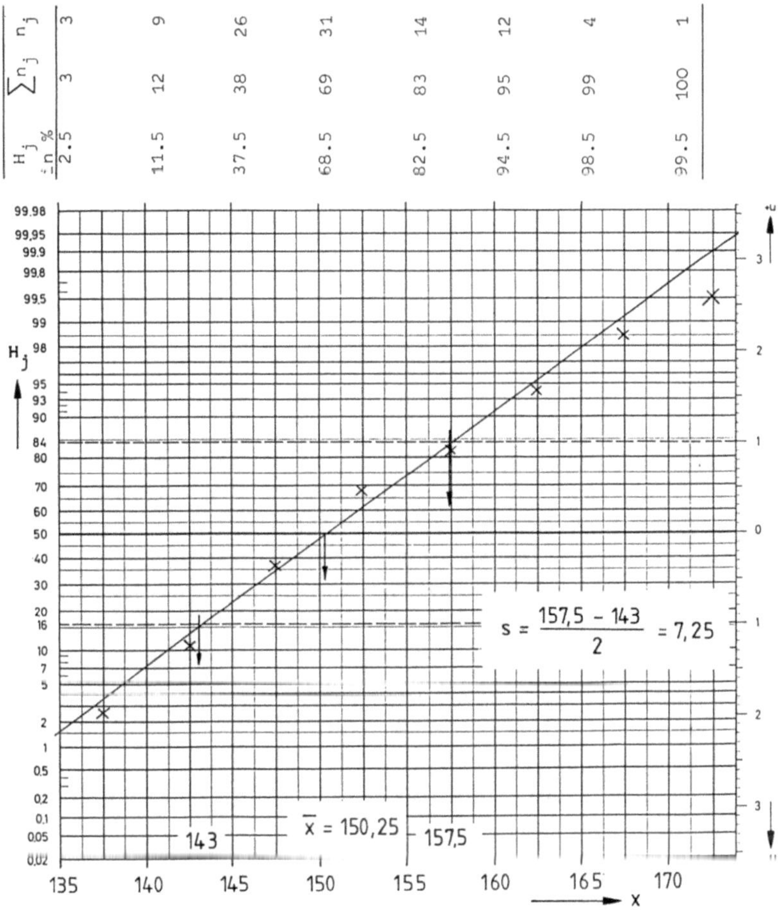

Fig. 12.4 b. Sum frequency in the PG for the random samples from the tolerance model

13 Assessment of Completed Batches and Production Batches

13.1 General

The basic philosophy for the assessment of finished batches or batches still in production was given in Sect. 5.7 in conjunction with Fig. 5.6.

In the following section it will be shown that with arithmetically combined dimensional chains, small error proportions can be completely harmless and larger error proportion often do not exhibit the effects which may be feared without taking account of the deviation progression principle.

It should not however be assumed from now on that high error proportions are always acceptable or permissible.

Basically if

—high conformity is produced, that means
—the individual values should deviate as little as possible from the desired value (mean value) and
—in no case should the limit values be exceeded by more than small error proportions.

High conformity never means the highest, technically achievable conformity. This would be the tolerance class 1 as applied to templates, gauges etc. A bicycle for example with bearings of this tolerance class would be too expensive; in any case it would not function substantially better nor last longer.

High conformity has its limits where they can be economically drawn.

The basic requirement that limit values are not significantly exceeded assumes that the tolerances

—must conform to functional and technical production standards, if not
—high costs must be paid both for production itself and/or for special measures such as sorting assembly or optional assembly or
—higher error proportion in the batches must be accepted, which many employees would not understand or
—tolerances must simply be expanded, as suggested in earlier sections, with the correct requirement that the new limit values either must not or must only occasionally be exceeded—and then by very little.

In addition with tolerances which belong to a statistically calculated dimensional chain, the requirement must be considered that the mean values must lie near the mean size. This can be even better fulfilled in cases where the tolerances are larger than the production spread range the smaller the AQL values are, Fig. 13.1. The AQL value (acceptable quality level) is an error proportion in percentage to which is given a given high acceptance probability.

While with values which belong to arithmetically calculated chains, AQL values over AQL = 1 can be thoroughly examined, this is not always possible with statistically calculated dimensional chains as in Fig. 13.1—in conjunction with Fig. 10.10. Further details on this in Sect. 13.3.

From Fig. 13.1 the following concept is important: If with the normal distribution the error proportion is $p = \mathrm{AQL}$, then with AQL = 10 the error proportion in a rectangular distribution is larger; with other examples the error proportion is $p_{\mathrm{RD}} = 0$. Both derive from Fig. 8.24 and are not incorrect. If a batch with an RD has exactly the same parameters μ and σ as a batch with ND, then for both batches always the same acceptance probability obtains. Finally both batches also have exactly the same effects on the closing value of the dimensional chain to which they belong.

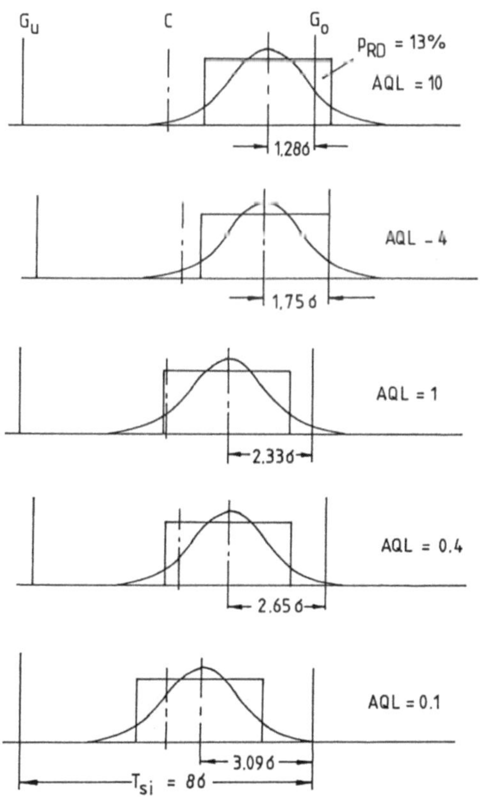

Fig. 13.1. Positions of normally distributed and rectangularly distributed production batches in the tolerance zone which with normal distribution is $p = \mathrm{AQL}$

13.2 Random Sample Dimension Tests

The data mentioned in Sect. 5.7

—sample range n
—acceptance factor k_A (index A often left out)
—a point of the OC, $P_a(p)$ and
—a further point of the OC

are generally used so that first two points of the OC, for example p_{90} and p_{10} are established; both the other data n and k_A are then calculated or simply specified graphically

- with Fig. 13.2, if the standard deviation of the batch is known, which is seldom the case, or
- with Fig. 13.3, if the standard deviation of the batch is unknown and thus must be estimated from the sample.

In both cases the same acceptance factor k will be derived; only the sample range is different

$$n_s = \left(1 + \frac{k^2}{2}\right) n_\sigma$$

In both cases the analysis of the batches is as follows:

—extraction of the random sample of the range n_σ or n_s,
—determination of the dimensional series for the relevant size,
—evaluation of the dimensional series by $\bar{x}$ and s (by the first-named method, the σ method, the s is only necessary for checking $s \approx \sigma$).
—formation of the test sizes.

$$Z = \bar{x} \pm k\sigma, \qquad (13.1)$$

or

$$Z = \bar{x} \pm ks. \qquad (13.2)$$

—decision about the batch:
—test size Z in the tolerance zone, limit values included, then acceptance of the batch.
—test size Z not in the tolerance zone, then rejection of the batch.

The effectiveness of a test demonstration is determined by its operating characteristics; as an example in Fig. 13.4a an OC for the test demonstration $n_\sigma - k = 6 - 1.8$ is represented and in Fig. 13.4b for $n_\sigma - k = 8 - 1.3$.

Exercise 13.1

Given: A store batch with drive shafts. For the shaft unit (Fig. 8.4 and Fig. 9.4) the size is

$$L = 59.5^{+0.15}_{-0.15}\ \text{mm}.$$

$p_{90} = 1\%$ and $p_{10} = 6\%$ are compatible.

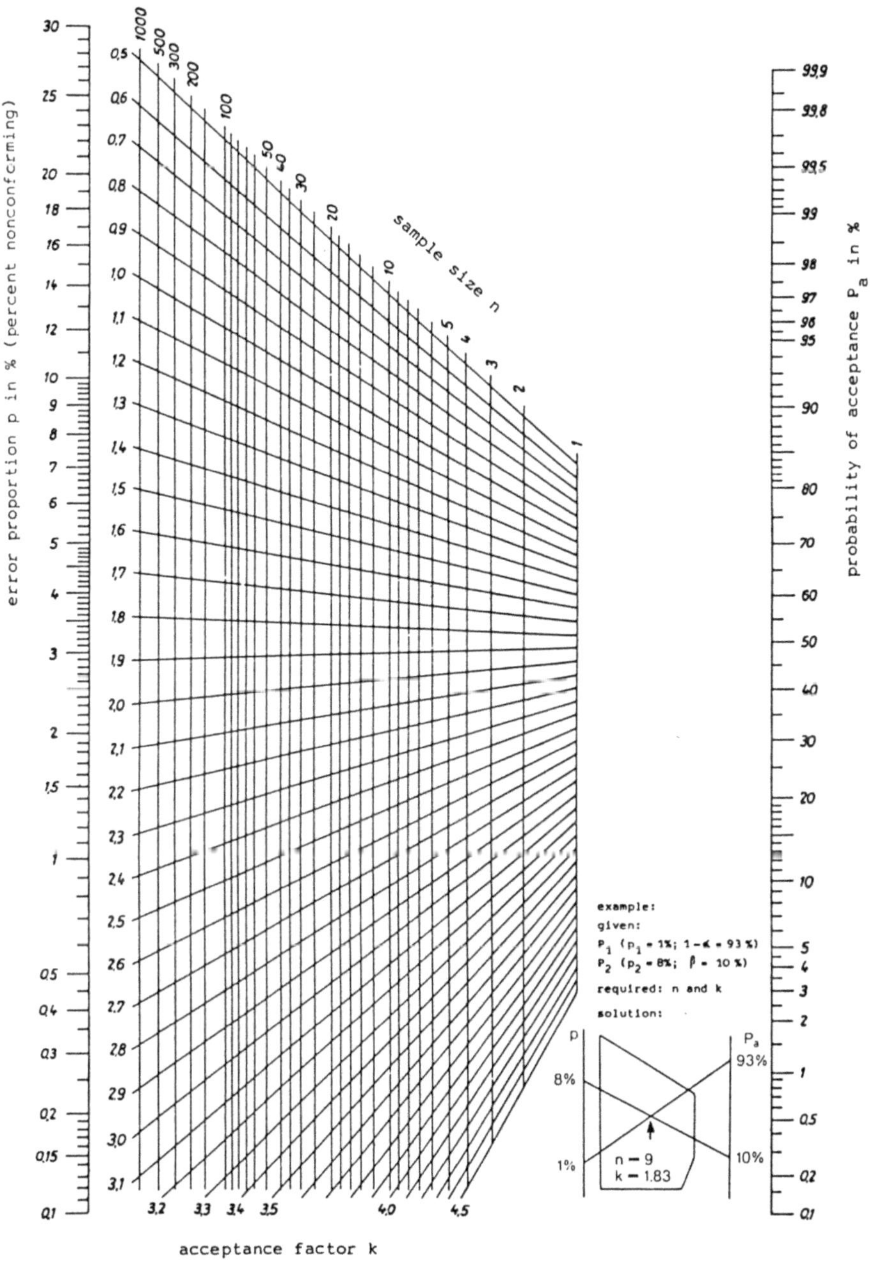

Fig. 13.2. Nomogram for determination of $(\bar{x}, \sigma)$-sample plans for measured testing of given limit values. From Wilrich, P.-Th: Nomogram for determination of sample plans for measured testing with a one-sided given tolerance limit. Quality and reliability 15 (1970) pp. 61–65 and 181–187, Munich–Vienna: Carl Hanser Verlag. Printed with kind permission of the author

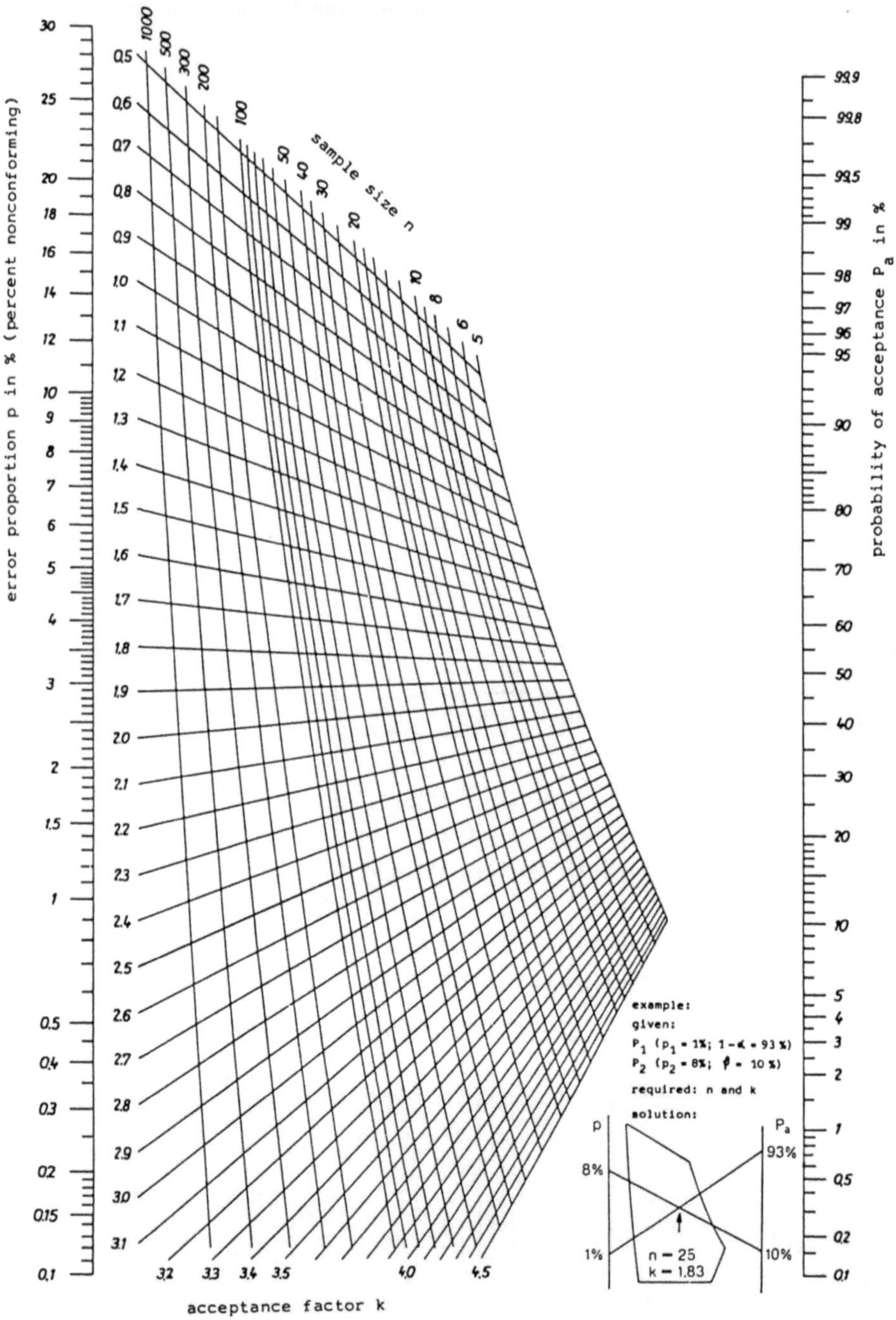

Fig. 13.3. Nomogram for determination of $(\bar{x}, s)$-sample plans for measured testing of given limit values. From Wilrich, P.-Th: Nomogram for determination of sample plans for measured testing with a one-sided given tolerance limit. Quality and reliability 15 (1970) pp. 61–65 and 181–187, Munich–Vienna: Carl Hanser Verlag. Printed with kind permission of the author

Required: $n_s - k$.
Solution: With Fig. 13.3 the test demonstration is given

$$n_s - k = 35 - 1.95.$$

The sample testing described can only be used when the same specification from continuous store batches is analysed over a long time period. If all batches (or the majority) are accepted then it can be stated with high probability that all batches (majority) are good with $p \lesssim \text{AQL}$.

On individual points about the sample test for quantitative characteristics, reference should be made to [4] or the given literature.

Here only the question which is not handled in sufficient detail in the literature will be discussed, what happens with a batch which is rejected. The answer can be for example

—sending back the batch to the supplier with the requirement to replace with a conforming batch,
—sorting the whole batch at the cost of the supplier and sending back the non-conforming items with the requirement to replace with conforming items,
—using all items in the batch, if necessary with a price reduction.

The first two solutions are very costly and—as shown in Chap. 8—often not efficient (Figs. 8.17 and 8.18) or in some cases senseless (Fig. 8.20).

Thus it is usually better to use the third answer and to test whether the batch can be used unsorted. This test is based on the answers to the following connected questions:

—How many members k does the dimensional chain have, to which this value belongs?
—How would the dimensional chain be calculated, arithmetically or statistically and if statistically under what assumptions?
—Is the tolerance of the value of the rejected batch relatively small or relatively large compared to the tolerances for the other members of the dimensional chain?
—Does the test size Z lie slightly or considerably outside the tolerance zone?
—What is known about the other batches which will be assembled into the dimensional chain? If values belong to this which come from the same production facility, can very much be known, if for example quality control charts are presented.

From the answers to these questions a decision can be made as to whether to deliver this batch to assembly unsorted.

If this is not possible, it is recommended that the batch should be subject to further tests by raising the sample already taken to $n = 200$ (or $n = 500$), after the evaluation of which according to $\bar{x}$ and s a properly safe analysis can be made of the position of the batch in the tolerance zone and whether the error proportion is acceptable.

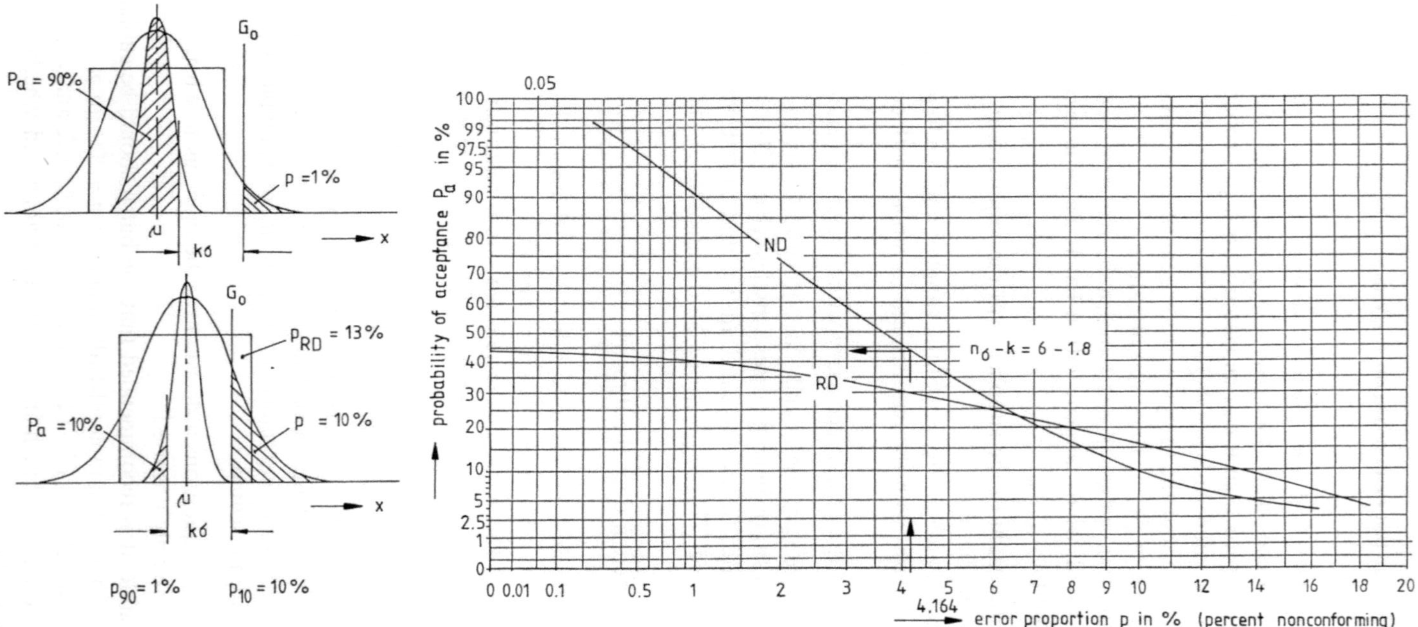

a

Fig. 13.4 a. Operational characterstics for the test $n_\sigma - k = 6 - 1.8$ with normally distributed and rectangularly distributed batches

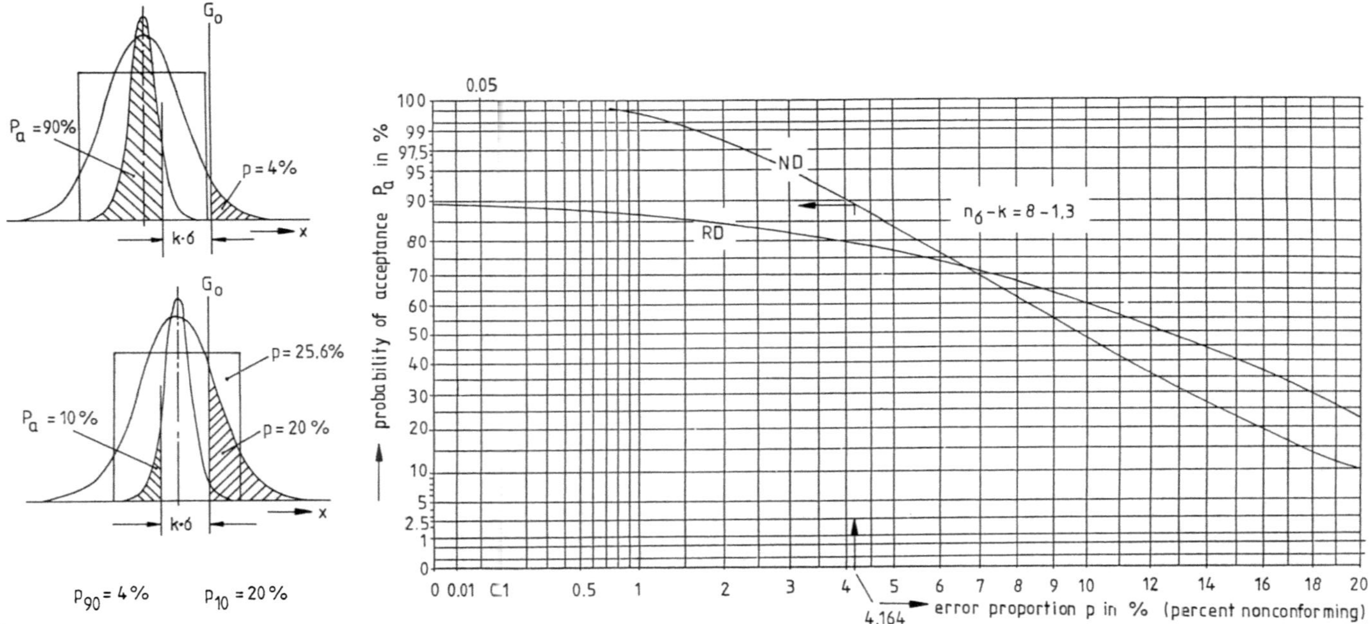

b

Fig. 13.4 b. Operatonal characteristics for the test $n_\sigma - k = 8 - 1.3$ with normally distributed and rectangularly distributed batches

Only when this analysis requires it should the whole batch be sorted.

It would be completely senseless to thoroughly sort the rejected batch in order to use, for example $N = 10\,000$ sorted parts and decide that the error proportion of the total $\Delta p = 0.5\%$ lies over the AQL value, when with the unsorted assembly no account has been taken of the function of the dimensional chain.

To conclude this section it should be pointed out that in the literature the sample test of quantitative characteristics is only recommended when it is shown that the completed batch has an (approximately) normal distribution.

In fact completed batches frequently have the shape of a mixed distribution of the first type, sometimes even the shape of a mixed distribution type 2 (see Chap. 6).

The sample test can also be used with deviations from the ND shape. In each case it is tested whether the test size

$$Z = \bar{x} \pm ks$$

lies in the tolerance zone or not. This test result depends only on the position of the distribution and its spread, not on the distribution shape of the individual values. Sometimes the OC are linked to the assumption of the presence of a normal distribution, for which they were calculated. If the distribution shape deviates from a normal distribution, then the OC which would have been calculated assuming the presence of a normal distribution are no longer valid. For example with a rectangular distribution of the individual values and the test demonstration $n_\sigma - k = 6 - 1.8$ the OC in Fig. 13.4a lie deeper and flatter than the OC for a normal distribution. The same applies for the test demonstration in Fig. 13.4b.

This should however have no influence on the decision about the acceptance or rejection of a batch; each abnormally distributed batch behaves in a dimensional chain as a normally distributed batch with the same parameters μ and σ.

13.3 Selection of AQL Values for Random Sample Tests of Dimensions (Variable Testing)

Statements about aspects of the selection of AQL values for sizes are not found in the literature. Since however, sizes are usually tested if they are members of a dimensional chain, it is left to conformity to principles according to which the characteristics of the size totals arise, to produce representative AQL values. These principles are the central limiting values, according to which size totals with a growing number of individual sizes quickly tend towards a normal distribution. Among other things according to the deviation progression principle the variance of size totals is always the sum of the individual variances,

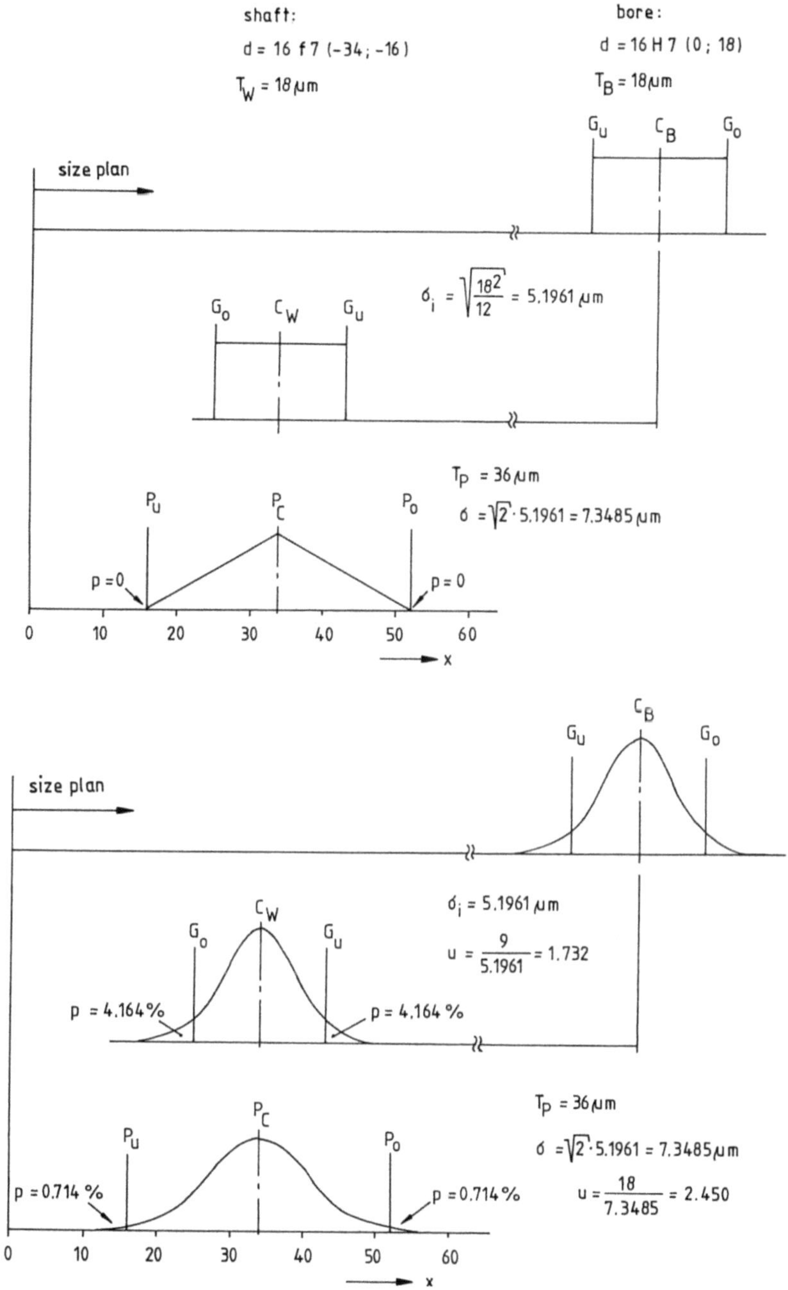

Fig. 13.5. Size plans for a two-member dimensional chain with rectangular distributions and with normal distributions in the individual batches

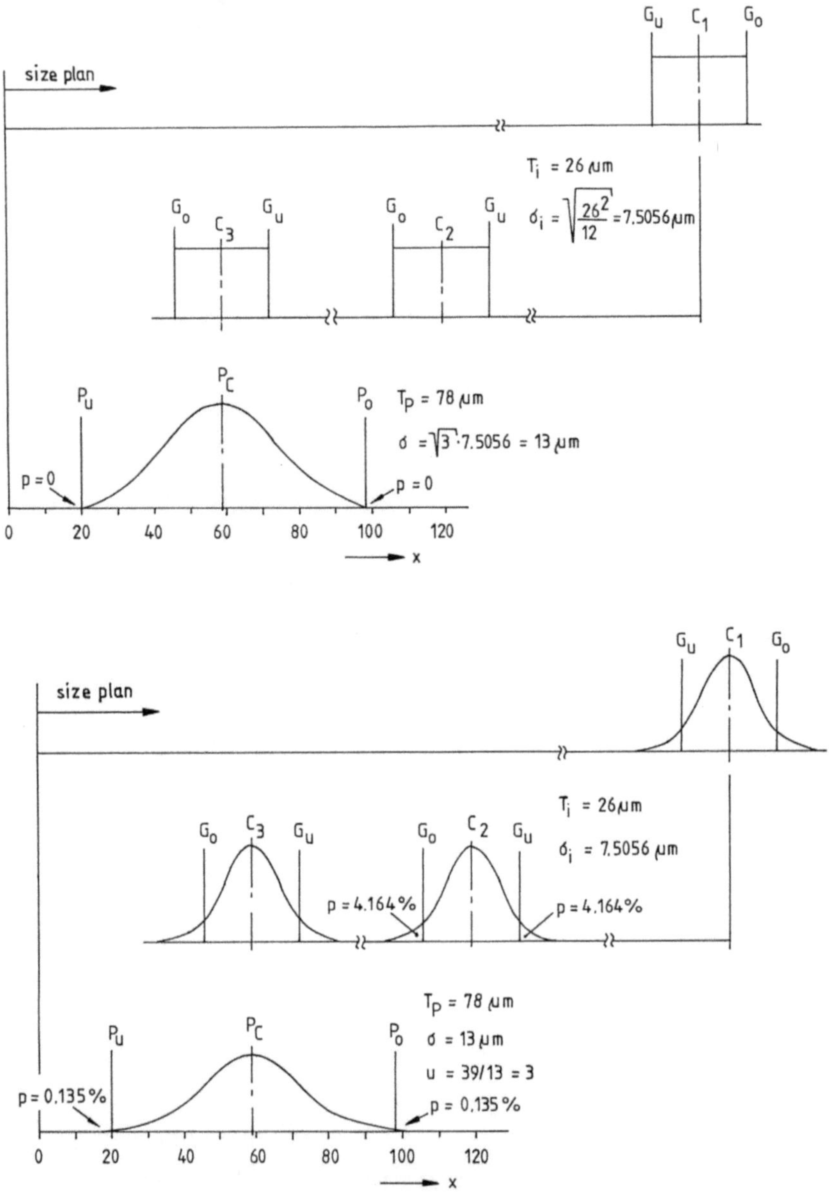

Fig. 13.6. Size plans for a three-member dimensional chain with equally large individual tolerances

and is indeed independent of the distribution shape of the individual size batch. What goes for the size totals goes equally well for the closing values.

In Fig. 13.5 above there is a size plan for the bearing with a nominal diameter $d = 16H7/f7$ represented with information about the tolerance zones and the additional assumptions and representation that since the batches of bores and shafts show rectangular distributions, they fill their tolerance zones. This occurs in practice for example if two-sided equally high error proportions are sorted from original wide normal distributions. In this way all size values in both batches fulfill the specification, errors are not present and both batches are useable without constraint. The standard deviation of both batches is as Fig. 4.9

$$\sigma_i = \sqrt{\frac{T^2}{12}}$$

with the numerical information in Fig. 13.5.

After assembly the fits (clearances) have a triangular distribution which exactly fills the tolerance on fit zone. The error proportion in the clearances is exactly nil below the lower fit P_u and above the upper fit P_o.

Since—as is often mentioned—the deviation progression principle is valid independent of the distribution shape, even optionally distributed batches of bores and shafts must be useable, in that they have the same parameters μ and σ as the rectangular distributions in Fig. 13.5 above. And these must hold true for an exact normal distribution, although this then has a two-sided error proportion of $p = 4.164\%$, Fig. 13.5 below. The tolerance on fit here with $\sigma = 7.3485$ µm has the same standard deviation as the triangular distribution above, however in the example below, the clearances are normally distributed with an error proportion of $p = 0.714\%$ below the lower tolerance on fit P_u and above the upper tolerance on fit P_o. This error proportion is small and technically representative, since the function of all assembled running fits is not in question.

The "error comparison" is somewhat clearer with a three-member dimensional chain. In Fig. 13.6 above it is stated that all three batches show useable rectangular distributions. All closing values lie in their (arithmetically calculated) tolerance zone with the tolerance $T_p = 3T_i = 78$ µm. In the lower size plan of Fig. 13.6 the individual batches are exactly normally distributed with the same parameters as above; each batch has an error proportion of $p_{tot} = 8.328\%$. However the normally distributed clearance gain the clearance tolerance zone with their 6σ range.

In Fig. 13.7 the same is represented for a four-member dimensional chain. The effects of rectangular distribution and normal distribution with the same parameters are almost identical on the clearance distribution.

At this stage the objection could be raised that the examples up to now are not generally valid, since all individual tolerances are the same size. This objection is justified but the effects of tolerances of different sizes are insignificant compared with those of equal size, Fig. 13.8. The reader can compare the error proportions in the clearances of Fig. 13.8 and those of Fig. 13.6.

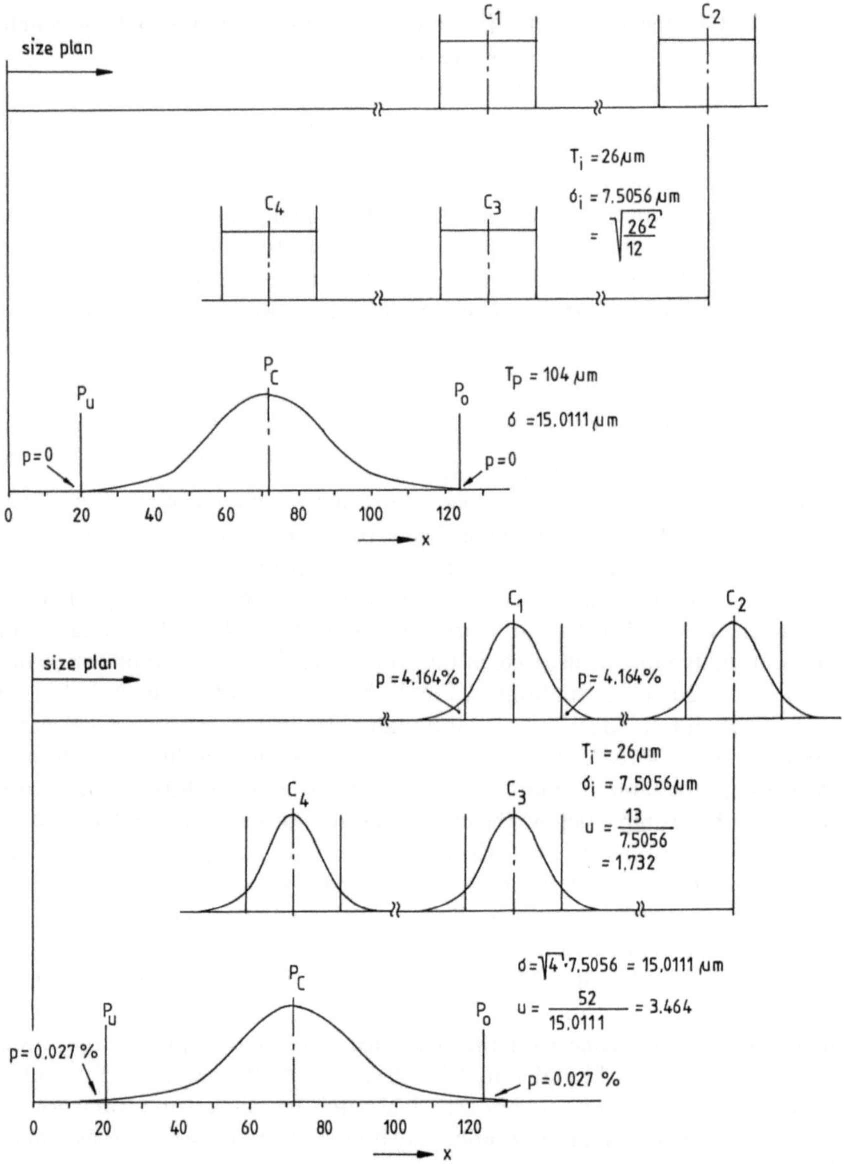

Fig. 13.7. Size plans for a four-member dimensional chain with equal individual tolerances

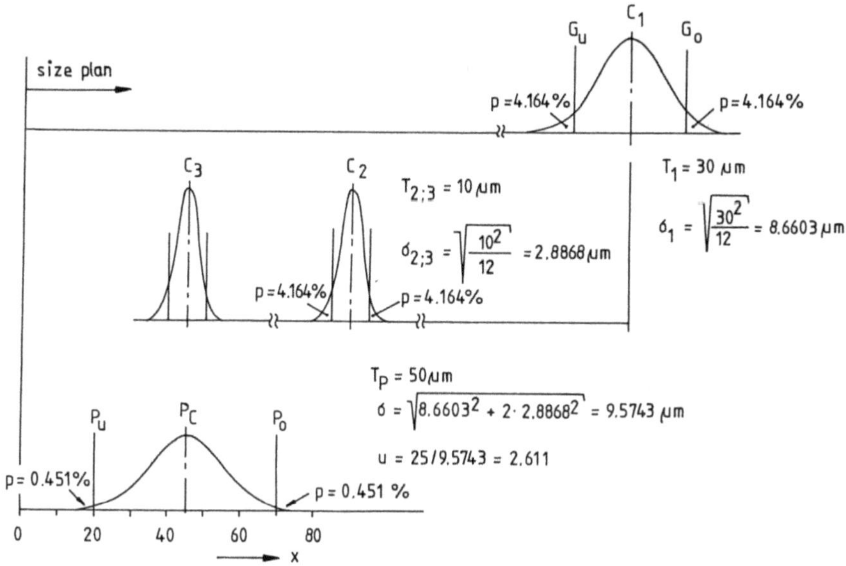

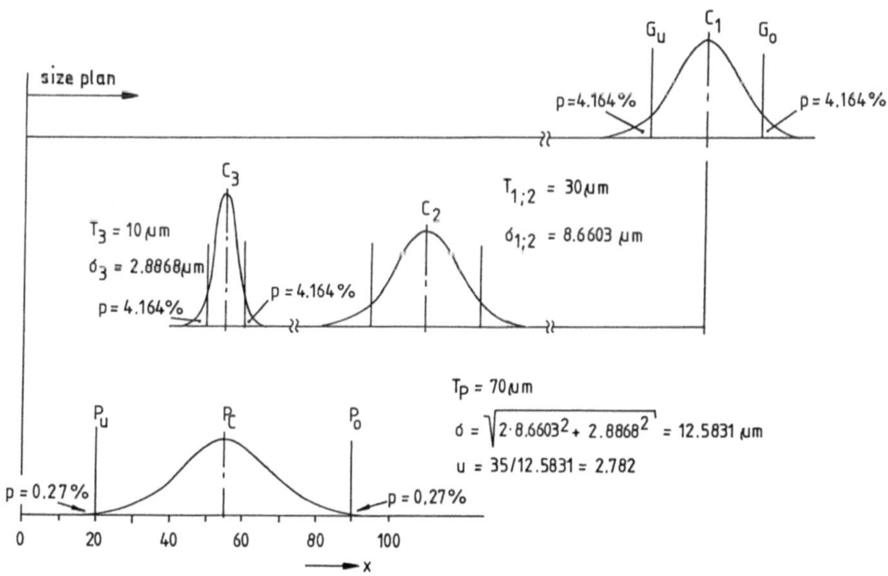

Fig. 13.8. Size plans for a three-member dimensional chain with unequal individual tolerances, however of the same order of size

From all examples up to now it can be concluded that with variable testing of sizes an acceptable quality level of AQL = 4.00 is representative, if it is a matter of sizes which belong to two to four-member dimensional chains. With sizes which belong to dimensional chains with $k > 4$ size members, higher AQL values could be considered; instead of that however a statistical calculation is suggested. For variable testing with AQL = 4 it is stated that—as usual—a one-sided test is made and indeed should be consecutive against the upper limit value G_o and the lower limit value G_u.

It will be seen that with the testing of completed batches with AQL = 4 and according to which factual assumptions without exception can be made over a long time period, the actual error proportions on average are clearly smaller than $p = 4.164\%$, so that the examples in the Figs. 13.5 to 13.8 with these reject proportions represent only the extreme worst cases of all individual size batches. The testing of individual size batches with AQL = 4 must therefore guarantee with very high probability constantly safe overall values.

x_j	N_j	
19	1	
20	1	
21	1	
22	1	
23	1	5
24	1	
25	1	
26	2	
27	3	
28	4	
29	4	20
30	5	
31	7	
32	9	
33	9	49
34	11	
35	13	
36	15	
37	17	105
38	20	
39	22	
40	24	171
41	27	
42	29	
43	31	258
44	33	
45	35	
46	37	
47	38	
48	39	
49	40	
50	40	
51	40	
52	39	
symmetrically continued		
81	1	

Fig. 13.9. Simulation of sorted batches taken from a model normal distribution

Parameters of the ND-model

$N = 1000$

$\mu = 50$

$\sigma_N = 9{,}992597$

	values sorted one-sidely					
remainig N	995	980	951	895	829	742
mean μ	50.145729	50.494898	51.057298	52.030168	53.062726	54.354447
standard deviation σ	9.802875	9.457584	9.003662	8.385555	7.836134	7.254498
proportion $\sigma_{sorted}/\sigma_{unsorted}$	0.981014	0.946459	0.901033	0.839177	0.784194	0.725987
distance $u = \frac{\mu - G_u}{\sigma}$	2.718	2.220	1.951	1.733	1.603	1.496
apparent error proportion p in %	0.328	1.321	2.553	4.155	5.447	6.733

	values sorted two-sidely					
remainig N	990	960	902	790	658	484
mean μ	50	50	50	50	50	50
standard deviation σ	9.609654	8.894169	7.914594	6.501606	5.158579	3.632384
proportion $\sigma_{sorted}/\sigma_{unsorted}$	0.961677	0.890076	0.792046	0.650643	0.516240	0.363508
distance $u = \frac{\mu - G}{\sigma}$	2.758	2.305	2.085	1.923	1.850	1.789
apparent error proportion at every side p in %	0.291	1.055	1.853	2.724	3.216	3.681

Fig. 13.9. (Continued)

Occasionally there will be a warning that with variable testing in particular incorrect decisions can occur if the assumption of the presence of a normal distribution is infringed so that the formerly normally distributed batch has been sorted by the supplier one-sidedly or two-sidedly at the ends.

To clarify these questions, in Fig. 13.9 different proportions have been sorted from a model normal distribution given in Fig. 4.8, both one-sidedly and two-sidedly. In addition the parameters μ and σ of the remaining distribution are calculated. As expected in all cases the standard deviation of the remaining distribution becomes smaller, Fig. 13.10. Further the apparent error proportions of the fictitious normal distributions outside the sort limits are calculated. The fictitious normal distributions are each normal distributions with the same parameters μ and σ as those in the actual remaining distributions.

The course of the apparent error proportions above the sorted error proportion is shown in Fig. 13.11. The lower curve is valid for the case where both sided equal proportions are sorted from the (exact) normal distribution. For further details see Fig. 13.12 for the case of the two-sided sort from $p=10.5\%$. If the sort limits are limit values, the remaining batch must be error-free. If this fact is however unknown and the presence of a normal distribution is suggested, the apparent error proportion includes both sides $p=2.72\%$.

Since the remaining distribution in every respect behaves as does the fictitious normal distribution, the same acceptance probability stands for the remaining distribution as that calculated assuming the presence of a normal distribution—according to test result—or can be read from [4]. This means that there is a very high acceptance probability for the testing of this remaining

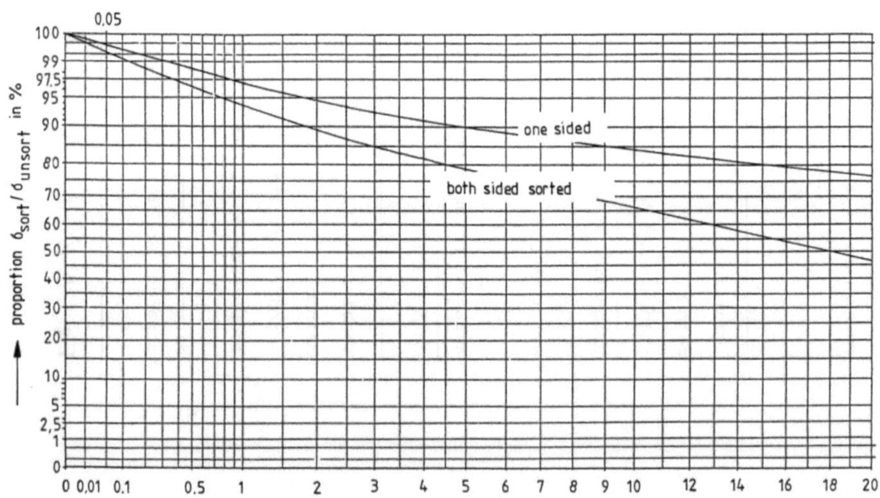

Fig. 13.10. Relative reduction in the standard deviation with sorted batches which are normal before sorting

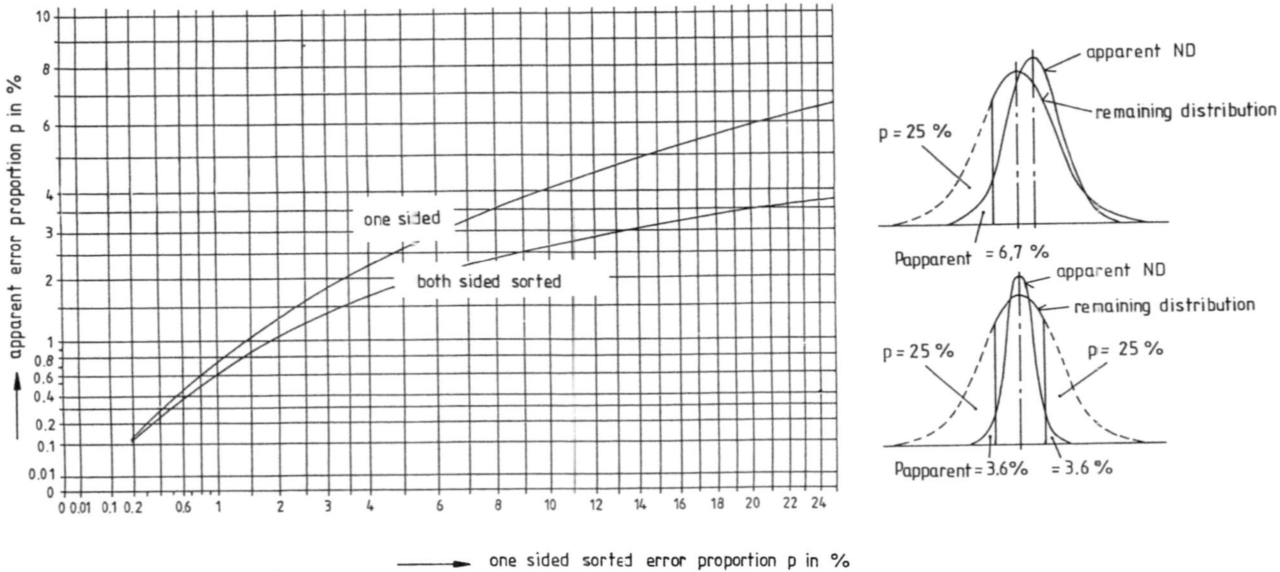

Fig. 13.11. Apparent error proportion with one-sided and two-sided sorted batches under the assumption of the presence of a (fictitious) normal distribution; the sorted batches were normal before the sort

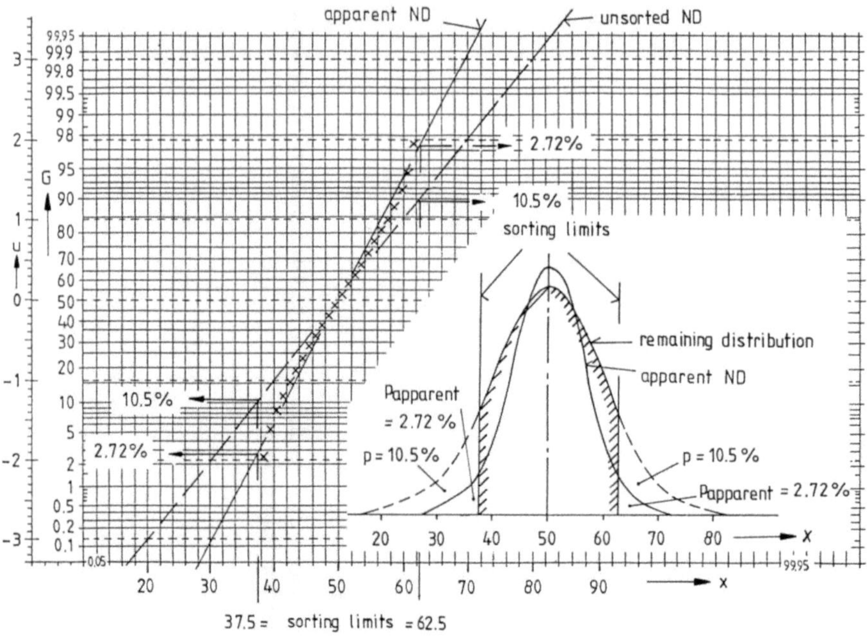

Fig. 13.12. Unsorted ND and fictitious ND after two-sided sort of $p = 10.5\%$

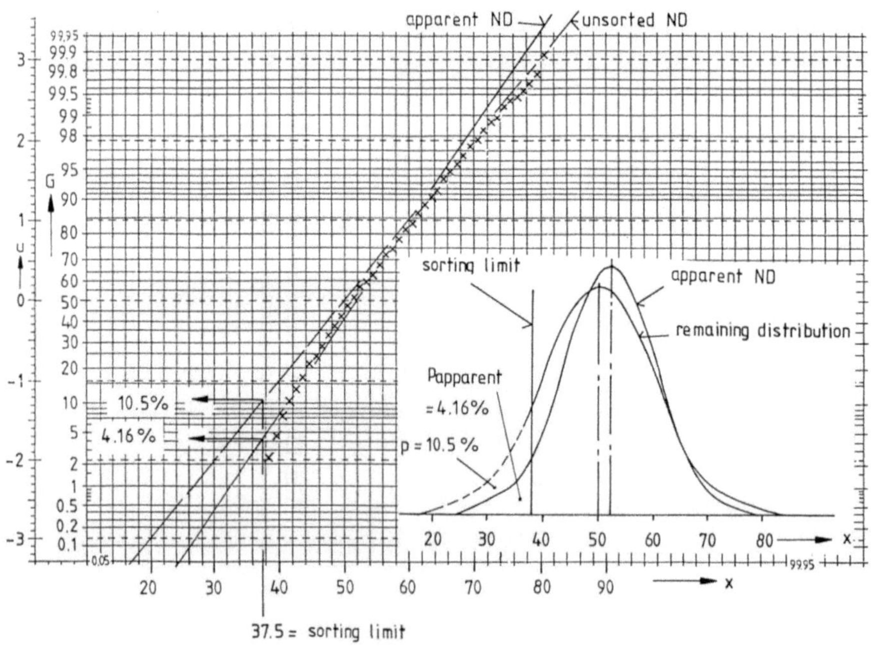

Fig. 13.13. Unsorted ND and fictitious ND after one-sided sort of $p = 10.5\%$

distribution against AQL = 4 as suggested at the beginning of this chapter. This is also valid for all other, two-sided similarly sorted batches. The lower curve in Fig. 13.11 is asymptotic against $P_{\text{apparent}} = 4.164\%$; the more the two-sided sorted error proportion nears 50%, the better the remaining distribution fulfills the criterion "rectangular distribution".

The high acceptance probability of this kind of sorted batch can only be right for the buyer if the batches have been made useable through sorting; they behave in the dimensional chain as the ordered fictitious normal distribution.

The circumstances are somewhat different if the first normally distributed batches are sorted one-sidedly at the suppliers. Then the mean value of the fictitious normal distribution moves away from the sort limit, but the standard deviation is larger than with two-sided sorting, Fig. 13.10, with the result that the apparent error proportion is larger, upper curve in Fig. 13.11. For further explanation of one-sided sorting of $p = 10.5\%$, see Fig. 13.13.

If one-sidedly sorted batches are tested against AQL = 4, then the acceptance probability is high if up to $p \approx 10\%$ is sorted. With error proportions sorted to a higher level on one side this will normally indicate rejection although the remaining batches are error-free and therefore useable.

While by the acceptance of a batch a decision has been made about the batch, rejection of a batch means only the deferment of a decision, until further information is presented about the batch. Usually the supplier is first required to express an opinion about the batch rejection and if it concerns a sorted batch the supplier will disclose this in his own interest. Then the buyer can restrict himself to testing this fact in an appropriate way, in order to decide about the acceptance of this batch in the case of a positive result.

Here the question presents itself, why the supplier did not inform the buyer at delivery of the sort test. Two reason are possible. Firstly it is possible that the supplier has not seen the connection and is of the opinion that his error-free sorted batch must be accepted with safety. Secondly it is possible that the supplier has seen the connection partly but not completely. He knows that the buyer analyses his stock batches by the variable method and that according to [4] the presence of a normal distribution is assumed. So he is afraid that his remaining distributed and error-free batch is however irregular somehow and he shies away from exposing his sort process in the hope that his batch will despite this be accepted; a justified hope, as long as he has sorted both sides to the same extent or one side less than $p \approx 10\%$ and as long as the buyer tests according to convention with AQL = 4.

According to all the discussions up to now, the supplier is well advised to expose his sort testing in every case; he risks no disadvantage. Also it is in the interest of the buyer to be informed about this at the time of delivery of the batch if it has been sorted. He can then restrict his input testing to an appropriate check of this information.

At the end of Sect. 13.2 a statement was made that variable testing of sizes which belong to a dimensional chain must be used independently of distribution shape. Against this thesis, considerations can be raised for the case where the completed batches are tested in advance, and which deviate in the extreme from the shape of the normal distribution, and where in one point of the distribution a considerably higher error proportion will lie than in the (fictitious) normal distribution with numerically identical parameters μ and σ.

To investigate such a case with simultaneous choice of an acceptable quality limit of AQL = 4 an (associated) mixed distribution would be chosen from a ND, with $N_1 = 1\,008$ and one of the range of smaller ND_2 with $N_2 = 0.1$ N, and with a standard deviation $\sigma_2 = 2\sigma_1$, Fig. 13.14. The distribution function of this MD is shown in Fig. 13.15. Above $\mu + 3$ lies a value proportion of

	ND_1	ND_2				rounded for folding:
x_j	N_{j1}	N_{j2}	N_j	$\sum N_j$	G(x) in %	N_j
1	1		1	1	0.0902	1
2	5		5	6	0.5411	5
3	15	parameters of ND_2	15	21	1.894	15
4	40		40	61	5.501	40
5	103	N_2 = 100.8	103	164	14.791	103
6	150		150	314	28.319	150
7	190	μ_2 = 26	190	504	45.455	190
8	190		190	694	62.590	190
9	150	σ_2 = 4	150	844	76.118	150
10	103		103	947	85.408	103
11	40		40	987	89.015	40
12	15		15	1002	90.368	15
13	5	0.1	5.1	1007.1	90.828	5
14	1		1	1008.1	90.918	1
15		0.5	0.5	1008.6	90.963	1
16						
17	parameters of ND_1	1.5	1.5	1010.1	91.098	2
18						
19	N_1 = 1008	4	4	1014.1	91.459	4
20						
21	μ_1 = 7.5	10.3	10.3	1024.4	92.388	10
22						
23	σ_1 = 2	15	15	1039.4	93.741	15
24						
25		19	19	1058.4	95.454	19
26						
27		19	19	1077.4	97.168	19
28						
29		15	15	1092.4	98.521	15
30						
31		10.3	10.3	1102.7	99.450	10
32						
33		4	4	1106.7	99.811	4
34						
35		1.5	1.5	1108.2	99.945	2
36						
37		0.5	0.5	1108.7	99.991	
38						
39		0.1	0.1	1108.8	100.000	

parameters of the mixed distribution:			
	N = 1108,8	lot size	N = 1109
	μ = 9.1818	mean	μ = 9.1749
	σ = 5.7772	standard deviation	σ = 5.7547

Fig. 13.14. Generation of an extremely sloping mixed distribution from two normal distributions ND_1 and ND_2

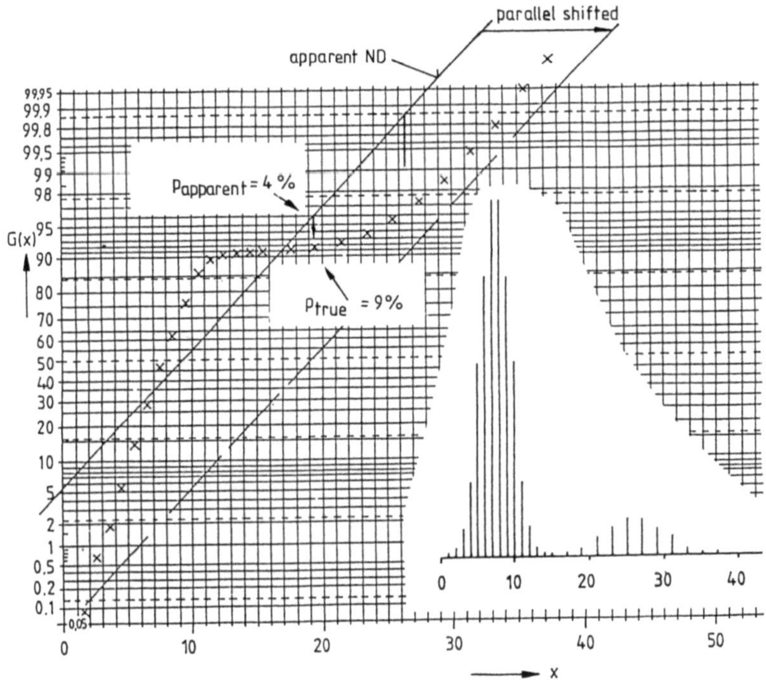

Fig. 13.15. Distribution function of an extremely sloping mixed distribution and ordered fictitious normal distribution, as Fig. 6.14b

$p \approx 4.5\%$. However this MD is not wider than the ND with the same parameters (fictitious ND) and its parallels in Fig. 13.15; only because of its sharpness is the value proportion of the MD so relatively large in a point.

Further it is evident from Fig. 13.15 that with a batch test according to AQL = 4 and an apparent error proportion of $p = 4\%$ the true error proportion $p \approx 9\%$.

In order to clarify the question of how this true error proportion effects the overall value of a two-member dimensional chain, two of this kind of mixed distributions were folded. To simplify the folding, all occupation numbers of the MD were rounded to whole numbers, last column in Fig. 13.14. The fold process (in addition) and the full fold product are shown in Fig. 13.16; the distribution function of the fold product and the ordered fictitious ND are represented in Fig. 13.17.

If two normally distributed batches are assembled with $p = 4\%$ in the same effective direction (reduction of clearance), then the error proportion expected in the clearances is $p = 0.664\%$, Fig. 13.18, upper size plan. If the two batches are only fictitiously normally distributed and in truth show sloping mixed distributions with an error proportion of $p \approx 9\%$, then the error proportion in the clearances is $p \approx 3\%$, Fig. 13.17 and lower size plan in Fig. 13.18.

Fold process (in addition) of two extremely mixed distributions with the parameters N = 1109, μ = 9.1749 and σ = 5.7547

x	g(x)	1	2	3	4	5	6	7	8
		1	5	15	40	103	150	190	190
1	1	2 1	3 5	4 15	5 40	6 103	7 150	8 190	9 190
2	5	3 5	4 25	5 75	6 200	7 515	8 750	9 950	10 950
3	15	4 15	5 75	6 225	7 600	8 1545	9 2250	10 2850	11 2850
4	40	5 40	6 200	7 600	8 1600	9 4120	10 6000	11 7600	12 7600
5	103	6 103	7 515	8 1545	9 4120	10 10609	11 15450	12 19570	13 19570
6	150	7 150	8 750	9 2250	10 6000				
7	190	8 190	9 950	10 2850					
8	190	9 190	10 950	etc.					

fold product:

x_j	N_j	$\sum N_j$	G(x) in %	x_j	N_j	$\sum N_j$	G(x) in %
2	1	1	0.00008	41	4668	1213309	98.65
3	10	11	0.00089	42	3244	1216553	98.92
4	55	66	0.00537	43	1914	1218467	99.07
5	230	296	0.0241	44	1578	1220045	99.20
6	831	1127	0.0916	45	552	1220597	99.24
7	2530	3657	0.2973	46	1037	1221634	99.3294
8	6570	10227	0.8315	47	68	1221702	99.3350
9	15020	25247	2.05	48	1138	1222840	99.4275
10	30509	55756	4.53	49	4	1222844	99.4278
11	53506	109262	8.88	50	1331	1224175	99.5361
12	82450	191712	15.59	51			
13	111660	303372	24.67	52	1412	1225587	99.6509
14	133600	436972	35.53	53			
15	142120	579092	47.09	54	1327	1226914	99.7588
16	133602	712694	57.94	55			
17	111670	824354	67.03	56	1110	1228024	99.8490
18	82484	906848	73.73	57			
19	53606	960454	78.09	58	817	1228841	99.9154
20	30783	991237	80.60	59			
21	15520	1006757	81.86	60	528	1229369	99.9584
22	7502	1014259	82.47	61			
23	3930	1018189	82.79	62	296	1229665	99.9824
24	3045	1021234	83.04	63			
25	3346	1024580	83.31	64	140	1229805	99.9938
26	4803	1029383	83.70	65			
27	6362	1035745	84.22	66	56	1229861	99.9984
28	8869	1044614	84.94	67			
29	10896	1055510	85.82	68	16	1229877	99.9997
30	13555	1069065	86.92	69			
31	15254	1084319	88.16	70	4	1229881	100.0000
32	16948	1101267	89.54				
33	17618	1118885	90.97				
34	17630	1136515	92.41				
35	16944	1153459	93.79				
36	15290	1168749	95.03				
37	13554	1182303	96.13				
38	10980	1193283	97.02				
39	8858	1202141	97.74				
40	6500	1208641	98,27				

parameters of the fold product:

$N = 1229881 = 1109^2$

$\mu = 18.3498 = 2 \cdot 9.1749$

$\sigma = 8.1384 = \sqrt{2} \cdot 5.7547$

Fig. 13.16. Fold process (in addition) and fold product from two extremely sloping mixed distributions with the same direction of slope

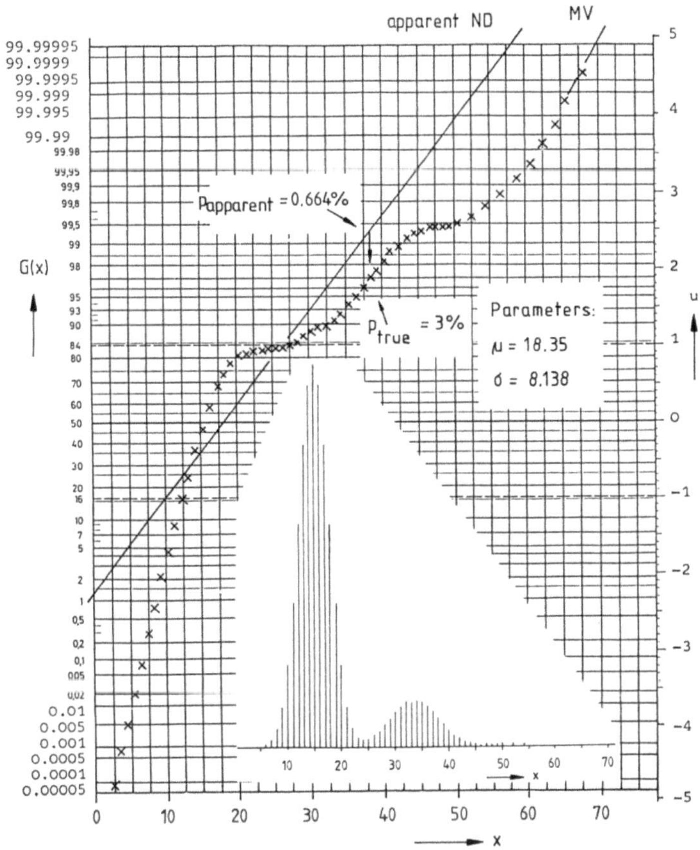

Fig. 13.17. Distribution function of the fold product of two extremely sloping mixed distributions which are folded in the same direction of slope

It is very unlikely that a completed batch would show an extremely sloping mixed distribution. It is also unlikely that both independently produced size batches would show the same extremely sloping mixed distribution. Should this case however occur, the error proportion in the clearances would only come to $p \approx 3\%$, which—depending on the individual case—would hardly affect the function of these bearings.

By folding the extremely sloping mixed distributions in Fig. 13.16 shown in Fig. 13.17 a similar direction is shown in the slopes of both distributions. If two extremely sloping distributions with the same effctive direction of the apparent error proportion (reduction in clearance) but dissimilar direction of slope are folded, Fig. 13.19, then the fold product is symmetrical, however the true error proportion in the clearances is $p \approx 3\%$, Fig. 13.20. Illustration of the "same effective direction" and the "dissimilar direction of slope" is given in Fig. 13.21.

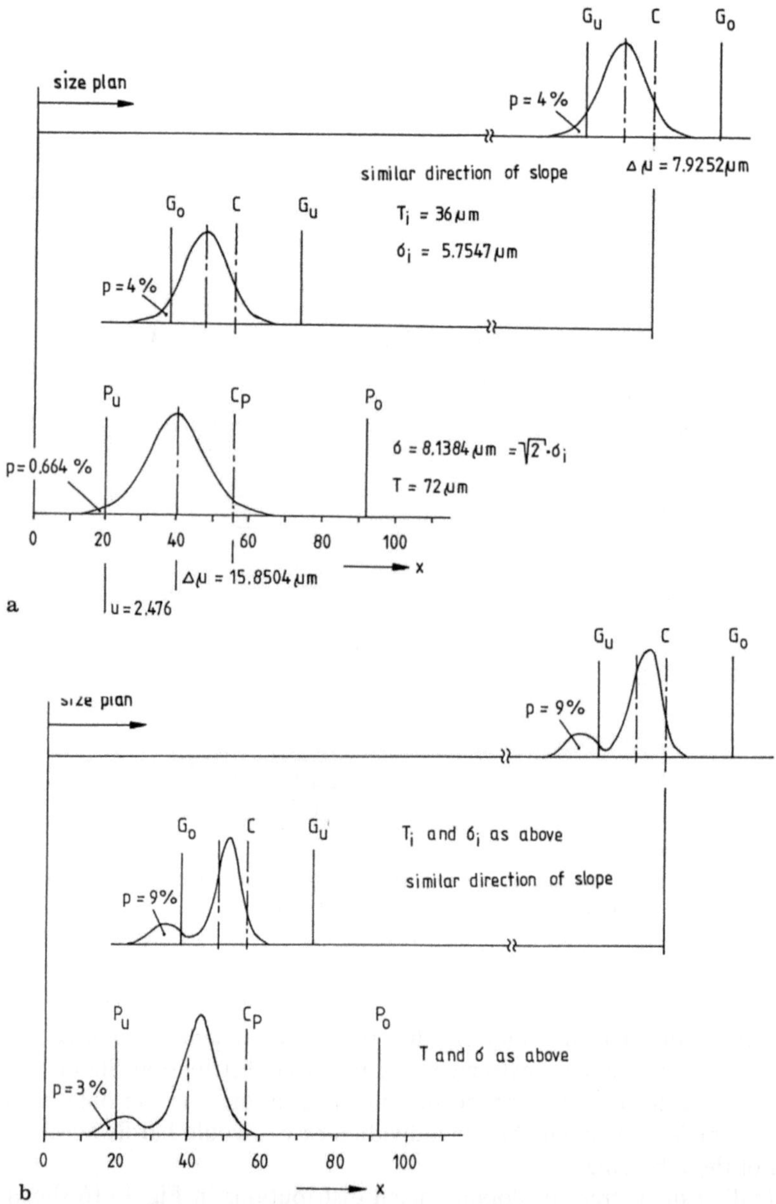

Fig. 13.18 a, b. Size plans for a two-member dimensional chain, **a** batches with fictitious ND, **b** batches with true MD

fold process (in addition) of two extremly mixed distributions

parameters: N = 1109; μ = 26.8251; σ = 5.7547

x	q(x)	1	2	3	4	5	6	7	8	9	10	11	12	13	14	15	16	17	18	19	20	21	22
		2		4		10		15		19		19		15		10		4		2		1	1
1	1	2 2		4 4		6 10		8		10 19		12 19		14 15		16 10		18 4		20 2		22 1	23 1
2	5	3 10		5 20		7 50		9 75		11 95		13 95		15 75		17 50		19 20		21 10		23 5	24 5
3	15	4 30		6 60		8 150		10 225		12 285		14 285		16 225		18 150		20 60		22 30			
4	40	5 80		7 160		9 400		11 600		13 760		15 760		17 600		19 400		21 160					
5	103	6 206		8 412		10 1030		12 1545		14 1975		16 1975		18 1545		20 1030							
6	150	7 300		9 600		11 1500		13 2250		15 2850		17 2850		19 2250									
7	190	8 380		10 760		12 1900		14 2850		16 3610		18 3610											
8	190	9 380		11 760		13 1900		15 2850		17 3610													
9	150	10 300		12 600		14 1500		16 2250 etc.															

parameters:

N = 1109

μ = 9.1749

σ = 5.7547

parameters of the fold product:

N = 1229881 = 1109^2

μ = 36 = 26.8251 + 9.1749

σ = 8.1384 = $\sqrt{2}\cdot 5.7547$

fold product:

x_j	$\sum N_j$	G(x) in %	N_i	$\sum N_i$	G(x) in %	x_i
2	2	0.00018	2	1229881	100.00000	70
3	12	0.00106	10	1229879	99.99984	69
4	46	0.00407	34	1229869	99.99902	68
5	146	0.01292	100	1229835	99.99626	67
6	422	0.03735	276	1229735	99.98813	66
7	932	0.08249	510	1229459	99.96569	65
8	1889	0.167	957	1228949	99.924	64
9	3344	0.296	1455	1227992	99.846	63
10	5678	0.503	2334	1226537	99.278	62
11	8839	0.782	3161	1224203	99.538	61
12	13268	1.174	4429	1221042	99.281	60
13	18715	1.656	5447	1216613	98.921	59
14	25492	2.256	6777	1211166	98.478	58
15	33119	2.931	7627	1204389	97.927	57
16	41593	3.681	8474	1196762	97.307	56
17	50402	4.461	8809	1188288	96.618	55
18	59219	5.241	8817	1179479	95.902	54
19	67691	5.991	8472	1170662	95.185	53
20	75344	6.668	7653	1162190	94.496	52
21	82121	7.268	6777	1154537	93.874	51
22	87640	7.751	5519	1147760	93.323	50
23	92075	8.149	4435	1142241	92.874	49
24	95420	8.445	3345	1137806	92.514	48
25	97849	8.660	2429	1134461	92.242	47
26	99952	8.846	2103	1132032	92.044	46
27	101890	9.018	1938	1129929	91.873	45
28	105663	9.352	3773	1127991	91.715	44
29	112699	9.974	7036	1124218	91.409	43
30	128796	11.399	16097	1117182	90.837	42
31	159442	14.111	30646	1101085	89.528	41
32	214112	18.950	34670	1070439	87.036	40
33	296579	26.249	82467	1015769	82.590	39
34	409573	36.249	112994	933302	75.886	38
35	543174	44.165	133601	820308	66.698	37
36	686707	55.835	143533			

Fig. 13.19. Fold process (extended) and fold product from two extremely sloping mixed distributions with dissimilar direction of slope; the fold product is symmetrical

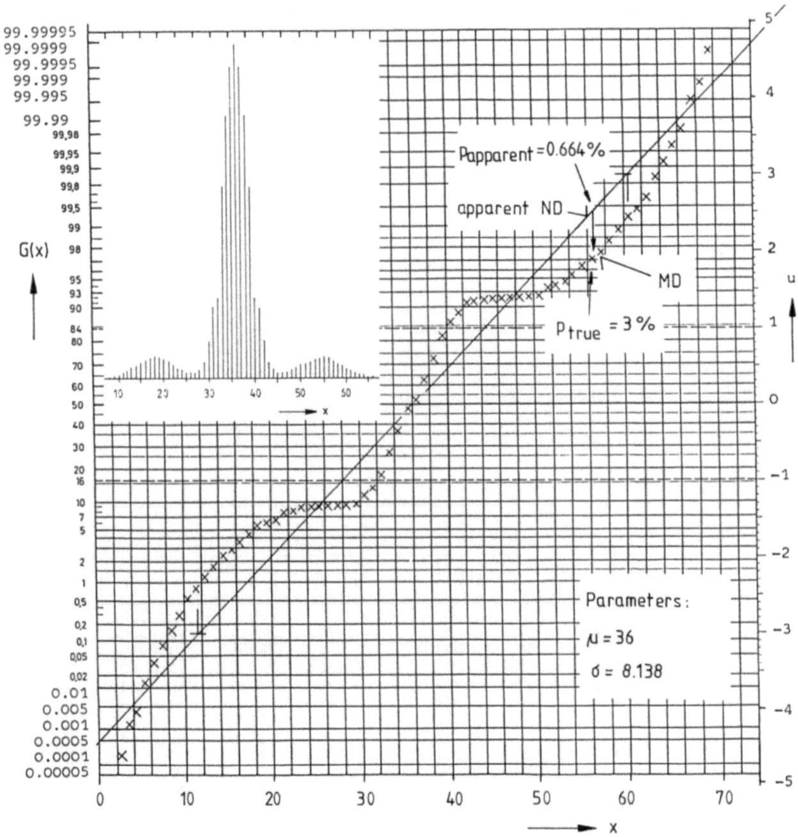

Fig. 13.20. Distribution function of the fold product from two extremely sloping mixed distributions which were folded in dissimilar direction of slope

Extreme deviations from the ND shape in individual size batches are completely unproblematical, if these belong to multiple member dimensional chains. In Fig. 13.22 are four normally distributed (or fictitious normally distributed) batches with an error proportion of $p = 4\%$ ($=$ AQL) added with the result that the error proportion in the clearances—purely mathematically—is $p = 0.023\%$. If all four batches should in truth show an extremely sloping mixed distribution with an error proportion of $p \approx 9\%$, then the true error proportion in the clearances is $p \approx 0.4\%$, Fig. 13.23 and lower size plan in Fig. 13.22.

Note: the differences with the parameters σ in Fig. 13.22 and in Fig. 13.23 require that for the repeated folding of the fold product as in Fig. 13.16, the class occupation numbers are reduced, through which the parameters will slightly alter.

All statements up to now about the selection of AQL values have concentrated on the case of the presence of arithmetically calculated dimensional

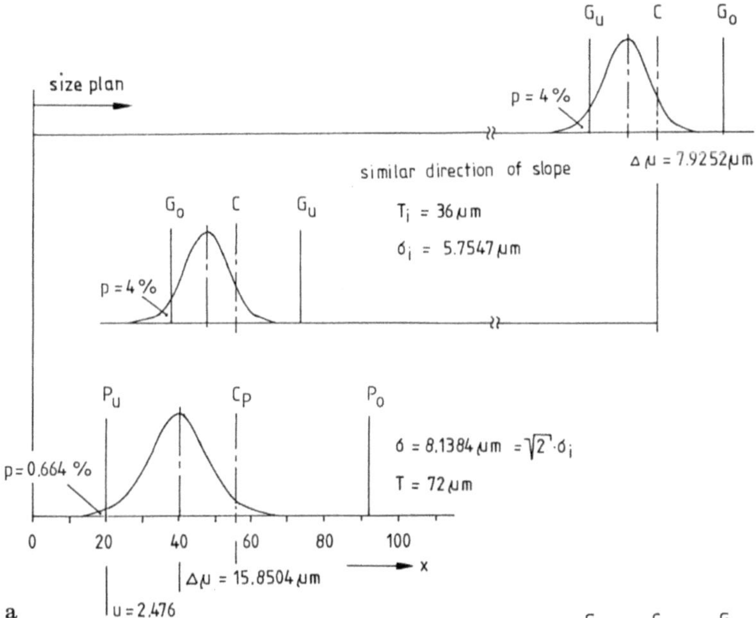

a

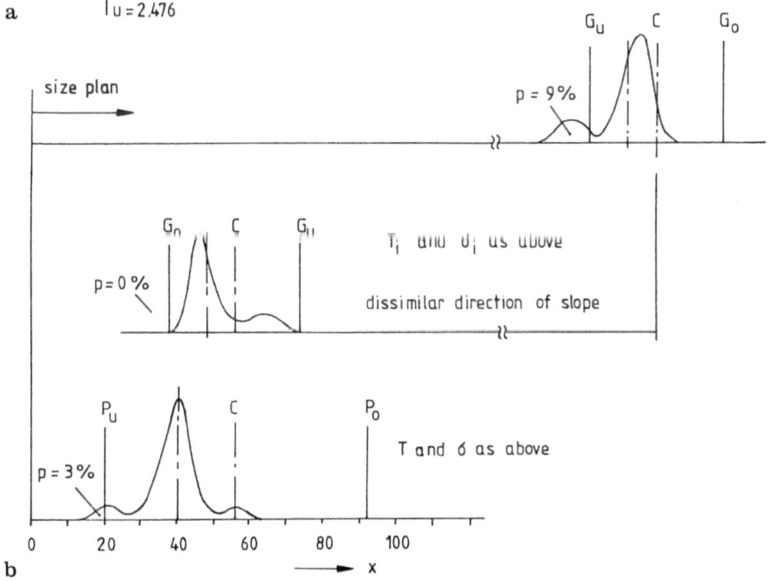

b

Fig. 13.21 a, b. Size plans for two-member dimensional chains. **a** batches with fictitious ND, **b** batches with true MD

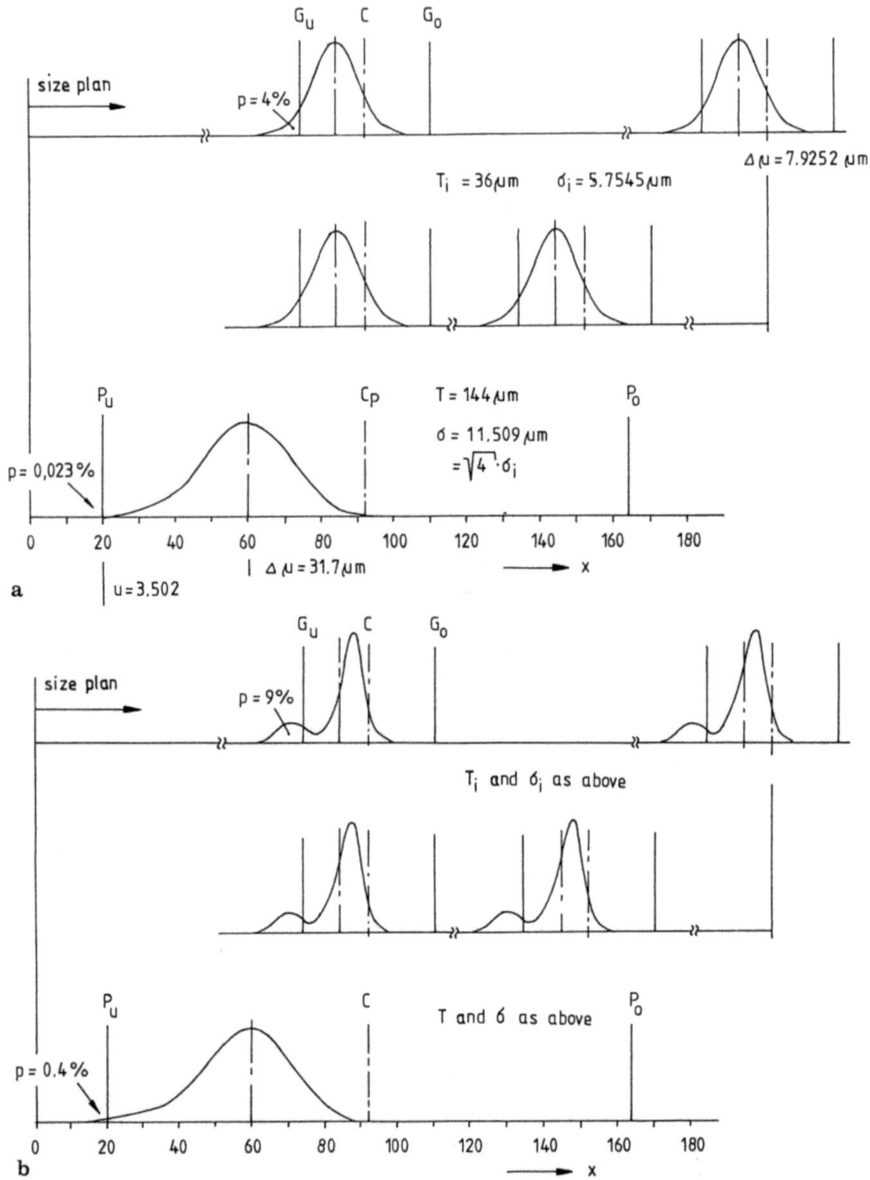

Fig. 13.22 a, b. Size plans for a two-member dimensional chains. **a** batches with fictitious ND, **b** batches with true MD

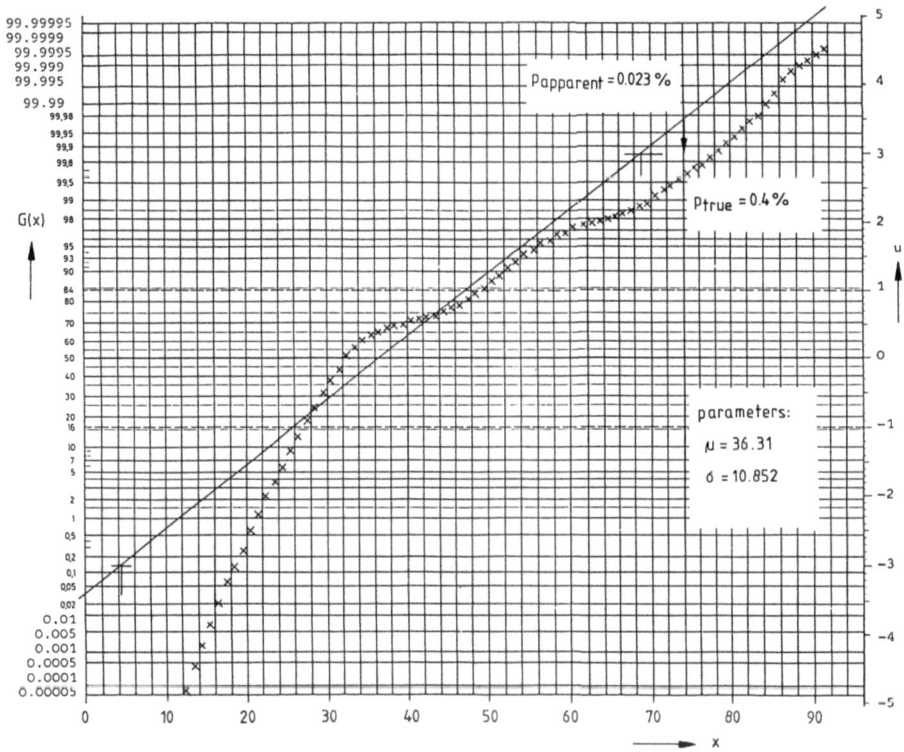

Fig. 13.23. Distribution function of the fold product from four extremely sloping mixed distributions, which were folded in the same direction of slope

chains. In the following we shall investigate which AQL values ae representative for statistically tolerated sizes.

The quadratic tolerance calculation in Chap. 9 has the advantage that the individual tolerances can be maximally widened with respect to the arithmetical calculation. The quadratic tolerance calculation lies at the basis of the assumption that all individual size batches are either exactly normally distributed or optionally distributed, in each case however with their mean values near the middle of the tolerance zone and the 6σ range maximally as large as the respective tolerance zones.

These assumptions are not supervised without great cost in production control and acceptance testing; thus statistical tolerance calculation should preferably be used under the assumption of the presence of rectangular distributions which fill their tolerance zones.

In the case where arithmetically calculated tolerances are tested with AQL = 4, there are no problems, if the assumption of the presence of rectangular distributions which fill the tolerance zones is fulfilled, Line 1) in Fig. 13.24. If however the apparent error proportion of $p = 4.164\%$ of the fictitious normal

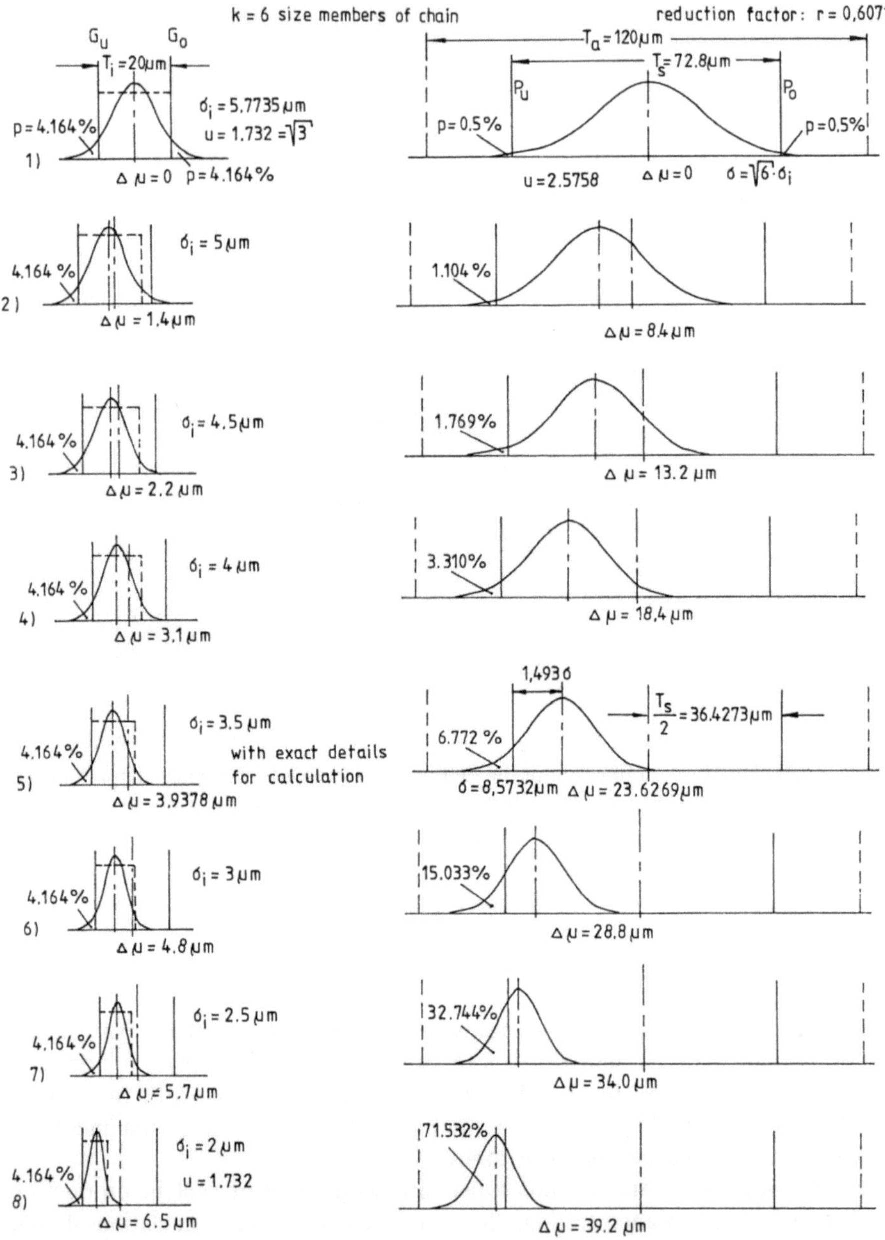

Fig. 13.24. Statistically calculated linear dimensional chains with $k = 6$ members. The calculation follows under the assumption that the individual distributions are rectangularly distributed, which fill the respective tolerance zones, or normal distributions with a limit-exceeding proportion both sides of $p = 4.164\%$, line 1). The lines 2) to 8) contain examples where the individual distributions do not fill the tolerance zones

distribution presents as only one-sided at full height, then the statistical tolerance zone of the closing values of the exceeded proportion are so much greater, the narrower the distributions become, or—put another way—the more the mean values of the individual actual distributions deviate from the centre of the tolerance zone, lines 2) to 8) in Fig. 13.24.

For the better understanding the illustration of the analysis, in line 5) all calculation sizes are given with up to four decimal places. The analysis follows the following steps:

—for $\sigma_i = 3.5\,\mu m$ and $p_{one} = 4.164\%$, the mean value is $\mu - G_u = u\sigma = \sqrt{3} \cdot 3.5 = 6.0622\,\mu m$ distant from the lower limit value,

—thus the mean deviation is

$$\Delta\mu = \frac{T_i}{2} - (\mu - G_u) = 10 - 6.0622 = 3.9378\,\mu m,$$

—for the case where all $k = 6$ size batches deviate about the same mean deviation of $\Delta\mu = 3.9378\,\mu m$ with respect to the tolerance zone centre in the same effective direction, a mean deviation of $\Delta\mu = 6 \cdot 3.9378 = 23.6269\,\mu m$ results for the closing values.

—so that the interval of the mean value of normally distributed closing values becomes

$$\frac{T_s}{2} - \Delta\mu = 36.4273 - 23.6269 = 12.8004\,\mu m$$

and this is in σ-units

$$u = \frac{T_s/2 - \Delta\mu}{\sigma} = \frac{12.8004}{\sqrt{6} \cdot 3.5} = 1.4930$$

—according to u-tables the error proportion for this is $p = 6.772\%$.

This error proportion is only expected in the overall values when randomly all $k = 6$ size batches

—show on one side the same error proportion $p = 4.164\%$,

—have the same standard deviation of $\sigma_i = 3.5\,mm$, and

—deviate in the same effective direction with respect to the centre of the tolerance zone, which is however unlikely.

It is especially unlikely that all size batches are taken according to AQL = 4 and yet at the same time show an error proportion of $p = 4.164\%$. In fact the middle error proportion in AQL = 4 tested stock batches according to their factual acceptance without exception over a long time period amounts to $p = AQL/2 = 2\%$, as is shown in Fig. 13.25. For this case and $\sigma_i = 3.5\,mm$, line 3), the error proportion amounts in unfavourable cases to only $p = 1.328\%$.

Only with even narrower distributions, lines 4) to 6), do higher error proportions result in unfavourable cases. If this kind of small spread is observed

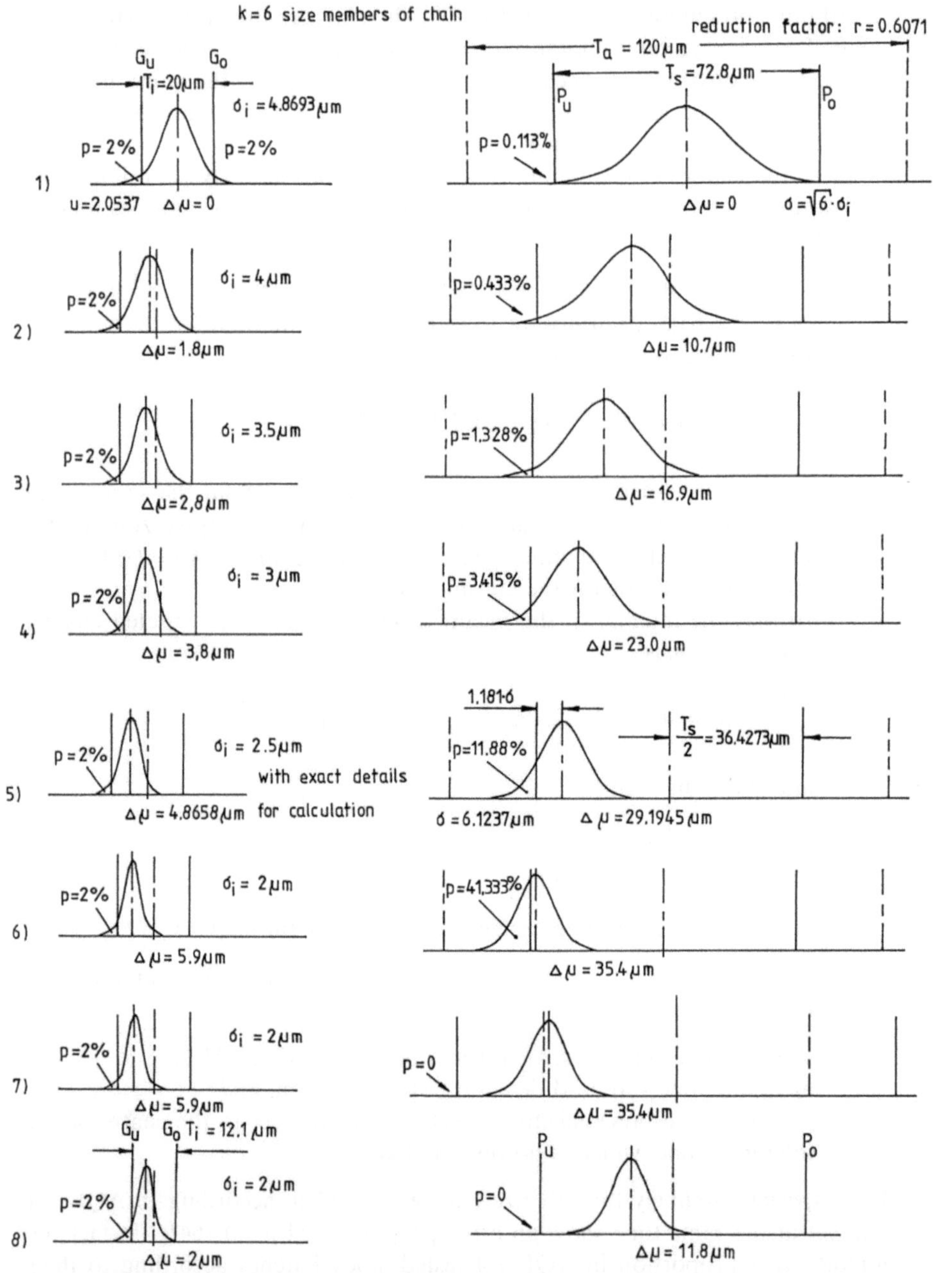

Fig. 13.25. Statistically calculated dimensional chains with $k = 6$ members. The calculation follows under the assumptions made in Fig. 13.24. It is stated that the actual error proportion in all individual batches is $p = 2\%$, in line 1), two-sided, in lines 2) to 8) one-sided

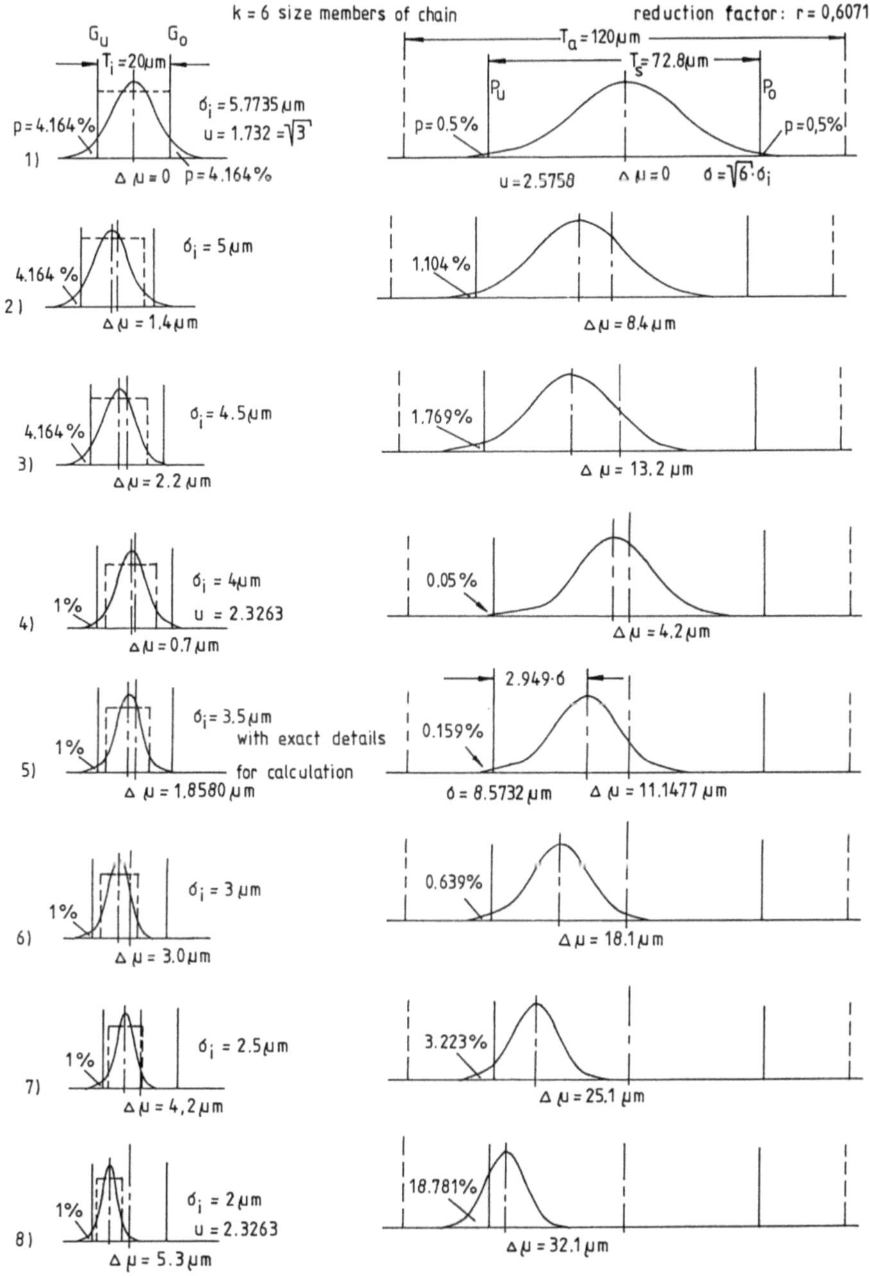

Fig. 13.26. Statistical calculated linear dimensional chains with $k = 6$ members. The calculation follows under the assumptions given in Fig. 13.24. In lines 4) to 8) are stated the actual error proportion (ND present) or the apparent error proportion (RD present) in all batches $p = 1\%$

in the size batches, the necessity for statistical tolerance calculations is questionable and a return may be made to arithmetical calculation in which either

— the closing tolerance is widened again to $T_a = 120\,\mu m$, line 7) in Fig. 13.25 or
— retaining the total tolerance the individual tolerances are narrowed according to the arithmetical calculations, line 8) in Fig. 13.25.

Furthermore there exists the possibility with the AQL values to go down for example to AQL = 2.5; if then the actual error proportion p in all batches is $p \approx 1\%$, there is hardly a problem, lines 4) to 6) in Fig. 13.26, except when the distributions are again so narrow that the necessity of statistical tolerance calculation is questioned, lines 7) and 8) in Fig. 13.26. Then the dimensional chain can be returned without consideration of the arithmetical calculation of tolerances.

13.4 Quality Control Charts for Dimensions

While final production batches are analysed by sample testing, in production control it is a matter of quality c.c. which instantly control the usually normally distributed production from a process according to position and as far as possible spread.

If the pocess has a trend, then the acceptable error proportion (AQL) will only be achieved or exceeded for a proportionately small time period. However it is advisable even here to operate first with AQL values about AQL = 1, especially when statistically tolerated sizes are produced and the parts are unmixed (e.g. on palettes) for assembly.

A quality c.c. (QCC) is a cartesian coordinate system, in that over time t the characteristic x or derived characteristics ($\bar{x}, \tilde{x}, s, R$) are applied.

From time to time (for example half-hourly) samples of the range $n \geqq 1$ are taken from the process and the test result is entered directly or after statistical evaluation of the respective sample in the QCC. A QCC also contains intervention limits in addition to the specified limit values, whose position is mathematically or graphically determined.

A QCC can be planned, calculated, designed, executed, evaluated, documented and destroyed.

Figure 13.27 shows diagramatically the representation of an original value QCC with intervention limits and limit values with three possible production positions.

Process analysis

Before a QCC can be calculated, a process analysis must be undertaken. In this an estimate of the process spread is made under the most realistic production

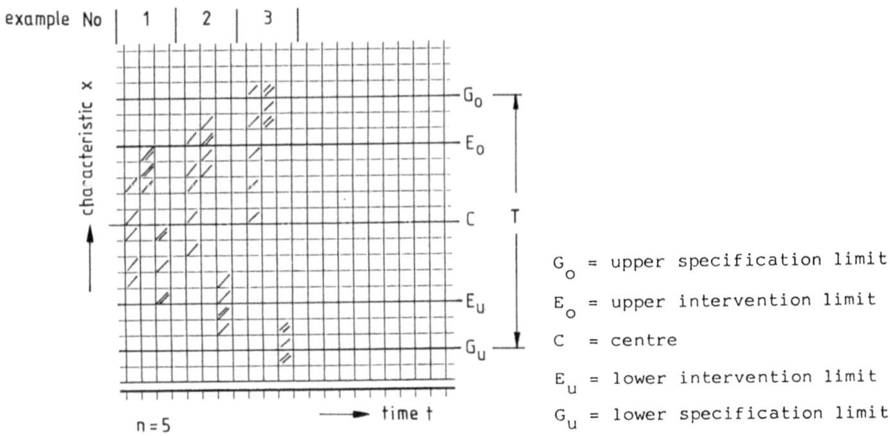

example No	production position	measure
1	all original values between E_u and E_o	nothing
2	one or more original values between E and G	intervention
3	one or more original values outside T	intervention and alarm; sorting and rework

Fig. 13.27. Original value QCC shown diagramatically with three cases of different production positions

conditions; the standard deviation $\hat{\sigma}$ of the process is estimated without the (controllable) methodical deviations of the mean value through incorrect settings (after tool changes) or other disruptive influences and without the systematic deviations such as trends.

The estimate of the (instant) standard deviation $\hat{\sigma}$ is possible through different methods, which are explained further in Chap. 14. In particular there is the root of the mean variance

$$\hat{\sigma} = \sqrt{\overline{s^2}}$$

to consider. With the (instant) standard deviation $\hat{\sigma}$

—the process capability can be analyzed and
—the quality control chart can be calculated.

The process capability is given when

$$c_p = \frac{G_o - G_u}{6 \cdot \hat{\sigma}} \gtrsim 1 \tag{13.3}$$

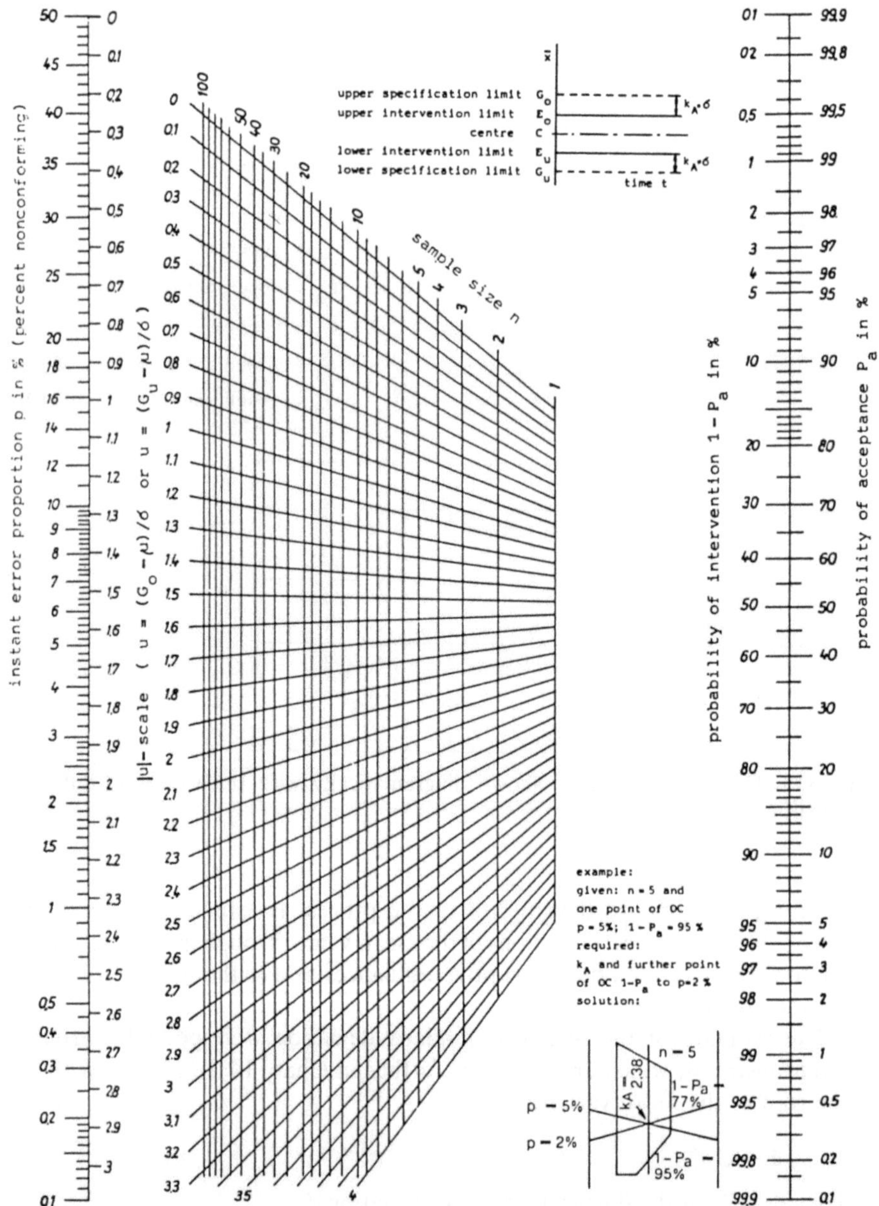

Fig. 13.28. Nomogram for determination of the limiting factors k_A and the OC for mean value QCC with given limit values. From: Wilrich, P.-Th.: Quality control charts, with given limit values. Quality and reliability 24 (1979), pages 260–271. Munich–Vienna: Carl Hanser Verlag. Printed with kind permission of the author

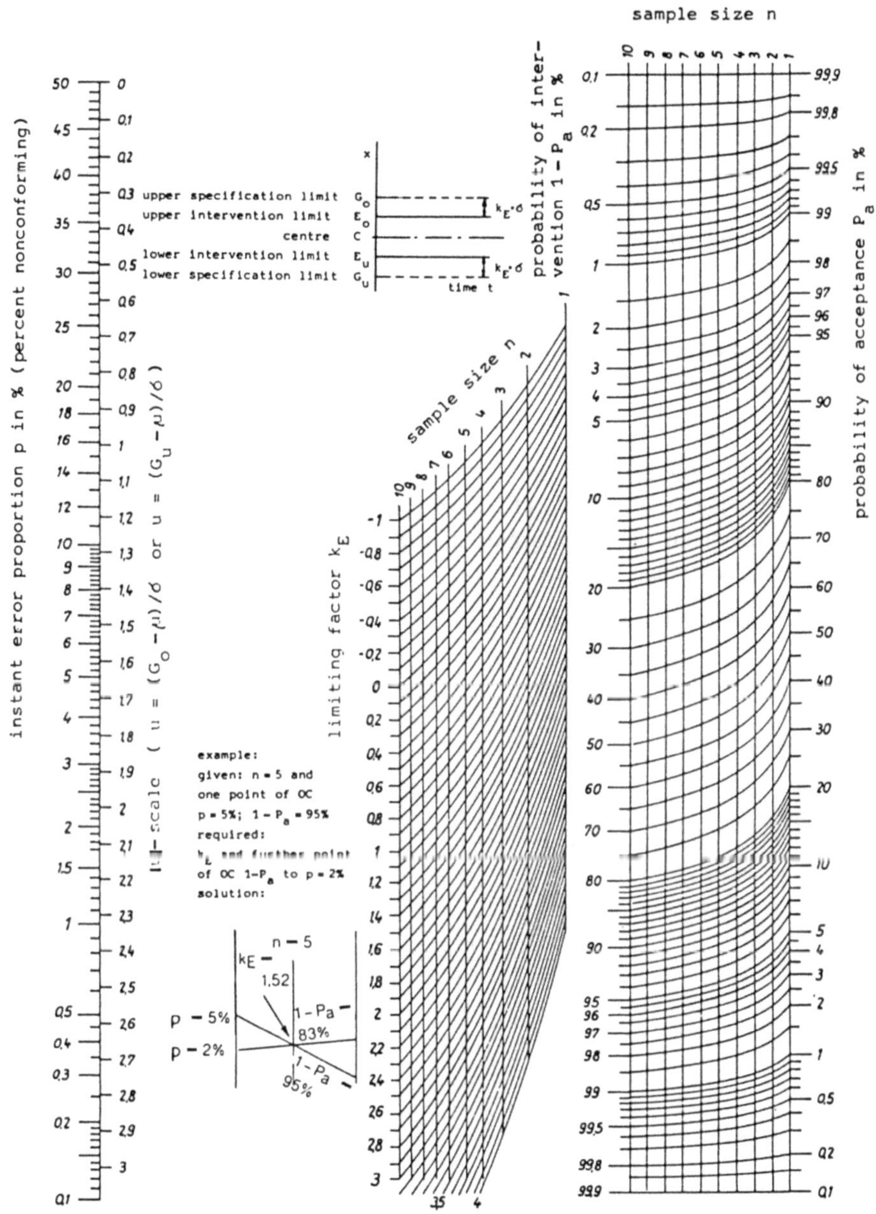

Fig. 13.29. Nomogram for the determination of the limiting factors k_E and the OC for the original value QCC with given limit values. From: Wilrich, P.-Th.: Quality control charts with given limit values. Quality and reliability 24 (1979), pages 260–271. Munich–Vienna: Carl Hanser Verlag. Printed with kind permission of the author

or, put another way,

$$T \gtrsim 6 \cdot \hat{\sigma} \tag{13.4}$$

Only then—apart from exceptions—can a QCC be sensibly calculated, designed and executed. Further points on the analysis of process capability in Chap. 14.

As long as the limit conditions of the process remain constant, it can frequently be assumed that the (instant) standard deviation also remains constant—during the application of the QCC—in the future.

As regards the four characteristics of the sample test—on grounds of practicality—first established in the QCC are

—the sample range n (preferably $n = 5$) and
—a point of the OC (for example $p_{90} = 1\%$).

The two other characteristics

—the limiting factor k and
—a further point of the OC (and thus the whole OC) are calculated or taken from the nomograms in Fig. 13.28 (mean value-QCC) or 13.29 (original value-QCC).

The intervention limits are at intervals

$$E_o = G_o - k\sigma \tag{13.5}$$

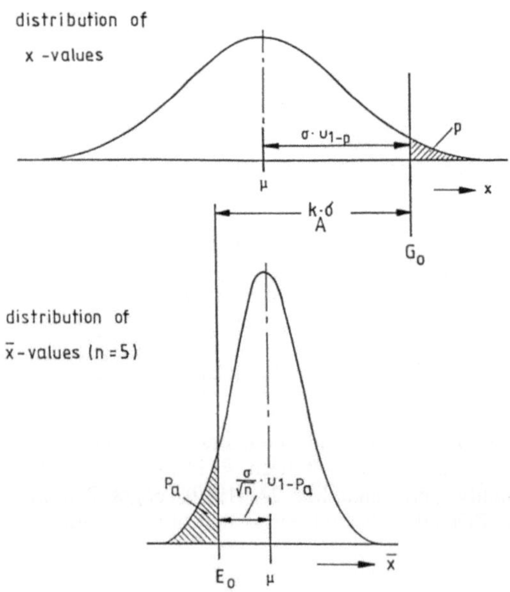

Fig. 13.30. Explanation of the production of limiting factors k_A for the mean value chart

and

$$E_u = G_u + k\sigma \tag{13.6}$$

to the limit values inserted in the QCC.

Mathematical Determination of the Limiting Factors of the Mean Value QCC

According to Fig. 13.30 the inteval $k\sigma$ between the intervention limits and the limit value is

$$k\sigma = \sigma u_{1-p} + \frac{\sigma}{\sqrt{n}} u_{1-P_a}. \tag{13.7}$$

Thus the limiting factor of the mean value chart is

$$k_A = u_{1-p} + \frac{u_{1-P_a}}{\sqrt{n}}. \tag{13.8}$$

Note: The index A stands for arithmetical mean value.

Mathematical Determination of the Limiting Factors of the Original Value QCC

According to Fig. 13.31 the interval $k\sigma$ between the intervention limits and the limit value is

$$k\sigma = \sigma u_{1-p} \quad \sigma u_{\sqrt[n]{P_a}}. \tag{13.9}$$

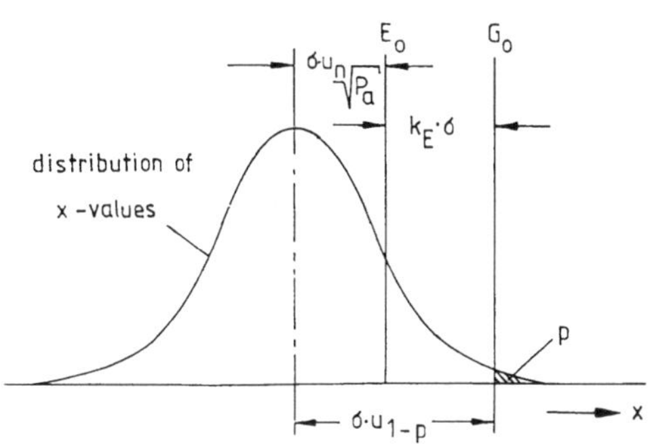

Fig. 13.31. Explanation of the production of the limiting factor k_E of the original value chart; the probability that an x-value lies below E_0 is $\sqrt[n]{P_a}$; the probability that n x-values simultaneously lie below E_0 is $(\sqrt[n]{P_a})^n = P_a$

p or P_a in %	u_{1-p}	u_{1-P_a}	$u_{\sqrt[n]{P_a}}$ n = 1	2	3	4	5
0.5	2.576						
1	2.326		-2.326	-1.282	-0.787	-0.478	-0.258
2	2.054						
3	1.881						
4	1.751						
5	1.645	1.645	-1.645	-0.760	-0.336	-0.068	0.124
6	1.555						
7	1.476						
8	1.405						
9	1.341						
10	1.282	1.282	-1.282	-0.476	-0.090	0.157	0.334
50		0.000	0.000	0.545	0.819	0.998	1.129
90		-1.282	1.282	1.633	1.819	1.944	2.037
95		-1.645	1.645	1.955	2.122	2.234	2.319
99		-2.326	2.326	2.575	2.712	2.806	2.877

Fig. 13.32. Extract from a table of the standardised normal distribution with the preferred QCC choice of *p*-, P_a-, and *u*-values

Thus the limiting factor of the original value chart is

$$k_E = u_{1-p} - u_{\sqrt[n]{P_a}}. \tag{13.10}$$

Note: the index *E* stands for extreme value.

In Fig. 13.32 are summarised some preferred required *u*-values from the ND table.

Exercise 13.2

Given: For the collar diameter of a turned part, the size $d = 15.10^{+0.1}_{-0.1}$ is given. For the purpose of a process analysis, half-hourly samples of the range $n = 5$ are taken and measured; this gives the following values:

i	1	2	3	4	5
x	15.08	15.10	15.08	15.12	15.14
	15.09	15.05	15.09	15.08	15.12
	15.06	15.11	15.08	15.10	15.13
	15.07	15.05	15.07	15.09	15.15
	15.04	15.09	15.08	15.13	15.13
$\bar{x}$ =	15.068	15.080	15.080	15.104	15.134
s^2 =	$3.7 \cdot 10^{-4}$	$.8 \cdot 10^{-4}$	$0.5 \cdot 10^{-4}$	$4.3 \cdot 10^{-4}$	$1.3 \cdot 10^{-4}$
i	6	7	8	9	10
x	15.10	15.12	15.13	15.13	15.17
	15.12	15.11	15.12	15.15	15.17
	15.15	15.15	15.13	15.15	15.12
	15.11	15.12	15.14	15.13	15.14
	15.13	15.14	15.13	15.12	15.14
$\bar{x}$ =	15.122	15.128	15.130	15.136	15.148
s^2 =	$3.7 \cdot 10^{-4}$	$2.7 \cdot 10^{-4}$	$0.5 \cdot 10^{-4}$	$1.8 \cdot 10^{-4}$	$4.7 \cdot 10^{-4}$

$$\overline{s^2} = 3.12 \cdot 10^{-4}$$

$$\sqrt{\overline{s^2}} = \hat{\sigma} = 0.01766 \text{ mm}$$

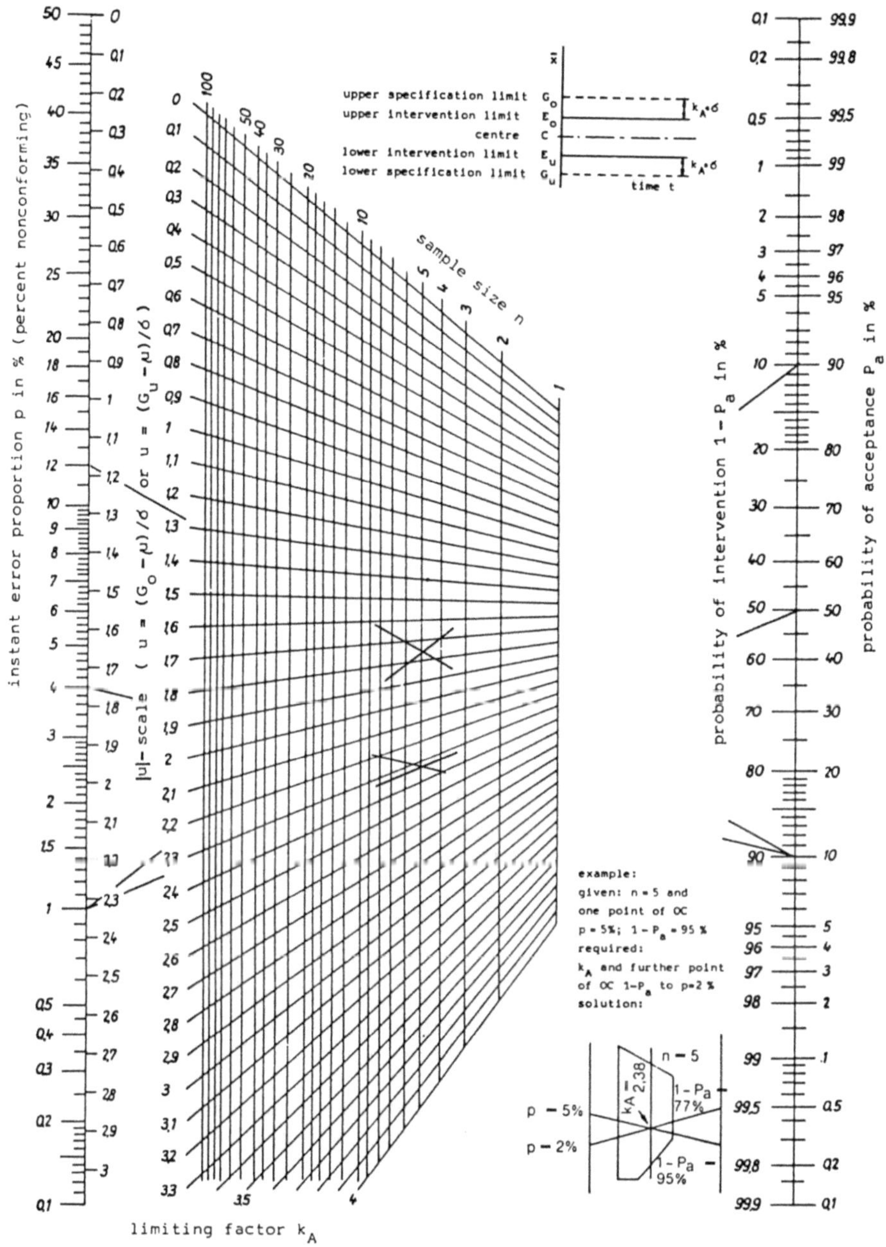

Fig. 13.33. Nomogram for the determination of the limiting factor k_A and the OC's for the mean value QCC with given limit values (from Wilrich) with solution of Exercise 13.2. From: see Fig. 13.29

Required:
1. Estimate of the (instant) standard deviation of the process.
2. Analysis of the process capability.
3. Calculation and design of a mean value QCC for $n = 5$ and $p_{90} = 1\%$ ($P_a = 90\%$ for $p = 1\%$).
4. Apply the chart with the above $\bar{x}$-values.
5. Determination and representation of the OC.
6. Analysis of the solution.
7. Renewed calculation with design and application of the chart for $p_{50} = 1\%$ ($P_a = 50\%$ for $p = 1\%$).

Solution:
1. As calculated in the value table

$$\hat{\sigma} = \sqrt{s^2} = 0.017\,66\,\text{mm}.$$

2. Thus the process capability is

$$c_P = \frac{G_o - G_u}{6\sigma} = \frac{0.2}{6 \cdot 0.017\,66} = 1.887 \quad \text{or}$$

$$T = 11.32 \cdot \sigma.$$

3. According to Fig. 13.33 for $n = 5$ and $p_{90} = 1\%$ the limiting factor $k_A = 1.76$. Mathematically it is

$$k_A = u_{1-p} + \frac{u_{1-P_a}}{\sqrt{n}} = 2.326 - \frac{1.282}{\sqrt{5}} = 1.753.$$

Thus the upper intervention limit lies at

$$E_o = G_o - k_A\sigma = 15.20 - 1.753 \cdot 0.017\,66 = 15.169\,\text{mm}$$

and involves

$$E_u = 15.031\,\text{mm}.$$

The $\bar{x}$-chart is designed in Fig. 13.34.

4. The $\bar{x}$-values from the given samples are inserted in Fig. 13.34.
5. From Fig. 13.33 the following value pairs $P_a(p)$ are taken for the OC:

p in %	0.5	1	2	4	6	8	10	20	12
P_a in %	96.7	90	75	50	32	22	14	2	10

The OC is shown in Fig. 13.35.

6. Analysis of the exercise and its solution:
   - —The tolerance is approximately correctly with $T = 0.2$ mm; in practice a QCC would not usually be necessary for this. The example was chosen because of its clarity.
   - —The given nm $= 5 \cdot 10 = 50$ values are not sufficient for the estimate of the process spread; in practice nm $\geqq 100$ values should be used as a basis.
   - —For $p_{90} = 1\%$ the slope of the OC is very small, thus for short times error ratios of several percent cannot be excluded. On the other hand, the existing clearance space for the trend is not completely required. For $T = 11.32\sigma$ a more sensitive QCC can be chosen.
7. For $p_{50} = 1\%$ is $k_A = 2.326$ and thus

$$E_o = 15.20 - 2.326 \cdot 0.017\,66 = 15.159\,\text{mm}$$

and

$$E_u = 15.041\,\text{mm}.$$

The chart is designed and executed as in Fig. 13.34; the OC is drawn in

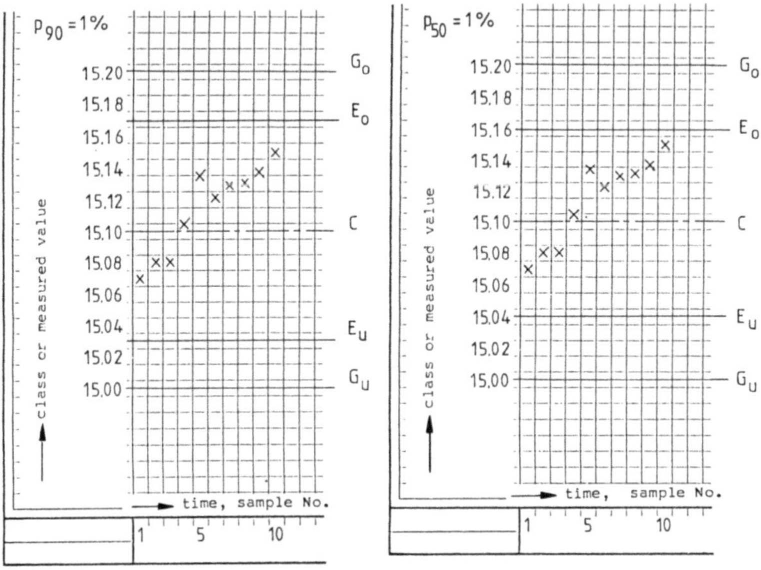

Fig. 13.34. Mean value quality control chart for Exercise 13.2. This and all further quality control charts accord with the German Quality Association—impression 18–178. With kind permission of the Deutsche Gesellschaft für Qualität e.V. Frankfurt

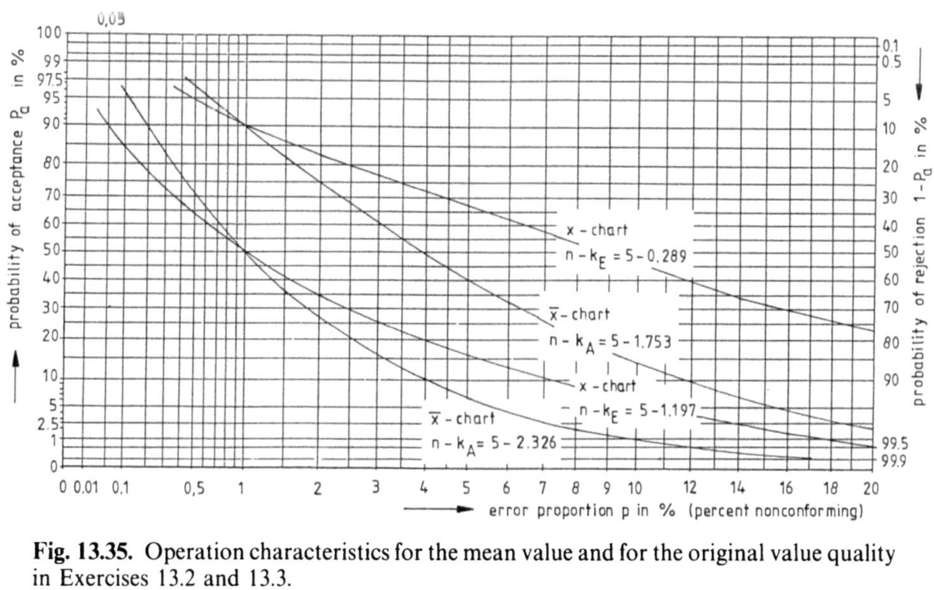

Fig. 13.35. Operation characteristics for the mean value and for the original value quality in Exercises 13.2 and 13.3.

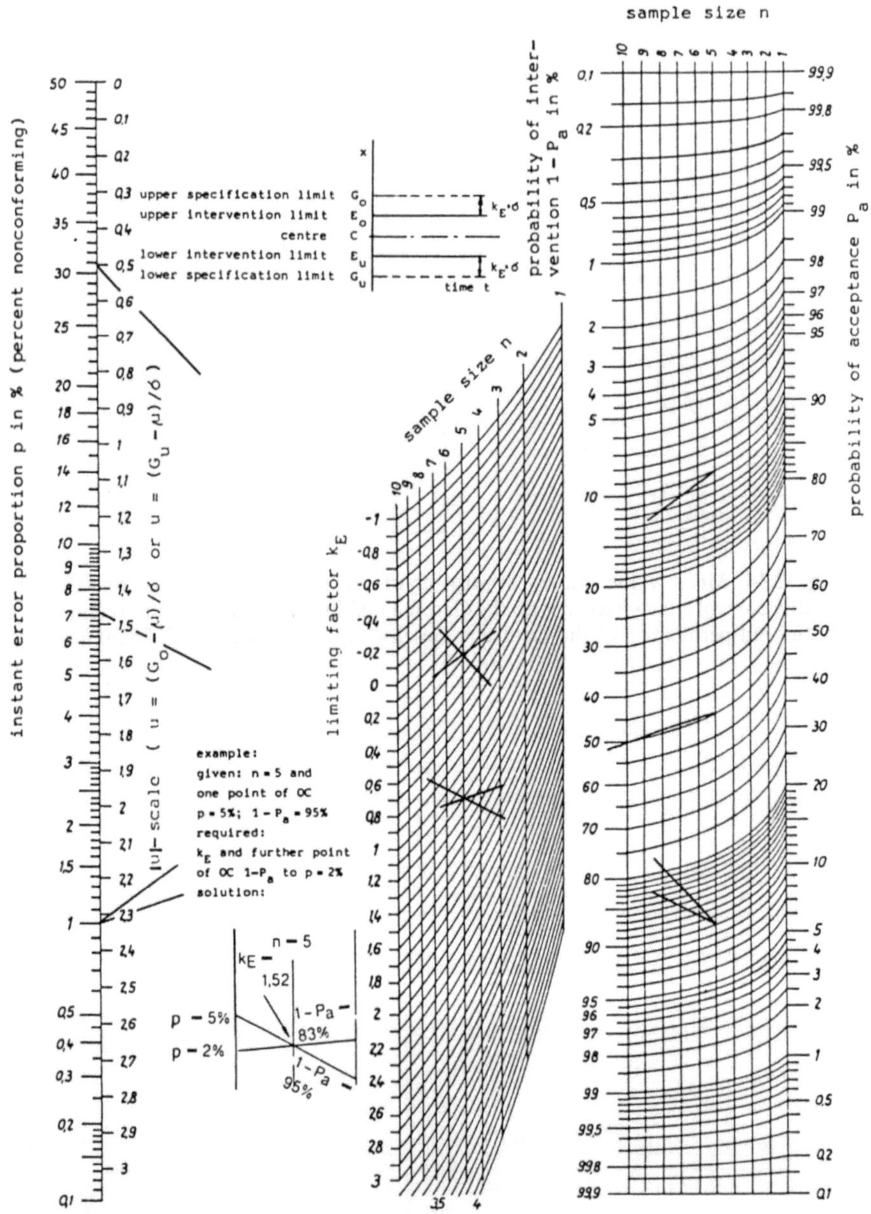

Fig. 13.36. Nomogram for the determination of the limiting factors k_E and the OC for the original value QCC with given limit values (from Wilrich) with solutions for the Exercise 13.3. From: see Fig. 13.29

Fig. 13.35. Even now there is sufficient clearance space for the trend and $p_{10} = 1\%$ could be considered for a basis for the calculation of the $\bar{x}$-chart

Exercise 13.3

Given: Data for Exercise 13.2.

Required: Calculation, design and application of two original value quality c.c. with representation of the operation characteristic for $n = 5$ and

1. $p_{90} = 1\%$ ($P_a = 90\%$ for $p = 1\%$),
2. $p_{50} = 1\%$ ($P_a = 50\%$ for $p = 1\%$).

Solution: 1. According to Fig. 13.36 is $k_E = 0.28$. Mathematically this is

$$k_E = u_{1-p} - u_{n_\surd P_a} = 2.326 - 2.037 = 0.289,$$

so that

$$E_o = G_o - k_E\sigma = 15.20 - 0.289 \cdot 0.01766 = 15.195\,\text{mm}$$

and involves

$$E_u = 15.005\,\text{mm}.$$

The chart is represented and applied with the given original values in Fig. 13.37; the OC is shown in Fig. 13.35.

2. According to Fig. 13.36 is $k_E = 1.2$. Mathematically this is

$$k_E = 2.326 - 1.129 = 1.197,$$

so that

$$E_o = 15.20 - 1.197 \cdot 0.01766 = 15.179\,\text{mm}$$

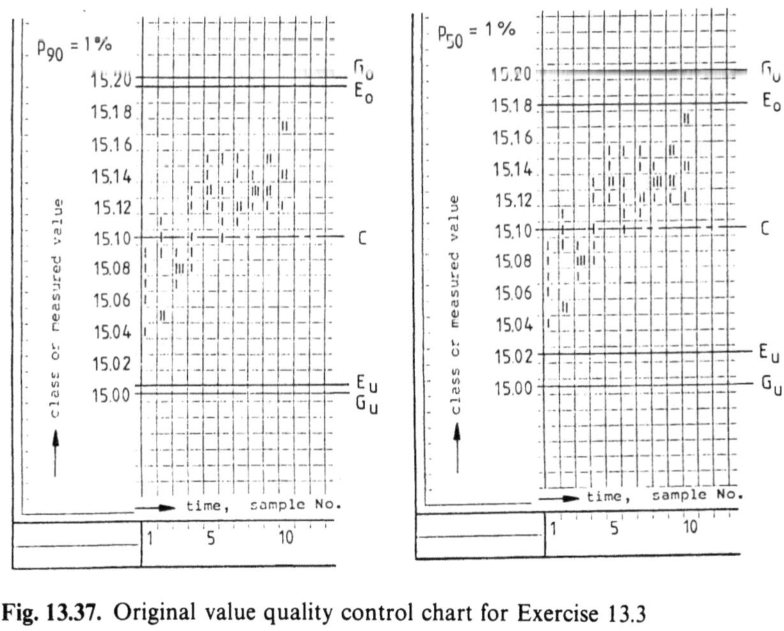

Fig. 13.37. Original value quality control chart for Exercise 13.3

and involves

$E_u = 15.021$ mm.

This is represented and applied too with the given original values in Fig. 13.37; the OC is represented in Fig. 13.35.

The question should also be asked as to which total should be corrected in the setting of the process, if one of the intervention limits in the QCC is exceeded.

The answer to this question should not only be left to the "hunch and experience" of the machine operator, although his experience is of great use and is often better than any theory.

As a simple rule:

— With $\bar{x}$-charts ($n \geqq 5$) the mean value $\bar{x}$ of the last sample test which led to intervention should be taken and corrected to the total $|\bar{x} - C|$.
— With original value cards ($n \geqq 5$) the mean value of the last sample test which led to intervention should be formed and even corrected to the total $|\bar{x} - C|$.

Further points on process correction are included in Chap. 15.

At the beginning of this section the frequency of sample taking was discussed, and it was said that samples are taken "from time to time" (for example half-hourly). At which intervals samples should be taken in the individual case depends on the number of parts produced. Many firms say that about 5% of production is taken in samples. This means that with an automatic robot which produces 100 parts per hour, sample testing should occur every hour. In addition the frequency of sample taking depends also on experience with the process and the occurrence of relevant disruptions.

Where the process shows a trend, then the periodicity of sample removal depends on the steepness of the trend and the size of the tolerance zone with respect to the (instant) spread. If the trend amounts to $\Delta\mu = 0.02\ \sigma$/part and 100 parts are produced every hour, then the steepness can also be given with $\Delta\mu = 2\ \sigma/h$. If the tolerance $T = 10\sigma$, for this case a half-hourly sample removal is necessary and measurable.

If however the steepness of the trend is larger than for example $\Delta\mu = 0.02\sigma$/part, then the necessary intervals between sample removal must be made considerably smaller. With a steepness of trend of $\Delta\mu \geqq 0.1\ \sigma$/part the distribution position drifts with exactly or less than 10 parts about $\Delta\mu = 1\sigma$ in the direction of the upper or the lower intervention limit. In this case a periodically executed quality c.c. must change to a continually executed quality c.c. in that every third or every second part or every part should be tested. More information about this in Chap. 15.

14 Process Analysis and Process Capability

14.1 Process Analysis

Process analysis is the painstaking observation of a process with the objective of distinguishing between as many important influences or groups of influences as possible and describing them as accurately as possible. In particular the non-correctable, instantaneous process spread should be determined.

The following description of process analysis uses an NC lathe as an example and can be applied in a similar way to other metal cutting or shaping processes.

For process analysis $n \approx 200$ consecutively produced parts are each removed. Thus each intervention in the process such as re-setting the machine or a tool change (type, size or period)—for example between part i and part $i+1$—will be recorded.

During the sample removal the process parameters such as tool material, depth of cut, cutting speed, feed and coolant lubrication are held constant and recorded. Also the machine should be kept in controlled condition (constant oil temperature).

After or even during the production of the sample, measurements can be made of the relevant size or sizes of the $n \approx 200$ parts. Care should be taken that measurement is always taken at the same coordinates of the part, for example the diameter of a bearing of a shaft should always be at the right or the left end or at both ends but separately recorded, in case it is proposed to determine the deviation from the cylinder shape.

Process analysis in the narrower sense is the evaluation of one or more size series in order to describe the influences separately. These influences can be divided into the following groups:

1. The trend in terms of
   - —direction,
   - —steepness (average steepness), and
   - —change of steepness in
     - —random or
     - —systematic manner.
2. The other random or systematical influences on the mean value of the production distribution.

Especially important: The limiting characteristics groups 1 and 2 are correctable by targeted control of the process.

3. The random influences on the deviations of the individual values from the respective mean value.

   These influences are divisible into the "5 M's", namely

   —material,
   —machine,
   —method,
   —measurement process,
   —men and women,

   but these can neither be clearly separated one from another nor corrected through the control of the machine.

All three groups of limiting characteristics cause the total spread of the process, which can, for example, be calculated through evaluation of the dimensional series according to the total standard deviation s_x.

The actual process spread is the total spread reduced by the influences in groups 1 and 2: this is also called

—instant process spread or
—remaining spread

and can be estimated in different ways—some better than others. In Fig. 14.1 five possibilities are given and illustrated.

With processes which show steep trends because of rapid tool wear and thus frequent tool changes, the possibilities 1 to 4 are often not applicable; that leaves only the difference method, which can be used in all cases.

possibility of estimation	estimation calculated by	estimated value	requirement	note
1	spread of mean values	$\hat{\sigma}_1 = \sqrt{n} \cdot s_{\bar{x}}$	sample size divided in subgroups of size n; no intervention within a subgroup	very sensitive against disruptions
2	mean of variances	$\hat{\sigma}_2 = \sqrt{\overline{s^2}}$		less sensitive against disruptions or trends; factors d_n and a_n by table 2
3	mean of ranges	$\hat{\sigma}_3 = \bar{R}/d_n$		
4	mean of standard deviation	$\hat{\sigma}_4 = \bar{s}/a_n$		
5	standard devi- deviation of differences	$\hat{\sigma}_5 = s_D/\sqrt{2}$	only differences without interventions between the parts	less sensitive against disruptions; unsensitive according trends

Fig. 14.1. Five different methods for estimating process spread σ

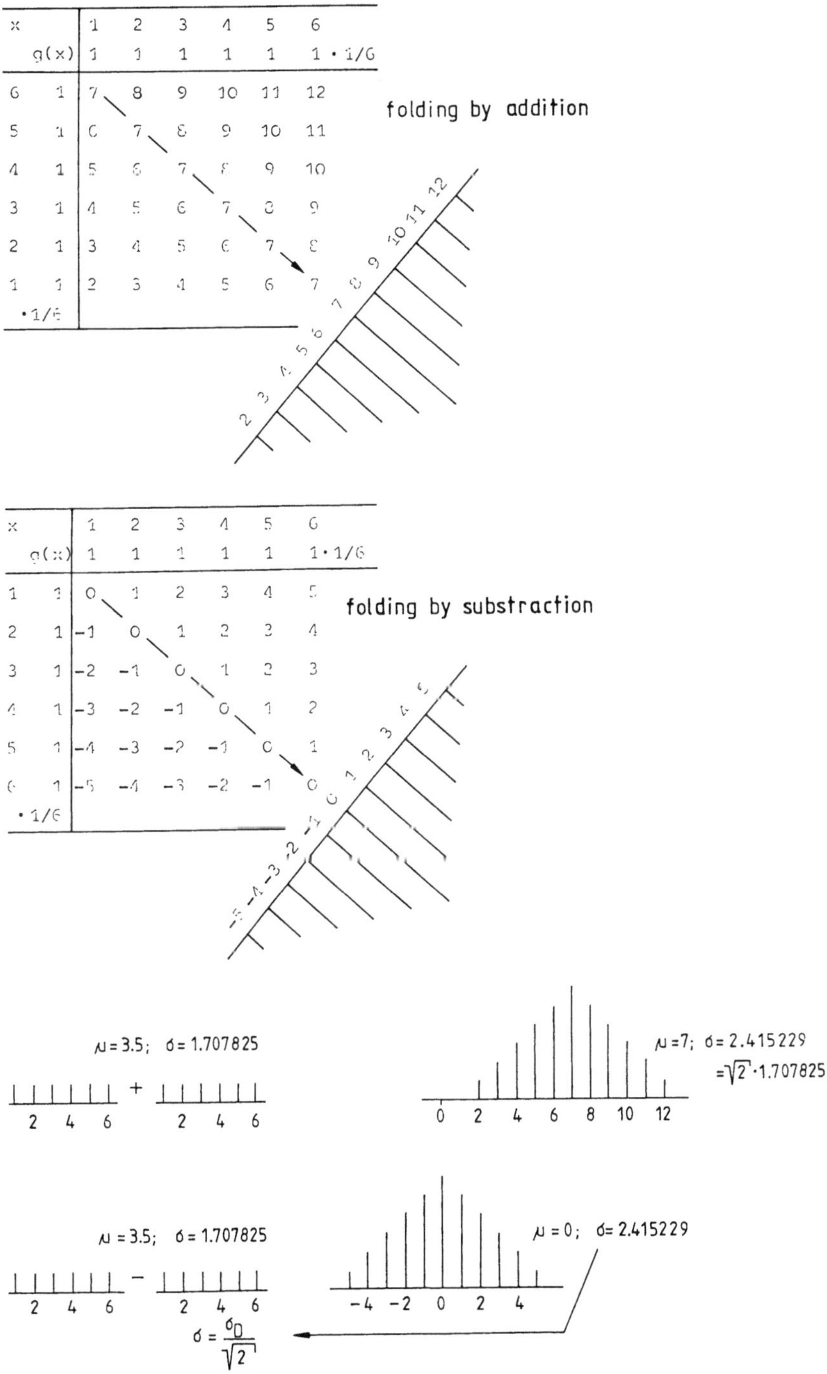

Fig. 14.2. Probability distributions for the sums and for the differences of two thrown dice

x	D	x	D	x	D	x	D	x	D	x	D	x	D	x	D	x	D	x	D
2	-3	3	-1	5	-1	2	-4	5	-1	3	1	5	1	6	4	1	-2	6	3
3	1	1	-2	1	-4	4	2	6	1	4	1	4	-1	1	-5	5	4	1	-5
6	3	6	5	4	3	2	-2	2	-4	4	0	6	2	1	0	1	-4	6	5
2	-4	5	-1	5	1	1	-1	4	2	5	1	5	-1	6	5	3	2	1	-5
3	1	4	-1	1	-4	4	3	5	1	4	-1	5	0	4	-2	6	3	2	1
6	3	5	1	3	2	4	0	5	0	1	-3	4	-1	2	-2	4	-2	2	0
5	-1	6	1	4	-1	6	2	3	-2	6	5	1	-3	3	1	3	1	4	2
5	0	2	-4	4	0	6	C	1	-2	4	-2	4	3	2	-1	6	3	2	-2
2	-3	2	0	4	0	5	-1	3	2	2	-2	4	0	3	1	5	-1	4	2
4	2	6	4	6	2	6	1	2	-1	4	2	2	-2	3	0	3	-2	5	1

D_j		n_j	
-5	///	3	42
-4	~~////~~ //	7	
-3	////	4	
-2	~~////~~ ~~////~~ ///	13	
-1	~~////~~ ~~////~~ ~~////~~	15	
0	~~////~~ ~~////~~ ///	13	
1	~~////~~ ~~////~~ ~~////~~ ///	18	45
2	~~////~~ ~~////~~ //	12	
3	~~////~~ ///	8	
4	///	3	
5	////	4	

evaluations of the differences:

$n = 100$

$\bar{D} = 0.02$

$s_D = 2.412007$

estimated value of the standard deviation of the population:

$$\hat{\sigma} = \frac{s_D}{\sqrt{2}} = 1.705547$$

x	
1	~~////~~ ~~////~~ ///
2	~~////~~ ~~////~~ ~~////~~ /
3	~~////~~ ~~////~~ ///
4	~~////~~ ~~////~~ ~~////~~ ~~////~~ ///
5	~~////~~ ~~////~~ ~~////~~ //
6	~~////~~ ~~////~~ ~~////~~ ///

evaluation of x-values:

$n = 100$

$\bar{x} = 3.69$

$s = 1.662$

Fig. 14.3. Results of 100 dice throws and their differences. The parameters of the equal distribution of a die are the mean value $\mu = 3.5$ and the standard deviation $\sigma = 1.707\,825$

The estimate possibilities 1 to 4 are easy to understand or can be understood simply through the examples given in Figs. 14.8 to 14.11 or 14.13 and 14.15 or 14.19 to 14.21; the difference method is somewhat more complicated and thus will be discussed in more detail.

Difference Method

This is best explained by the example of the symmetrical dice. If the probability distributions of two dice are folded, then a triangular distribution is formed, Fig. 14.2, and this is valid not only for the sum but also for the differences of the two dice.

With the sum distribution, the smallest sum is $2 = 1 + 1$ and the largest sum $12 = 6 + 6$; with the difference distribution the smallest difference is

$-5 = 1 - 6$ and the largest difference is $5 = 6 - 1$, Fig. 14.2. From the standard deviation of the differences (sums) the standard deviation of a die can be calculated.

This is valid not only for the probability distributions but also for frequency distributions. In order to show this, a die is thrown 100 times, Fig. 14.3. Theoretically each result should occur $16.\bar{6}$ times; the actual frequencies are shown on the slash list for the x values in Fig. 14.3. The sample characteristics $\bar{x} = 3.69$ and $s = 1.662$ lie randomly deviated near the parameters $\mu = 3.5$ and $\sigma = 1.707\,825$.

The differences D are also calculated and given in Fig. 14.3. The first difference was formed from $x_1 - x_{100} = 2 - 5 = -3$.

The slash list for the differences in Fig. 14.3 shows: 13 differences are nil; the remaining differences are as expected approximately half negative and half positive.

The statistical evaluation of the differences gives the characteristics in Fig. 14.3. The estimated value for the standard deviation of the dice is $\hat{\sigma} = 1.705\,547$ with the exact value near enough identical.

In order to show the functioning of the difference method (sum method) with the normal distribution, from an ND model with $N = 50$ values as in Fig. 14.4 a random sample was taken of $n = 50$ values—item after item—putting back allowed, Fig. 14.5. The evaluation according to the difference method and the sum method delivered randomly deviated estimated values for the standard

x_j	N_j	$\sum N_j$	G_j in %
1	1	1	2
2	1	2	4
3	3	5	10
4	6	11	22
5	9	20	40
6	10	30	60
7	9	39	78
8	6	45	90
9	3	48	96
10	1	49	98
11	1	50	100

parameters:

$N = 50$

$\mu = 6$

$\sigma = 2.009975$

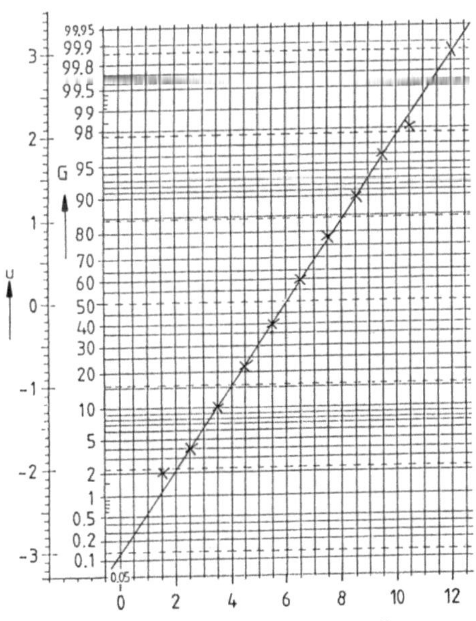

Fig. 14.4. Model distribution with nearly ND shape

1	2	3	4	5
i	x_i	D_i	S_i	$\bar{x}_i = S_i/2$
1	6			
2	11	5	17	8.5
3	5	-6	16	8
4	8	3	13	6.5
5	5	-3	13	6.5
6	8	3	13	6.5
7	7	-1	15	7.5
8	6	-1	13	6.5
9	1	-5	7	3.5
10	8	7	9	4.5
11	8	0	16	8
12	6	-2	14	7
13	6	0	12	6
14	6	0	12	6
15	6	0	12	6
16	8	2	14	7
17	6	-2	14	7
18	9	3	15	7.5
19	5	-4	14	7
20	7	2	12	6
21	3	-4	10	5
22	9	6	12	6
23	4	-5	13	6.5
24	5	1	9	4.5
25	5	0	10	5
26	5	0	10	5
27	6	1	11	5.5
28	6	0	12	6
29	4	-2	10	5
30	6	2	10	5
31	6	0	12	6
32	7	1	13	6.5
33	6	-1	13	6.5
34	4	-2	10	5
35	8	4	12	6
36	5	-3	13	6.5
37	3	-2	8	4
38	2	-1	5	2.5
39	6	4	8	4
40	7	1	13	6.5
41	10	3	17	8.5
42	8	-2	18	9
43	7	-1	15	7.5
44	4	-3	11	5.5
45	8	4	12	6
46	5	-3	13	6.5
47	6	1	11	5.5
48	3	-3	9	4.5
49	6	3	9	4.5
50	7	1	13	6.5

not applicable in cases of a disruption or a trend

evaluation:

according to the difference method:

D_j		n_j
-6	/	1
-5	//	2
-4	//	2
-3	/////	5
-2	//////	6
-1	/////	5
0	////////	8
1	//////	6
2	///	3
3	/////	5
4	///	3
5	/	1
6	/	1
7	/	1

$n = 49$

$\bar{D} = 0.020408$

$s_D = 2.933214$

$\hat{\sigma} = s_D/\sqrt{2} = 2.074095$

according to the sum method and the mean method:

S_j	$\bar{x}_j$		n_j
5	2.5	/	1
6	3		
7	3.5	/	1
8	4	//	2
9	4.5	////	4
10	5	//////	6
11	5.5	///	3
12	6	/////////	9
13	6.5	///////////	11
14	7	////	4
15	7.5	///	3
16	8	//	2
17	8.5	//	2
18	9	/	1

$n = 49$ | $n = 49$

$\bar{S} = 12.102041$ | $\bar{\bar{x}} = 6.051020$

$s_S = 2.663370$ | $s_{\bar{x}} = 1.331685$

$\hat{\sigma} = s_S/\sqrt{2} = 1.883287$ | $\hat{\sigma} = \sqrt{2} \cdot s_{\bar{x}} = 1.883287$

Fig. 14.5. Estimation of the standard deviation of the total population according to the difference method, the sum method and the mean value method (mean values of two consecutive values) in the example of a random sample of $n = 50$ values with the characteristics $\bar{x} = 6.06$, $s = 1.963036$ from an ND model with the parameters $\mu = 6$ and $\sigma = 2.009975$. The sum and mean value methods lead to identical estimated values

	undisrupted		slightly disrupted		strongly disrupted		very strongly disrupted		
1	2	3	4	5	6	7	8	9	
i	x_i	D_i	x_i	D_i	x_i	D_i	x_i	D_i	
1	6		6		6		6		
2	11	5	11	5	11	5	11	5	
3	5	-6	5	-6	5	-6	5	-6	
4	8	3	8	3	8	3	8	3	
5	5	-3	5	-3	5	-3	5	-3	
6	8	3	8	3	8	3	8	3	
7	7	-1	7	-1	7	-1	7	-1	
8	6	-1	6	-1	6	-1	6	-1	
9	1	-5	1	-5	1	-5	1	-5	
10	8	7	8	7	8	7	8	7	
11	8	0	8.5	0.5	9	1	10	2	
12	6	-2	7	-1.5	8	-1	10	0	
13	6(16)	0(10)	7.5	0.5	9	1	12	2	
14	6(16)	0(0)	8	0.5	10	1	14	2	linear
15	6	0(-10)	8.5	0.5	11	1	16	2	upwards
16	8	2	11	2.5	14	3	20	4	trend
17	6	-2	9.5	-1.5	13	-1	20	0	
18	9	3	13	3.5	17	4	25	5	
19	5	-4	9.5	-3.5	14	-3	23	-2	
20	7	2	12	2.5	17	3	27	4	
21	3	-4	8	-4	13	-4	23	-4	
22	9	6	13.5	5.5	18	5	27	4	
23	4	-5	8	-5.5	12	-6	20	-7	
24	5	1	8.5	0.5	12	0	19	-1	linear
25	5	0	8	-0.5	11	-1	17	-2	downwards
26	5	0	7.5	-0.5	10	-1	15	-2	trend
27	6	1	8	0.5	10	0	14	-1	
28	6	0	7.5	-0.5	9	-1	12	-2	
29	4	-2	5	-2.5	6	-3	8	-4	
30	6	2	6.5	1.5	7	1	8	0	
31	6	0	6	-0.5	6	-1	6	-2	
32	7	1	7	1	7	1	7	1	
33	6	-1	6	-1	6	1	6	-1	
34	4	-2	4	-2	4	-2	4	-2	
35	8	4	8	4	8	4	8	4	
36	5	-3	5	-3	5	-3	5	-3	
37	3	-2	3	-2	3	-2	3	-2	
38	2	-1	2	-1	2	-1	2	-1	
39	6	4	6	4	6	4	6	4	
40	7	1	7	1	7	1	7	1	
41	10	3	10	3	10	3	10	3	
42	8	-2	8	-2	8	-2	8	-2	
43	7	-1	7	-1	7	-1	7	-1	
44	4	-3	4	-3	4	-3	4	-3	
45	8	4	8	4	8	4	8	4	
46	5	-3	5	-3	5	-3	5	-3	
47	6	1	6	1	6	1	6	1	
48	3	-3	3	-3	3	-3	3	-3	
49	6	3	6	3	6	3	6	3	
50	7	1	7	1	7	1	7	1	

Fig. 14.6. Simulation of undisrupted and disrupted processes

undisrupted

D_j		n_j
-7		
-6	/	1
-5	//	2
-4	//	2
-3	/////	5
-2	//////	6
-1	/////	5
0	////////	8
1	//////	6
2	///	3
3	/////	5
4	///	3
5	/	1
6	/	1
7	/	1

n = 49

$\bar{D}$ = 0.020408

s_D = 2.933214

$\hat{\sigma}$ = 2.074095

evaluation with the single disruptions in lines i = 13 and i = 14

-10 /

0 //////

+10 /

otherwise as above

n = 49

$\bar{D}$ = 0.020408

s_D = 3.573571

$\hat{\sigma}$ = 2.526896

strongly disrupted

D_j		n_j
-7		
-6	//	2
-5	/	1
-4	/	1
-3	///////	7
-2	///	3
-1	///////////	11
0	//	2
1	/////////	9
2		
3	//////	6
4	////	4
5	//	2
6		
7	/	1

n = 49

$\bar{D}$ = 0.020408

s_D = 2.975524

$\hat{\sigma}$ = 2.104013

very strongly disrupted

D_j		n_j
-7	/	1
-6	/	1
-5	/	1
-4	//	2
-3	/////	5
-2	////////	8
-1	///////	7
0	///	3
1	////	4
2	////	4
3	////	4
4	//////	6
5	//	2
6		
7	/	1

n = 49

$\bar{D}$ = 0.020408

s_D = 3.152313

$\hat{\sigma}$ = 2.229022

slightly disrupted

D_j		n_j
-6	/	1
-5.5	/	1
-5	/	1
-4.5		
-4	/	1
-3.5	/	1
-3	/////	5
-2.5	/	1
-2	///	3
-1.5	//	2
-1	/////	5
-0.5	////	4
0		
0.5	//////	6
1	////	4
1.5	/	1
2		
2.5	//	2
3	////	4
3.5	/	1
4	///	3
4.5		
5	/	1
5.5	/	1
6		
6.5		
7	/	1

n = 49

$\bar{D}$ = 0.020408

s_D = 2.936763

$\hat{\sigma}$ = 2.076605

Fig. 14.7. Evaluation of different, simulated processes by the difference method

deviation σ of the ND model. Since the sum method for estimation of the standard deviation of the total population is identical to the mean value method (method 1 in Fig. 14.1) and fails in cases of a disruption or a trend, in the following text the difference method will be used.

The random sample from the ND model will be used in the following figures to simulate different, frequently arising process features.

Fig. 14.6 contains the undisrupted process, columns 2 and 3, as well as the simulation of a slight, a strong and a very strong disruption.

The evaluation of all four processes by the difference method in Fig. 14.7 can be followed by a calculator with $\bar{x}/s$-automatic facility.

	1	2	3	4	5	6	7	8	9	10
	6	8	8	8	3	5	6	5	10	5
	11	7	6	6	9	6	7	3	8	6
	5	6	6 (16)	9	4	6	6	2	7	3
	8	1	6 (16)	5	5	4	4	6	4	6
	5	8	6	7	5	6	8	7	8	7
$\bar{x}$	7.0	6.0	6.4(8.4)	7.0	5.2	5.4	6.2	4.6	7.4	5.4
R	6	7	2(10)	4	6	2	4	5	6	4
s	2.549510	2.915476	0.894427 (5.176872)	1.581139	2.280351	0.894427	1.483240	2.073644	2.190890	1.516575

sums:

$\sum \bar{x} = 60.6\ (64.6)$

$\sum R = 46\ (54)$

$\sum s = 18.379679\ (22.662124)$

$\sum s^2 = 37.900001\ (63.900005)$

characteristic values:

$\bar{\bar{x}} = 6.06\ (6,46)$

$s_x = 1.963026\ (2.779003)$

$s_{\bar{x}} = 0.909456\ (1.652069)$

estimated values:

$\hat{\sigma}_1 = \sqrt{5} \cdot s_{\bar{x}} = 2.033606\ (3.694140)$

$\hat{\sigma}_2 = \sqrt{\overline{s^2}} = 1.946792\ (2.527845)$

$\hat{\sigma}_3 = \bar{R}/d_n = 1.977644\ (2.321582)$

$\hat{\sigma}_4 = \bar{s}/a_n = 1.955285\ (2.410864)$

$\hat{\sigma}_5 = s_D/\sqrt{2} = 2.074095\ (2.526896)$

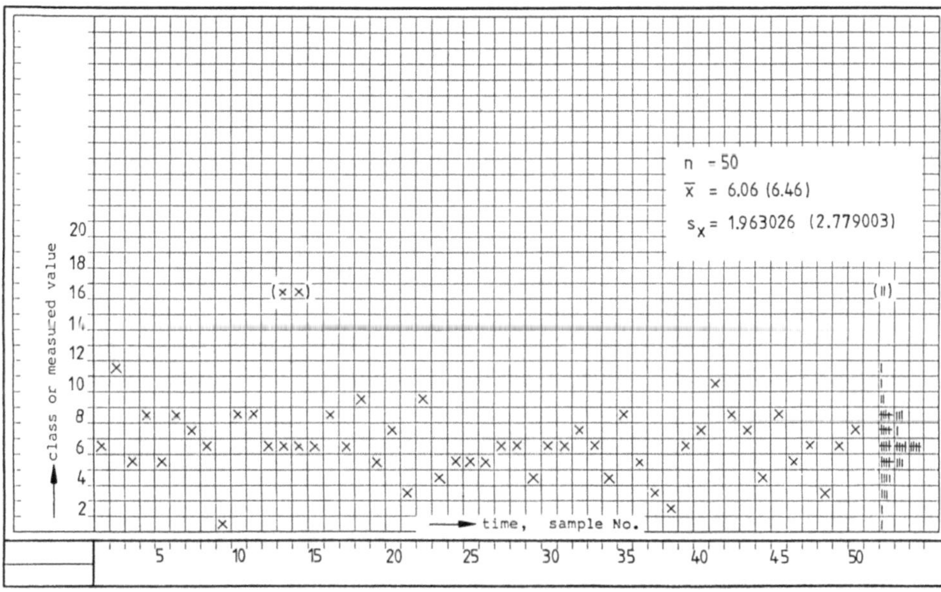

Fig. 14.8. Evaluation and representation of an undisrupted process. The five estimated values for the process spread $\hat{\sigma}$ are all approximately the same. This is an ideal diagram of undisrupted, normally distributed production. The additional single disruption (shown in brackets) during the 13th and 14th test have an effect on the total spread s_x and an especially strong effect on the estimated value $\hat{\sigma}_1$

	1	2	3	4	5	6	7	8	9	10
	6	8	8.5	11	8	7.5	6	5	10	5
	11	7	7	9.5	13.5	8	7	3	8	6
	5	6	7.5	13	8	7.5	6	2	7	3
	8	1	8	9.5	8.5	5	4	6	4	6
	5	8	8.5	12	8	6.5	8	7	8	7
$\bar{x}$	7.0	6.0	7.9	11	9.2	6.9	6.2	4.6	7.4	5.4
R	6	7	1.5	3.5	5.5	3	4	5	6	4
s	2.549510	2.915476	0.651920	1.541104	2.413504	1.193734	1.483240	2.073644	2.070890	1.516576

sums:

$\sum \bar{x} = 71.6$

$\sum s = 18.529598$

$\sum s^2 = 38.650007$

$\sum R = 45.5$

characteristic values:

$\bar{\bar{x}} = 7.16$

$s_x = 2.524250$

$s_{\bar{x}} = 1.871541$

estimated values:

$\hat{\sigma}_1 = \sqrt{5} \cdot s_{\bar{x}} = 4.184893$

$\hat{\sigma}_2 = \sqrt{\overline{s^2}} = 1.965961$

$\hat{\sigma}_3 = \bar{R}/d_n = 1.956148$

$\hat{\sigma}_4 = \bar{s}/a_n = 1.971234$

$\hat{\sigma}_5 = s_D/\sqrt{2} = 2.076605$

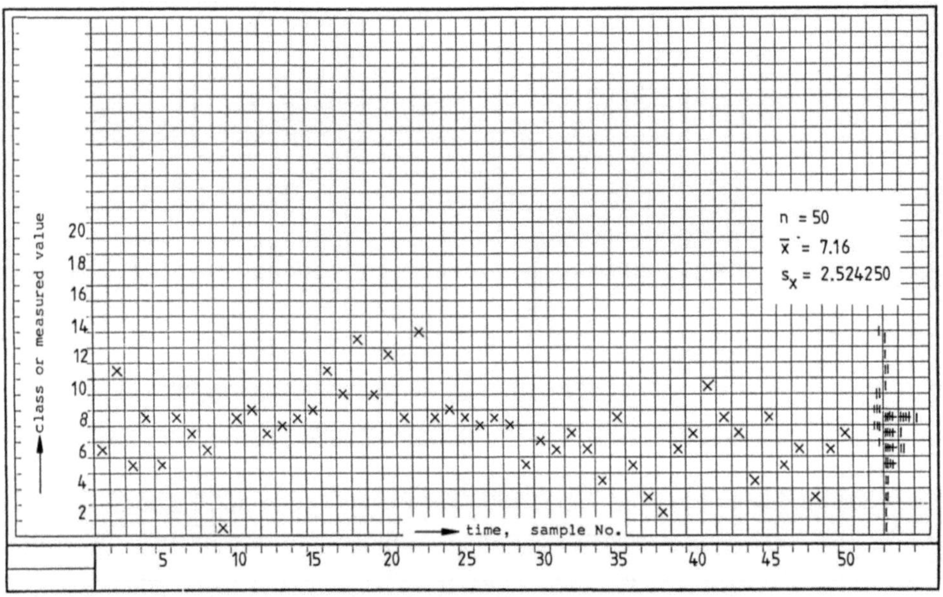

Fig. 14.9. Evaluation and representation of a disrupted process. The disruption affects strongly the total spread s_x and the estimated value $\hat{\sigma}_1$. The other estimated values are scarcely altered by the slight disruption

	1	2	3	4	5	6	7	8	9	10
	6	8	9	14	13	10	6	5	10	5
	11	7	8	13	18	10	7	3	8	6
	5	6	9	17	12	9	6	2	7	3
	8	1	10	14	12	6	4	6	4	6
	5	8	11	17	11	7	8	7	8	7
$\bar{x}$	7.0	6.0	9.4	15	13.2	8.4	6.2	4.6	7.4	5.4
R	6	7	3	4	7	4	4	5	6	4
s	2.549510	2.915476	1.140175	1.870829	2.774887	1.816590	1.483240	2.073644	2.190890	1.516575

sums:

$\sum \bar{x} = 82.6$

$\sum s = 20.331816$

$\sum s^2 = 44.399998$

$\sum R = 50$

characteristic values:

$\bar{\bar{x}} = 8.26$

$s_x = 3.778619$

$s_{\bar{x}} = 3.405943$

estimated values:

$\hat{\sigma}_1 = \sqrt{5} \cdot s_{\bar{x}} = 7.615919$

$\hat{\sigma}_2 = \sqrt{\overline{s^2}} = 2.107131$

$\hat{\sigma}_3 = \bar{R}/d_n = 2.149613$

$\hat{\sigma}_4 = \bar{s}/a_n = 2.162959$

$\hat{\sigma}_5 = s_D/\sqrt{2} = 2.104013$

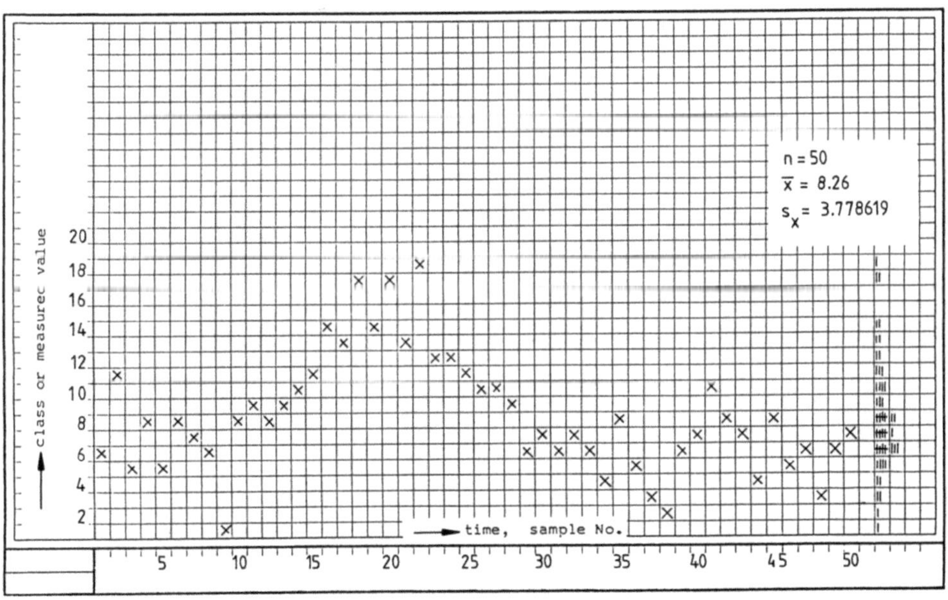

Fig. 14.10. Evaluation and representation of a strongly disrupted process. The disruption affects the total spread and the estimated value $\hat{\sigma}_1$. The other estimated values for the process spread are only slightly enlarged

1	2	3	4	5	6	7	8	9	10
6	8	10	20	23	15	6	5	10	5
11	7	10	20	27	14	7	3	8	6
5	6	12	25	20	12	6	2	7	3
8	1	14	23	19	8	4	6	4	6
5	8	16	27	17	8	8	7	8	7
$\bar{x}$ 7.0	6.0	12.4	23	21.2	11.4	6.2	4.6	7.4	5.4
R 6	7	6	7	10	7	4	5	6	4
s 2.549510	2.915476	2.607681	3.082207	3.898718	3.286335	1.483240	2.073644	2.190890	1.516575

sums:

$\sum \bar{x} = 104.6$

$\sum s = 25.604276$

$\sum s^2 = 70.900000$

$\sum R = 62$

characteristic values:

$\bar{\bar{x}} = 10.46$

$s_x = 6.800990$

$s_{\bar{x}} = 6.637972$

estimated values:

$\hat{\sigma}_1 = \sqrt{5} \cdot s_{\bar{x}} = 14.842957$

$\hat{\sigma}_2 = \sqrt{\overline{s^2}} = 2.662705$

$\hat{\sigma}_3 = \bar{R}/d_n = 2.665520$

$\hat{\sigma}_4 = \bar{s}/a_n = 2.723859$

$\hat{\sigma}_5 = s_D/\sqrt{2} = 2.229022$

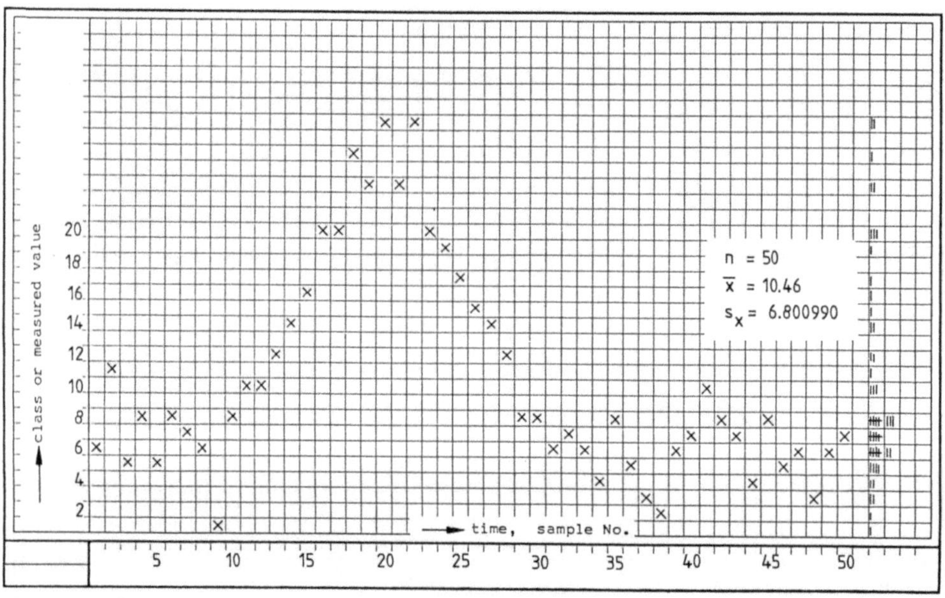

Fig. 14.11. Evaluation and representation of a strongly disrupted process. The disruption affects particularly the total spread and the estimated value $\hat{\sigma}_1$. The other estimated values are less strongly enlarged. The strongest method against the disruptive effect is the difference method. The estimated value $\hat{\sigma}_5 = 2.229$ lies only about 10% above the true value of the total population

	without disruption		disruption from i = 11 to i = 30	
1	2	3	4	5
i	x_i	D_i	x_i	D_i
1	6		6	
2	11	5	11	5
3	5	-6	5	-6
4	8	3	8	3
5	5	-3	5	-3
6	8	3	8	3
7	7	-1	7	-1
8	6	-1	6	-1
9	1	-5	1	-5
10	8	7	8	7
11	8	0	9	1
12	6	-2	8	-1
13	6	0	9	1
14	6	0	10	1
15	6	0	11	1
16	8	2	14	3
17	6	-2	13	-1
18	9	3	17	4
19	5	-4	14	-3
20	7	2	17	3
21	3	-4	14	-3
22	9	6	21	7
23	4	-5	17	-4
24	5	1	19	2
25	5	0	20	1
26	5	0	21	1
27	6	1	23	2
28	6	0	24	1
29	4	-2	23	-1
30	6	2	26	3
31	6	0	6	
32	7	1	7	1
33	6	-1	6	-1
34	4	-2	4	-2
35	8	4	8	4
36	5	-3	5	-3
37	3	-2	3	-2
38	2	-1	2	-1
39	6	4	6	4
40	7	1	7	1
41	10	3	10	3
42	8	-2	8	-2
43	7	-1	7	-1
44	4	-3	4	-3
45	8	4	8	4
46	5	-3	5	-3
47	6	1	6	1
48	3	-3	3	-3
49	6	3	6	3
50	7	1	7	1

evaluation by difference method:

D_1		n_j
-6	/	1
-5	/	1
-4	/	1
-3	///////	7
-2	///	3
-1	////////	8
0		
1	///////////	11
2	//	2
3	///////	7
4	////	4
5	/	1
6		
7	//	2

$n = 48$

$\bar{D} = 0.4375$

$s_D = 2.888675$

$\hat{\sigma} = s_D/\sqrt{2}$
$= 2.113313$

Fig. 14.12. Simulation of a process with disruption followed by resumption of a normal pattern

	1	2	3	4	5	6	7	8	9	10
	6	8	9	14	14	21	6	5	10	5
	11	7	8	13	21	23	7	3	8	6
	5	6	9	17	17	24	6	2	7	3
	8	1	10	14	19	23	4	6	4	6
	5	8	11	17	20	26	8	7	8	7
$\bar{x}$	7.0	6.0	9.4	15.0	18.2	23.4	6.2	4.6	7.4	5.4
R	6	7	3	4	7	5	4	5	6	4
s	2.549510	2.915476	1.140175	1.870829	2.774887	1.816590	1.483240	2.073644	2.190890	1.516575

sums:

$\sum \bar{x} = 102,6$

$\sum R = 51$

$\sum s = 20.331816$

$\sum s^2 = 44.399998$

characteristic values:

$\bar{\bar{x}} = 10.26$

$s_x = 6.416846$

$s_{\bar{x}} = 6.394477$

estimated values:

$\hat{\sigma}_1 = \sqrt{5} \cdot s_{\bar{x}} = 14.298485$

$\hat{\sigma}_2 = \sqrt{\overline{s^2}} = 2.107131$

$\hat{\sigma}_3 = \bar{R}/d_n = 2.192605$

$\hat{\sigma}_4 = \bar{s}/a_n = 2.162959$

$\hat{\sigma}_5 = s_D/\sqrt{2} = 2.113313$

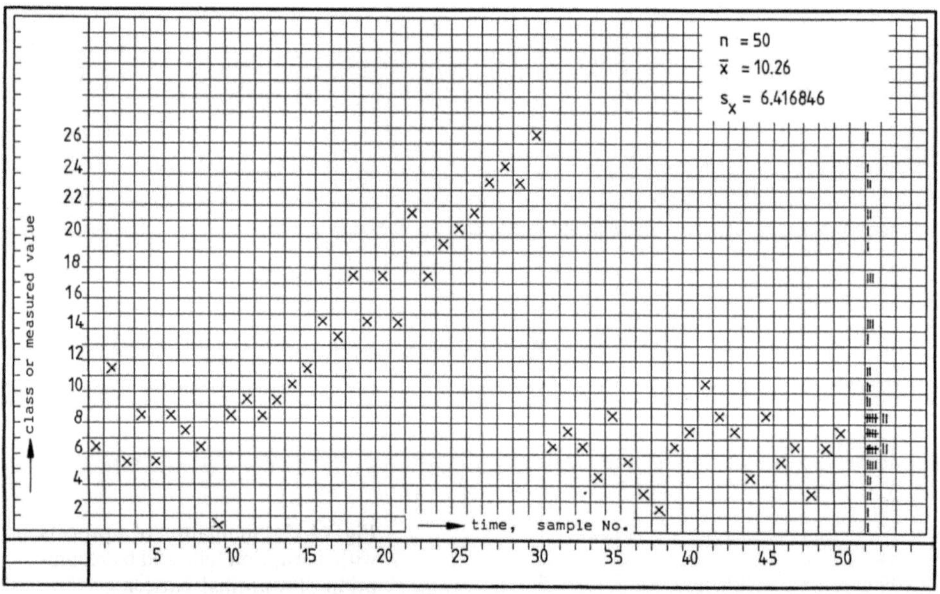

Fig. 14.13. Evaluation and representation of a process with one passing disruption. The effects on the estimated values are slight; only $\hat{\sigma}_1$ is strongly enlarged. The total spread s_x is more than three times as big as the process spread

	without disruption		with mean value fluctuations		
1	2	3	4	5	6
i	x_i	D_i	$\Delta\mu$	x_i	D_i
1	6			6	
2	11	5		11	5
3	5	-6	0	5	-6
4	8	3		8	3
5	5	-3		5	-3
6	8	3		10	
7	7	-1		9	-1
8	6	-1	2	8	-1
9	1	-5		3	-5
10	8	7		10	7
11	8	0		7	
12	6	-2		5	-2
13	6	0	-1	5	0
14	6	0		5	0
15	6	0		5	0
16	8	2		9	
17	6	-2		7	-2
18	9	3	1	10	3
19	5	-4		6	-4
20	7	2		8	2
21	3	-4		6	
22	9	6		12	6
23	4	-5	3	7	-5
24	5	1		8	1
25	5	0		8	0
26	5	0		2	
27	6	1		3	1
28	6	0	-3	3	0
29	4	-2		1	-2
30	6	2		3	2
31	6	0		6	
32	7	1		7	1
33	6	-1	0	6	-1
34	4	-2		4	-2
35	8	4		8	4
36	5	-3		3	
37	3	-2		1	-2
38	2	-1	-2	0	-1
39	6	4		4	4
40	7	1		5	1
41	10	3		14	
42	8	-2		12	-2
43	7	-1	4	11	-1
44	4	-3		8	-3
45	8	4		12	4
46	5	-3		1	
47	6	1		2	1
48	3	-3	-4	-1	-3
49	6	3		2	3
50	7	1		3	1

evaluation by difference method:

D_j		n_j
-6	/	1
-5	//	2
-4	/	1
-3	///	3
-2	//////	6
-1	/////	5
0	/////	5
1	//////	6
2	//	2
3	///	3
4	///	3
5	/	1
6	/	1
7	/	1

$n = 40$

$\bar{D} = 0.075$

$s_D = 3.024579$

$\hat{\sigma} = 2.138700$

Fig. 14.14. Simulation of a process with mean value fluctuations

	1	2	3	4	5	6	7	8	9	10
	6	10	7	9	6	2	6	3	14	1
	11	9	5	7	12	3	7	1	12	2
	5	8	5	10	7	3	6	0	11	-1
	8	3	5	6	8	1	4	4	8	2
	5	10	5	8	8	3	8	5	12	3
μ	0	2	-1	1	3	-3	0	-2	4	-4
$\bar{x}$	7.0	8.0	5.4	8.0	8.2	2.4	6.2	2.6	11.4	1.4
R	6	7	2	4	6	2	4	5	6	4
s	2.549510	2.915476	0.894427	1.581139	2.280351	0.894427	1.483240	2.073644	2.190890	1.516575

sums:

$\sum \bar{x} = 60{,}6$

$\sum R = 46$

$\sum s = 18.379679$

$\sum s^2 = 37.900001$

characteristic values:

$\bar{\bar{x}} = 6.06$

$s_x = 3.489986$

$s_{\bar{x}} = 3.145438$

estimated values:

$\hat{\sigma}_1 = \sqrt{5} \cdot s_{\bar{x}} = 7.033412$

$\hat{\sigma}_2 = \sqrt{\overline{s^2}} = 1.946792$

$\hat{\sigma}_3 = \bar{R}/d_n = 1.977644$

$\hat{\sigma}_4 = \bar{s}/a_n = 1.955285$

$\hat{\sigma}_5 = s_D/\sqrt{2} = 2.138700$

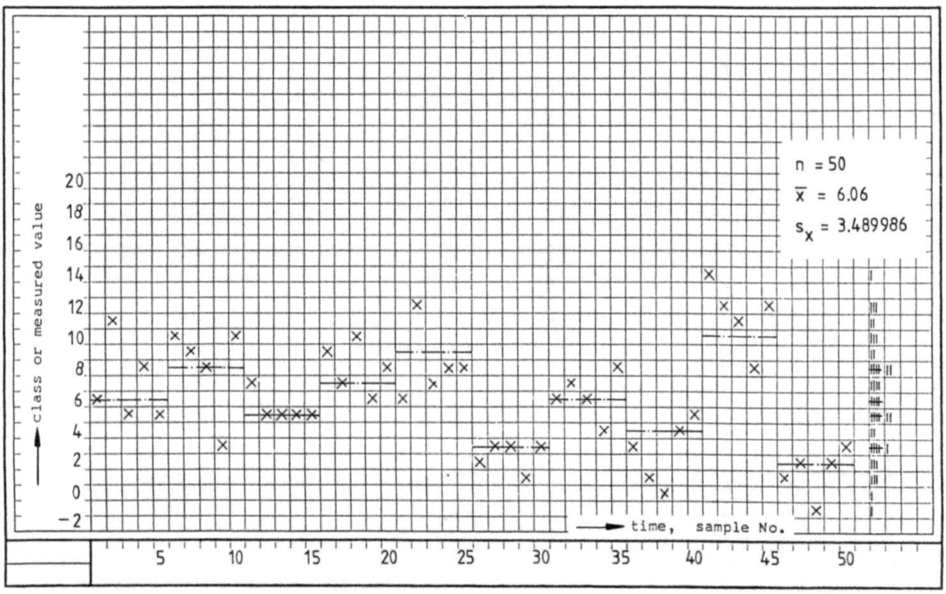

Fig. 14.15. Evaluation and representation of a process with mean value fluctuations; within the lower sample the process is normally distributed. The estimated values $\hat{\sigma}_2$ to $\hat{\sigma}_4$ are identical to those of undisrupted processes (Fig. 14.8). $\hat{\sigma}_5$ deviates somewhat because here only 40 differences are available

The calculation of the other estimated values for σ, the simulated processes and the representation of the processes with calculation of the total spread are given in Figs. 14.8 to 14.11 with commentary in diagram legends.

In Figs. 14.12 and 14.13 a process is simulated with a single disruption.

Finally in Figs. 14.14 and 14.15 the frequently occurring case is simulated where a process is analysed in such a way that samples are taken at greater intervals, for example five every 2 hours (long term investigation). Between the samples tool changes and new settings can take place.

Trend Processes

In the examples given up to now for simulated processes, cases have been shown where a limited disruption has taken place. Frequently processes show a more or less permanent trend.

In Fig. 14.16 a linear trend within a process which is temporarily without spread (left side) is represented diagrammatically. Then all differences are equal and the spread of the differences is nil. This leads to the assumption that with a linear trend process of differing differences, the spreads of the differences are caused exclusively or predominantly by temporary process spreads, Fig. 14.16 (right).

In Fig. 14.17 linear trend processes are simulated and in Fig. 14.18 evaluated according to the difference method. Already at the slash list stage for the differences it can be recognised that they show the same occupation numbers. In columns 2, 3 and 4 the slash lists are shifted by the trend to higher values with respect to column 1. Against this the estimated value $\hat{\sigma}$ for the spread of the total population (process spread) for the columns 1 to 4 are exactly the same size.

The further evaluation of the trend process according to the other estimated values and the total spread is shown in Figs. 14.19 to 14.21.

A weak trend, Fig. 14.19, affects the estimated value $\hat{\sigma}_1$ very strongly and the total spread strongly. The estimated values $\hat{\sigma}_2$ to $\hat{\sigma}_5$ are scarcely enlarged

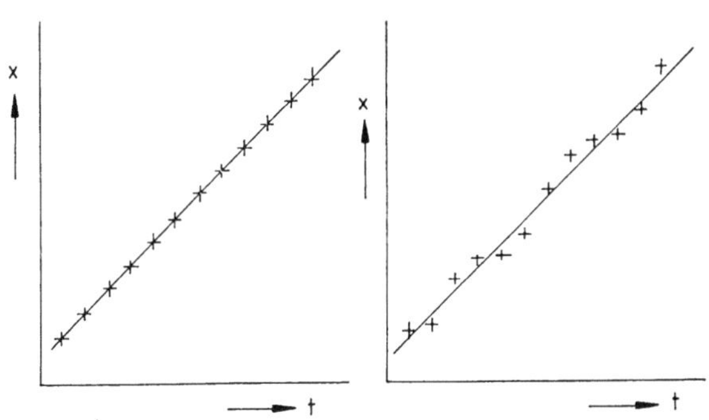

Fig. 14.16. Diagrammatic representation of a linear trend in an ideal process without spread (left) and a real process with spread (right)

	trend nil		trend slight		trend strong		trend very strong	
			$\Delta\mu = 0.25\sigma$/part		$\Delta\mu = 0.5\sigma$/part		$\Delta\mu = 1\sigma$/part	
1	2	3	4	5	6	7	8	9
i	x_i	D_i	x_i	D_i	x_i	D_i	x_i	D_i
1	6		6.5		7		8	
2	11	5	12	5.5	13	6	15	7
3	5	-6	6.5	-5.5	8	-5	11	-4
4	8	3	10	3.5	12	4	16	5
5	5	-3	7.5	-2.5	10	-2	15	-1
6	8	3	11	3.5	14	4	20	5
7	7	-1	10.5	-0.5	14	0	21	1
8	6	-1	10	-0.5	14	0	22	1
9	1	-5	5.5	-4.5	10	-4	19	-3
10	8	7	13	7.5	18	8	28	9
11	8	0	13.5	0.5	19	1	30	2
12	6	-2	12	-1.5	18	-1	30	0
13	6	0	12.5	0.5	19	1	32	2
14	6	0	13	0.5	20	1	34	2
15	6	0	13.5	0.5	21	1	36	2
16	8	2	16	2.5	24	3	40	4
17	6	-2	14.5	-1.5	23	-1	40	0
18	9	3	18	3.5	27	4	45	5
19	5	-4	14.5	-3.5	24	-3	43	-2
20	7	2	17	2.5	27	3	47	4
21	3	-4	13.5	-3.5	24	-3	45	-2
22	9	6	20	6.5	31	7	53	8
23	4	-5	15.5	-4.5	27	-4	50	-3
24	5	1	17	1.5	29	2	53	3
25	5	0	17.5	0.5	30	1	55	2
26	5	0	18	0.5	31	1	57	2
27	6	1	19.5	1.5	33	2	60	3
28	6	0	20	0.5	34	1	62	2
29	4	-2	18.5	-1.5	33	-1	62	0
30	6	2	21	2.5	36	3	66	4
31	6	0	21.5	0.5	37	1	68	2
32	7	1	23	1.5	39	2	71	3
33	6	-1	22.5	-0.5	39	0	72	1
34	4	-2	21	-1.5	38	-1	72	0
35	8	4	25.5	4.5	43	5	78	6
36	5	-3	23	-2.5	41	-2	77	-1
37	3	-2	21.5	-1.5	40	-1	77	0
38	2	-1	21	-0.5	40	0	78	1
39	6	4	25.5	4.5	45	5	84	6
40	7	1	27	1.5	47	2	87	3
41	10	3	30.5	3,5	51	4	92	5
42	8	-2	29	-1.5	50	-1	92	0
43	7	-1	28.5	-0.5	50	0	93	1
44	4	-3	26	-2.5	48	-2	92	-1
45	8	4	30.5	4.5	53	5	98	6
46	5	-3	28	-2.5	51	-2	97	-1
47	6	1	29.5	1.5	53	2	100	3
48	3	-3	27	-2.5	51	-2	99	-1
49	6	3	30.5	3.5	55	4	104	5
50	7	1	32	1.5	57	2	107	3

Fig. 14.17. Simulation of processes with linear trends; the signs of weaker, stronger, or very much stronger trends are relative details.

D_j	1 trend nil	n_j	2 trend slight	n_j	3 trend strong	n_j	4 trend very strong	n_j	D_i	5 mixed distribution from column 1 and 3	n_i
-6	/	1	/	1					-6	/	1
-5	//	2	//	2	/	1			-5	///	3
-4	//	2	//	2	//	2	/	1	-4	////	4
-3	/////	5	/////	5	//	2	//	2	-3	///////	7
-2	//////	6	//////	6	/////	5	//	2	-2	///////////	11
-1	/////	5	/////	5	//////	6	/////	5	-1	///////////	11
0	////////	8	////////	8	/////	5	//////	6	0	/////////////	13
1	//////	6	//////	6	////////	8	/////	5	1	//////////////	14
2	///	3	///	3	//////	6	////////	8	2	/////////	9
3	/////	5	/////	5	///	3	//////	6	3	////////	8
4	///	3	///	3	/////	5	///	3	4	////////	8
5	/	1	/	1	///	3	/////	5	5	////	4
6	/	1	/	1	/	1	///	3	6	//	2
7	/	1	/	1	/	1	/	1	7	//	2
8					/	1	/	1	8	/	1
9							/	1	9		

1	2	3	4	5
$\Delta\mu = 0$	$\Delta\mu = 0.25\,\sigma$/part	$\Delta\mu = 0.5\,\sigma$/part	$\Delta\mu = 1\,\sigma$/part	
$n = 49$	$n = 49$	$n = 49$	$n = 49$	$n = 98$
$\bar{D} = 0.020408$	$\bar{D} = 0.520408$	$\bar{D} = 1.020408$	$\bar{D} = 2.020408$	$\bar{D} = 0.520408$
$s_D = 2.933214$	$s_D = 2.933214$	$s_D = 2.933214$	$s_D = 2.933214$	$s_D = 2.961017$
$\hat{\sigma} = 2.074095$	$\hat{\sigma} = 2.074095$	$\hat{\sigma} = 2.074095$	$\hat{\sigma} = 2.074095$	$\hat{\sigma} = s_D/\sqrt{2} = 2.093755$

Fig. 14.18. Evaluation of linear trend processes and a non-linear trend according to the difference method

by the trend. This is different with a strong or very strong trend, Figs. 14.20 and 14.21. All estimated values $\hat{\sigma}$ are larger; only the estimated value $\hat{\sigma}_5$ is not influenced by the steepness of the trend.

Often trend processes are not linear. More often in practice after a tool change the trend is first very steep until the extreme sharpness of the tool is worn, thereafter the trend becomes flatter and later, if the tool is almost blunt the trend becomes steeper again.

This process is simulated in the extreme by a process first without trend and then during the second half of the time period with a strong trend. For the differences a mixed distribution results in column 5 of Fig. 14.18. It is a question of an MD from two equal distributions with the same standard deviation and a slight difference in mean value. With this type of mixed distribution the spread with respect to that of the components alters hardly at all. Thus with $\hat{\sigma}_5 = 2.093\,755$ the process spread is very precisely estimated; the estimated values $\hat{\sigma}_2$ to $\hat{\sigma}_4$ also lie correctly, Fig. 14.22.

All examples up to now have been to show the effects of disruptions or trends; therefore the assumption of limit values has been avoided. The effects of limit values are discussed in Chap. 15.

	1	2	3	4	5	6	7	8	9	10
	6.5	11	13.5	16	13.5	18	21.5	23	30.5	28
	12	10.5	12	14.5	20	19.5	23	21.5	29	29.5
	6.5	10	12.5	18	15.5	20	22.5	21	28.5	27
	10	5.5	13	14.5	17	18.5	21	25.5	26	30.5
	7.5	13	13.5	17	17.5	21	25.5	27	30.5	32
$\bar{x}$	8.5	10.0	12.9	16	16.7	19.4	22.7	23.6	28.9	29.4
R	5.5	7.5	1.5	3.5	6.5	3.0	4.0	6.0	4.5	5.0
s	2.423840	2.761340	0.651920	1.541104	2.413504	1.193734	1.753568	2.583602	1.850676	1.981161

sums:

$\sum \bar{x} = 188.1$

$\sum R = 47$

$\sum s = 19.154449$

$\sum s^2 = 40.650003$

characteristic values:

$\bar{\bar{x}} = 18.81$

$s_x = 7.256869$

$s_{\bar{x}} = 7.330067$

estimated values:

$\hat{\sigma}_1 = \sqrt{5}\ s_{\bar{x}} = 16.390529$

$\hat{\sigma}_2 = \sqrt{\overline{s^2}} = 2.016185$

$\hat{\sigma}_3 = \bar{R}/d_n = 2.020636$

$\hat{\sigma}_4 = \bar{s}/a_n = 2.037707$

$\hat{\sigma}_5 = s_D/\sqrt{2} = 2.074095$

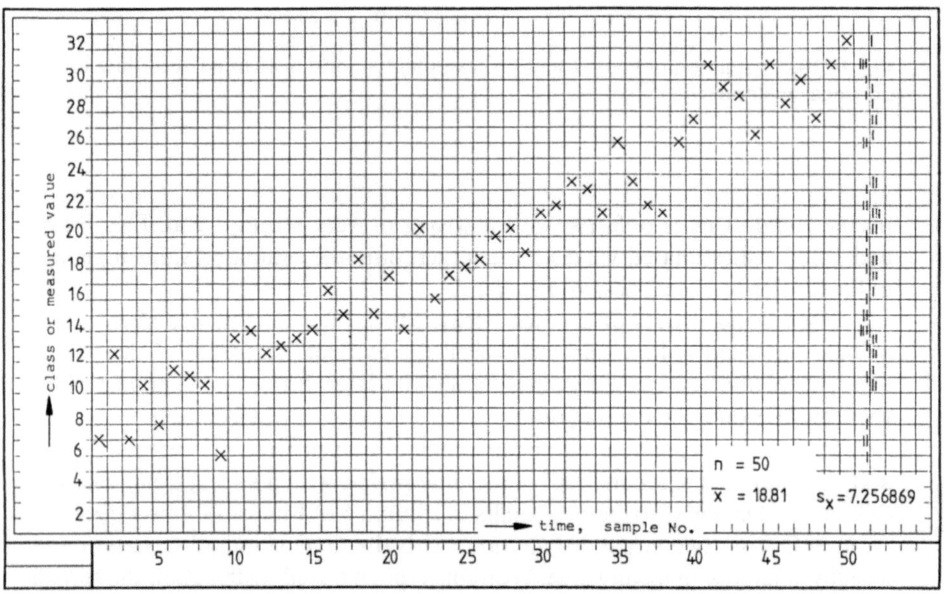

Fig. 14.19. Evaluation and representation of a process with a linear weak trend of $\Delta\mu = 0.25\sigma$/part

	1	2	3	4	5	6	7	8	9	10
	7	14	19	24	24	31	37	41	51	51
	13	14	18	23	31	33	39	40	50	53
	8	14	19	27	27	34	39	40	50	51
	12	10	20	24	29	33	38	45	48	55
	10	18	21	27	30	36	43	47	53	57
$\bar{x}$	10.0	14.0	19.4	25.0	28.2	33.4	39.2	42.6	50.4	53.4
R	6	8	3	4	7	5	6	6	5	6
s	2.549510	2.828427	1.140175	1.870829	2.774887	1.816590	2.280351	3.209361	1.816590	2.607681

sums:

$\sum \bar{x} = 315.6$

$\sum R = 56$

$\sum s = 22.894401$

$\sum s^2 = 55.899996$

characteristic values:

$\bar{\bar{x}} = 31.56$

$s_x = 14.412523$

$s_{\bar{x}} = 14.873332$

estimated values:

$\hat{\sigma}_1 = \sqrt{5} \cdot s_{\bar{x}} = 33.257781$

$\hat{\sigma}_2 = \sqrt{\overline{s^2}} = 2.364318$

$\hat{\sigma}_3 = \bar{R}/d_n = 2.407566$

$\hat{\sigma}_4 = \bar{s}/a_n = 2.435575$

$\hat{\sigma}_5 = s_D/\sqrt{2} = 2.074095$

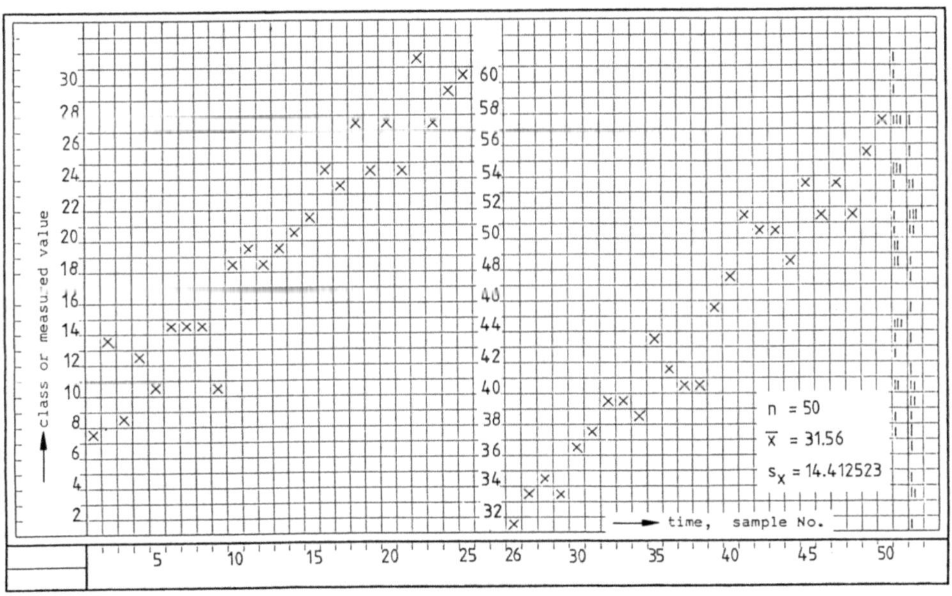

Fig. 14.20. Evaluation and representation of a process with linear strong trend of $\Delta\mu = 0.5\sigma/\text{part}$

Summary

The difference method is especially useful for the estimation of process spread, since it is particularly robust with respect to disruptions and trends. It is the only possible method if the process has a very steep trend or is frequently subject to intervention and thus for example five samples without intervention are quite unusual. In addition the difference method allows the mean steepness of the trend to be known; the mean value of the differences $\bar{D}$ is an estimated value for the mean steepness $\Delta\mu$ of the trend. Knowledge of the steepness of the trend is important for the selection of quality control charts and for the analysis of their effectiveness, Chap. 15.

Exercise 14.1

Given: Section 14.1
Required: Completion of the analysis of the characteristics and the estimated values in Figs. 14.8 to 14.22 with the aid of a calculator with $\bar{x}/s$-automatic facility.

14.2 Process Capability

In this section the information given already in Sect. 13.4 on process capability will be repeated and completed.

The concept of process capability is not standardised in the DIN standards or designs. In an ISO standard about concepts of quality assurance [45], different concepts are however defined. Three of these are quoted here, including the English and French original text:

	1	2	3	4	5	6	7	8	9	10
	8	20	30	40	45	57	68	77	92	97
	15	21	30	40	53	60	71	77	92	100
	11	22	32	45	50	62	72	78	93	99
	16	19	34	43	53	62	72	84	92	104
	15	28	36	47	55	66	78	87	98	107
$\bar{x}$	13.0	22.0	32.4	43.0	51.2	61.4	72.2	80.6	93.4	101.4
R	8	9	6	7	10	9	10	10	6	10
s	3.391165	3.535534	2.607681	3.082207	3.898718	3.286335	3.633180	4.615192	2.607681	4.037326

sums:

$\sum \bar{x} = 570.6$

$\sum R = 85$

$\sum s = 34.689519$

$\sum s^2 = 123.864550$

characteristic values:

$\bar{\bar{x}} = 57.06$

$s_x = 28.92$[illegible]834

$s_{\bar{x}} = 29.998970$

estimated values:

$\hat{\sigma}_1 = \sqrt{5}\, s_{\bar{x}} = 67.077974$

$\hat{\sigma}_2 = \sqrt{\overline{s^2}} = 3.519440$

$\hat{\sigma}_3 = \bar{R}/d_n = 3.654342$

$\hat{\sigma}_4 = \bar{s}/a_n = 3.690374$

$\hat{\sigma}_5 = s_D/\sqrt{2} = 2.074095$

a

Fig. 14.21 a. Evaluation of a process with linear, very strong trend of $\Delta\mu = 1\sigma$/part

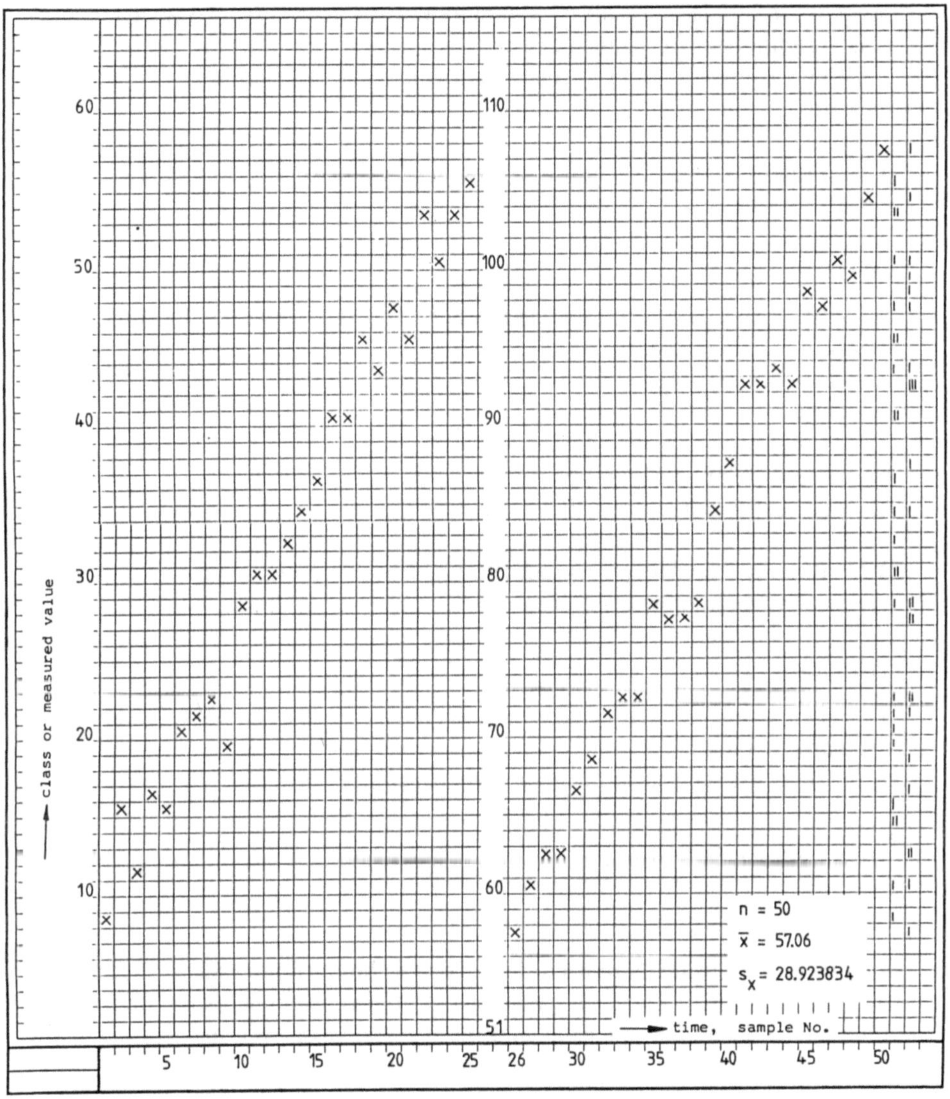

b

Fig. 14.21 b. Representation of a process with linear, very strong trend of $\Delta\mu = 1\sigma$/part

3.2.3 Process Capability: A measure of inherent process variability.

NOTES

1 Standard techniques for determining process capability have not achieved consensus at the present time. Some examples are the standard deviation or

range (or multiples thereof) or a component for acceptable assignable causes plus a component for chance causes.
2 When using the term "process capability", it is essential to state which measure is being used.

3.2.6 Process Capability Index (PCI): The value of the tolerance specified divided by the process capability.

NOTES

1 When using the term "process capability", it is essential to state which measure is being used.
2 This ratio is often used to classify a process into one of the following categories:
a) low relative capability: PCI < 6,
b) medium relative capability: $6 \leqq$ PCI $\leqq 8$,
c) high relative capability: PCI > 8.

3.2.7 Process Capability Fraction (PCF): The value of the process capability divided by the tolerance specified.

NOTES

1 When using the term "process capability", it is essential to state which measure is being used.
2 PCF is the reciprocal of the *process capability index* (3.2.6).

3.2.3 Aptitude du Processus: Mesure de la variabilité inhérente au processus.

NOTES

1 On ne s'est pas encore accordé sur des techniques normalisées permettant de déterminer l'aptitude du processus. Certains exemples sont fournis par l'écart-type, l'étendue (ou leurs multiples) ou la prise en considération d'une composante pour les causes assignables acceptables plus une composante pour les causes dues au hasard.
2 Quand on utilise le terme «aptitude du processus», il est essentiel de préciser quelle est la mesure appliquée.

3.2.6 Indice d'Aptitude du Processus (IAP): Valeur de la tolérance spécifiée divisée par l'aptitude du processus.

NOTES

1 Quand on utilise le term «aptitude du processus», il est essentiel de préciser quelle est la mesure utilisée.

2 On se sert souvent de cet indice pour classer un processus dans l'une des catégories suivantes:
a) aptitude relative faible: IAP < 6,
b) aptitude relative moyenne: $6 \leqq \text{IAP} \leqq 8$,
c) grande aptitude relative: IAP > 8.

3.2.7 Proportion d'Aptitude du Processus (PAP): Valeur de l'aptitude du processus divisée par la tolérance spécifiée.

NOTES

1 Quand on utilise le terme ≪aptitude du processus≫, il est essentiel de préciser quelle est la mesure utilisée.
2 La PAP est l'inverse de l'*indice d'aptitude du processus* (3.2.6).

Thus the process capability is the process spread and the tolerance based on the standard deviation is the process capability index. In summary there are the following three possibilities to meet the given tolerance requirements for the capability of a process, formulated:

$$1)\ \frac{T}{\hat{\sigma}} \geqq 6 \quad \text{or} \quad T \geqq 6\hat{\sigma},$$

$$2)\ c_p = \frac{T}{6\hat{\sigma}} \geqq 1 \quad \text{or} \quad c_p \geqq 100\%,$$

$$3)\ \frac{6\hat{\sigma}}{T} = \frac{1}{c_p} \leqq 1 \quad \text{or} \quad \frac{1}{c_p} \leqq 100\%.$$

In accordance with the ISO standard for the analysis of the established (relative) process capability the details in Fig. 14.23 are suggested.

Note 1. In the literature and in internal works standards occasionally the production position will be used in addition to determine process capability, for example by the index

$$c_{PK} = \frac{G_o - \bar{\bar{x}}}{3\hat{\sigma}} \geqq 1.$$

And if $c_{PK} < 1$, then the process capability is not given. Where the process capability, analysed according to c_P, is good or very good and however $c_{PK} < 1$, then the process is not "incapable" but the machine operator is not in the position to correct the process to the centre; and in the case where the machine control is automatic, the automatic facility does not function or is wrongly programmed. With an effective process control $c_{PK} < 1$ over a long time period should not occur; more on this in Chap. 15.

If the process position can be corrected, then the analysis of a process by the c_{PK} value should usually be superfluous.

	1	2	....	9	10	11	12	....	19	20
	6	8		10	5	7	14		51	51
	11	7		8	6	13	14		50	53
	5	6		7	3	8	14		50	51
	8	1		4	6	12	10		48	55
	5	8		8	7	10	18		53	57
$\bar{x}$	7.0	6.0		7.4	5.4	10.0	14.0		50.4	53.4
R	6	7		6	4	6	8		5	6
s	2.549510	2.915476		2.190890	1.516575	2.549510	2.828427		1.816590	2.607681

sums:

$\sum \bar{x} = 60.6 + 315.6 = 376.2$

$\sum s = 18.379679 + 22.894401 = 41.274080$

$\sum s^2 = 37.900001 + 55.899996 = 93.799997$

$\sum R = 46 + 56 = 102$

characteristic values:

$\bar{\bar{x}} = 18.81$

$s_x = \sqrt{\frac{1}{99}(49\cdot 1.96^2 + 49\cdot 14.41^2)} = 10.233210$

$s_{\bar{x}} = 16.622176$

estimated values:

$\hat{\sigma}_1 = \sqrt{5}\, s_{\bar{x}} = 37.168316$

$\hat{\sigma}_2 = \sqrt{\overline{s^2}} = 2.165616$

$\hat{\sigma}_3 = \bar{R}/d_n = 2.192605$

$\hat{\sigma}_4 = \bar{s}/a_n = 2.195430$

$\hat{\sigma}_5 = s_D/\sqrt{2} = 2.093755$

Fig. 14.22. Evaluation of a process which runs to up to 50 parts without trend and after this up to 100 parts with a linear trend with the slope $\Delta\mu = 0.5\sigma$/part. The representation of this process results from the consecutive series of Figs. 14.8 and 14.20

process capability	analysis
$c_p = \frac{T}{6\cdot\hat{\sigma}} < 1$ or $T < 6\cdot\hat{\sigma}$	process capability is not given
$c_p = \frac{T}{6\cdot\hat{\sigma}} \approx 1$ or $T \approx 6\cdot\hat{\sigma}$	process capability is given, if a few error proportion can be accepted
$c_p = \frac{T}{6\cdot\hat{\sigma}} \geq 1.3\bar{3}$ or $T \geq 8\cdot\hat{\sigma}$	process capability is good
$c_p = \frac{T}{6\cdot\hat{\sigma}} \geq 1.6\bar{6}$ or $T \geq 10\cdot\hat{\sigma}$	process capability is very good

$\hat{\sigma}$ = estimated value for the instant standard deviation, figure 14.1

Fig. 14.23. The (relative) process capability and its analysis; this analysis assumes that the process control is optimal

In any case the c_{PK} value—instead of the c_P value—can be useful in cases where only a limit value is given; this is however not usually the case with sizes except for shape and position deviations.

Note 2. Alongside the concept of process capability there is also that of machine capability. Both concepts are often either distinguished from each other too much or mixed up one with another. Both (not standardised) concepts are roughly similar.

The investigation of machine capability of a lathe follows for example as in [46] the precisely established conditions below:

—material (C45 i.e. unalloied steel with 0.45%C),
—part diameter (e.g. 25 to 40 mm),
—clamp length (e.g. 35 mm),
—measured length (e.g. 16 mm),
—cutting speed ($v = 100^{+20}_{-20}$ m/min),
—cut depth ($a = 0.5$ mm),
—feed ($s = 0.1\ \text{mm}^{-1}$).

and various other conditions. The determination of machine capability has the preferred objective of comparing equal or similar machines (from different producers).

From a qualitative point of view it is usually justified to say that the machine with the best machine capability will show the best process capability in special applications. This cannot however always be true.

The investigation of process capability should answer the question as to whether the process is in a position to produce specific parts from specified materials and with specified measurements and above all within specified tolerances in the future without errors. Whether or not process capability is given, it is certainly possible that the same machine may have a quite different process capability when producing other parts with other measurements and other tolerances.

Finally it happens that with the same part having two relevant sizes

size 1 is given in process capability and
size 2 is not given.

In the case where—as in fact in all cases in which the process capability is not given—it is to be investigated whether a fundamental explanation can be found, with the objective of reducing the spread and thus improving the process capability. The limiting characteristics on the instant standard deviation σ are the already mentioned "5 M's", namely

—material,
—machine,
—method,
—measurement process,
—men and women.

In the individual case an investigation should be made as to whether one or more of these "5 M's" can be influenced in this sense. The first thing to clarify is whether the measurement process is in order or could be improved in terms of correctness and precision. Also the possibility often arises that the material could be improved, for example in the plastic manufacture through a preliminary treatment of the granulated material.

A very strong limiting characteristic is the method. To this belong the following components, which are briefly explained below:

—Clamping methods and clamping force. For example three-jaw chuck, strong bracing of the (hollow) part with strong deviations of the circular shape as a result; this "inner spread" shows in the total spread. Cure: another chuck, less bracing force with the last cut, thus lessening the inner spread.

—Cutting conditions. By whose alteration (v, a, s) the cutting force is also altered, thus producing less elastic deformation of the part and tool; in addition the vibration frequencies are changed which may have caused resonance in the part or the tool.

—Tool. Material can be changed, turning tool may be reinforced.

—Coolant lubrication. If there is not enough or too irregular large temperature differences can result inside and between parts, which is noticeable in the spread, in particular with large measurements.

With the machining of a steel ring for example, with a diameter of 200 mm, temperature differences of 10 K produce the following diameter changes:

$$\begin{aligned}\Delta D &= D_0 \alpha_{th} \Delta\vartheta \\ &= 0.2 \cdot 12 \cdot 10^{-6} \cdot 10 \,\frac{\text{mK}}{\text{K}} \\ &= 24 \cdot 10^{-6}\,\text{m} \\ &= 24\,\mu\text{m}.\end{aligned}$$

Where IT 7 is given for the ring, the tolerance amounts to $T = 46\,\mu\text{m}$ from DIN 7151; thus the diameter change as a result of thermal expansion is more than half of the tolerance.

Where it is not possible to improve process capability by these or other measures, there are still the following decision possibilities:

—change the machine or after a very careful tolerance analysis as in Chaps. 8 to 11,

—widen the tolerance and thus provide for the process capability, or

—accept a more or less large error proportion in the future, or

—at least partly sort.

The latter is problem free with measurement control, but with very expensive parts, the efficiency will reduce considerably if parts cannot be repaired.

15 Process Control

15.1 General Aspects

At the end of Chap. 13 it had already been shown that with the presence of a trend of a steepness of $\Delta\mu \geqq 0.1\ \sigma$/part process control can only be carried out successfully if a continuous QCC is used. This is not problematical if micro-processor controlled measuring machines are used, which store and statistically compute measurements from several measurement stations and can output the results in tabular form of several measurement stations or as QCC's for a particular, chosen measurement station or as graphical data for a fully graphic monitor, Fig. 15.1.

This comfortable combination of automatic measurements and computers can only be used with success, however, if they are relevantly programmed. A pre-requisite for this is a comprehensive knowledge of quality control chart techniques and the special cases which must be considered where trend processes are present.

At the current stage of development these luxury QCC's are indeed produced; process control by correction of the setting positions is however still largely manual. In isolated cases automatic controls are already available but their efficiency nonetheless still depends on relevant programming. Since because of the spread of processes the mean deviation of production distributions from the desired value is not exactly known, the question of optimal correction totals is a special problem.

A further and more important aspect is the fact that the steepness of the trends frequently cannot be known and thus is under estimated. To this end Fig. 15.61 shows a simulated controlled trend process with the slope $\Delta\mu = 0.1$/part. The trend will only become "visible" with the details of interventions, Fig. 15.62.

A further general aspect which is often not known is the fact that the (instant) process spread cannot be reduced by process control. With each intervention in the process, a new setting of the mean value of the distribution of individual values is assumed. Good process control can indeed reduce the total spread of the produced batch s_X; this is however always greater than the process spread σ; this is the reason for the (relative) process capability being $c_P \geqq 1.3\bar{3}\ (T \geqq 8\sigma)$. Bad process control can make the total spread of the produced batch considerably greater. The analysis of a process by its process capability assumes that the process control is optimal.

The data output of the QDS 803 is shown on a large fully graphical monitor screen It is clearly laid out for practical use and is restricted to information which is actually necessary for the control and optimisation of the production process

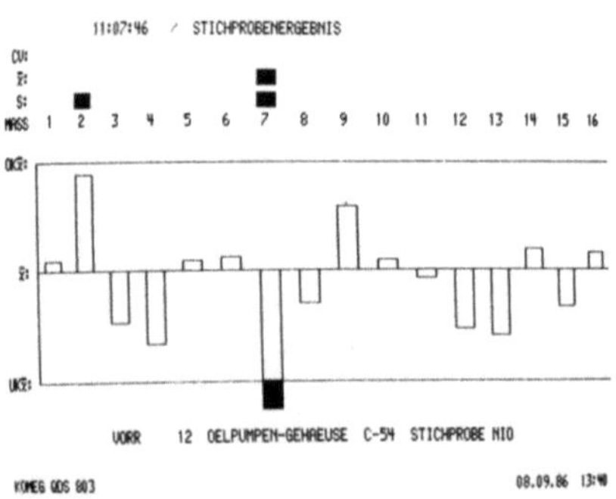

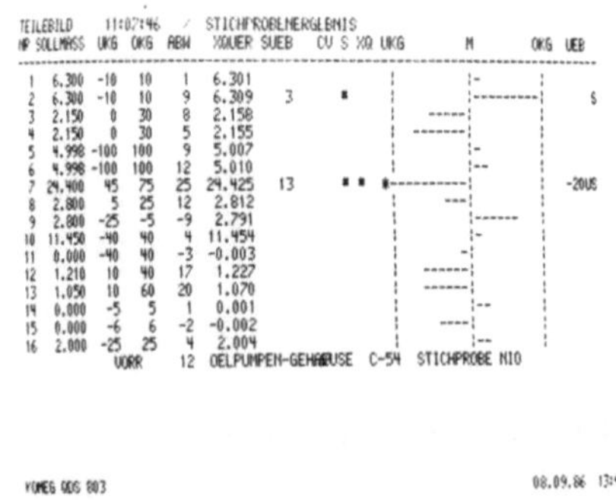

Graphical Result Screen

The graphical results screen supplies a quick overview of all the characteristics. To simplify understanding, size columns are shown as plus or minus deviations from the centre of the tolerance range

Excesses over individually inputted control limits are shown by inverse display of the section concerned

According to the chosen operation method, the columns show either the individual values of a part or the mean value of a sample With sample control, all tolerance limits are established by the control limits

Alphanumeric Result Screen

The result shown is the same as that on the column screen

All the desired, limit and actual values essential for decision-making are presented in clearly arranged tabular form. Where the control or tolerance limit is exceeded, the size and the direction of the non-conformity is specifically shown

A simplified bar chart next to the number series also helps to facilitate the overview

At the touch of a button the type of representation on screeh can be switched to the other type available

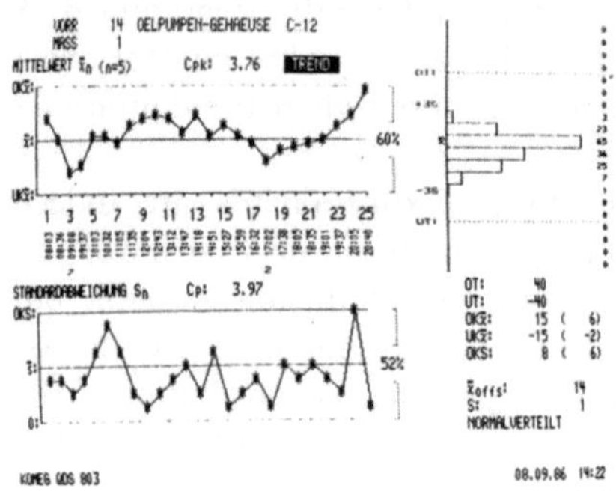

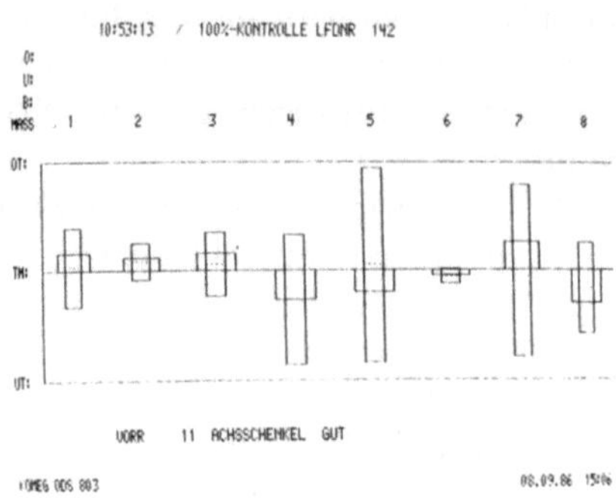

Quality Control Chart

The quality cc is illustrated in one complete screen of the process course (with time clock) and the current stage of process and quality

- Trend course of $\bar{x}_n$ within the control limits
- Course followed by the standard deviation against the control limit
- Histogram with optional class number up to 16
- Given and (in brackets behind) newly calculated control limits
- Last $\bar{x}$-value ($\bar{x}$ offset) in μ-deviation from the centre of the control range
- Last standard deviation
- Process capability index CpK and Cp as well as notes on the process behaviour, TREND, RUN or MIDDLE THIRD
- Marker for process intervention (below the time)

Combined Result Screen After Rolling Calculation

At each measurement column three sets of data are available

1 The size deviation of the last part (wide column)
2 The interval of current work uncertainty $\bar{x} \pm f \cdot S_n$ (narrow column) f optional: 1, 2 or 3
3 The position of the mean value $\bar{x}_n$ in the narrow column

If the work uncertainty leaves the tolerance bandwidth, then a warning is given with a direction indicator: excess upwards, excess downwards or two-sided excess through undue spread

Fig. 15.1. Data output from a quality data system of the firm KOMEG with the illustrations of the manufacturer. Various concepts are used in a different form in the text

The following discussions about process control (process guidance would be correct to the standard but in operational practice is little used) proceed under the assumption that the size tolerances belong to arithmetically calculated dimensional chains, because this is usually the case in practice. With statistically tolerated sizes, the intervention limits have to be at their most dense at the mean value in order to achieve control of the production distribution at the centre. This is a disadvantage of statistical tolerance calculation, which is however unimportant in comparison to the fact that through the statistical expansion of individual tolerances, the process capability can be significantly improved.

15.2 Process Control by Mean Value Quality Control Charts

If no trend is present or if a present trend is very small, the designed QCC's for mean values can be determined with the aid of the nomograms in Fig. 13.6. In Fig. 15.2 the operational characteristics for $\bar{x}$-charts for $n = 5$ and various limiting factors of $k_A = 1.0$ and 3.0 are represented.

As a rule of thumb it can be said that over a short time period on average for every error proportion calculated an acceptance probability of $P_a = 50\%$ can be assigned. For $k_A = 2.0$ this non-conformity proportion amounts to

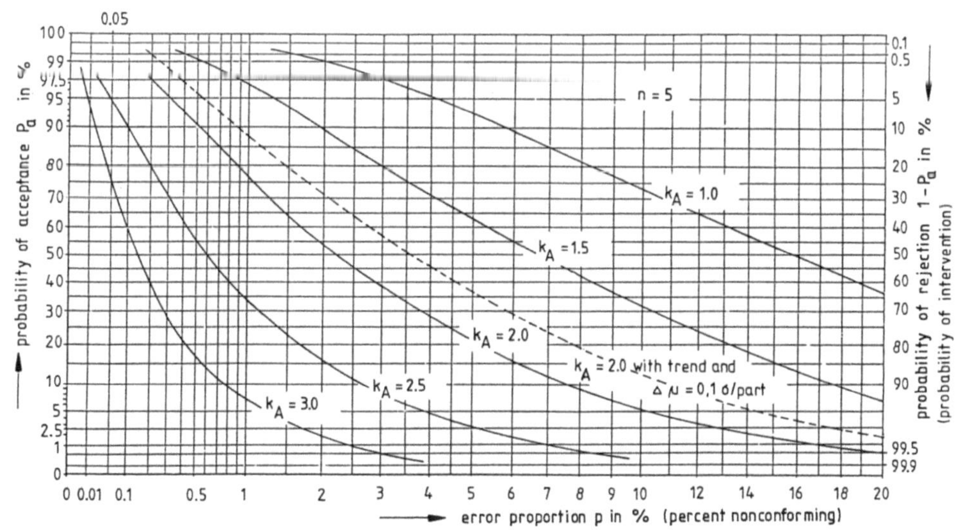

Fig. 15.2. Operational characteristics for the mean value quality control chart for $n = 5$ and various limiting factors k_A. The curves are valid for the case where there is no trend or a trend with only slight slope

$p_{50} = 2.275\%$, whereas applied to the whole produced batch the mean non-conformity proportion of the batch is smaller. A limiting factor of $k_A \geqq 2$ is an appropriate value for a sample range of $n = 5$.

This also covers analysis through process capability, Fig. 15.3. If the process capability is at $c_P = 1.0$ at the lower limit, then the acceptance probability is $P_a = 97.4464\%$ as long as the process is set exactly at the tolerance centre. With increasing process capability the clearance room for non-conformity settings and trends becomes greater, Fig. 15.3.

There are firms which set their employees internal works standards, independent of the respective process capability, of intervention limits for the mean value QCC's always at 99% limits or—according to American practice—established as $6\sigma_{\bar{x}}$ limits ("classical" quality control charts according to Shewhart). Also currently marketed measurement computers are programmed in this way. In Fig. 15.4 the resulting effects are represented. If the process capability lies a $c_P = 1.0$ at the lower limit, this way of proceeding is correct and necessary; with good ($c_P = 1.3\bar{3}$) or very good ($c_P = 1.6\bar{6}$) process capability, independent of the size of the tolerance zone, always the best possible quality will be produced with the result that "the individual values will deviate as little as possible from the mean value".

This consequent target-setting does however have the effect that no "clearance room" is available for non-conformity settings and trends. In

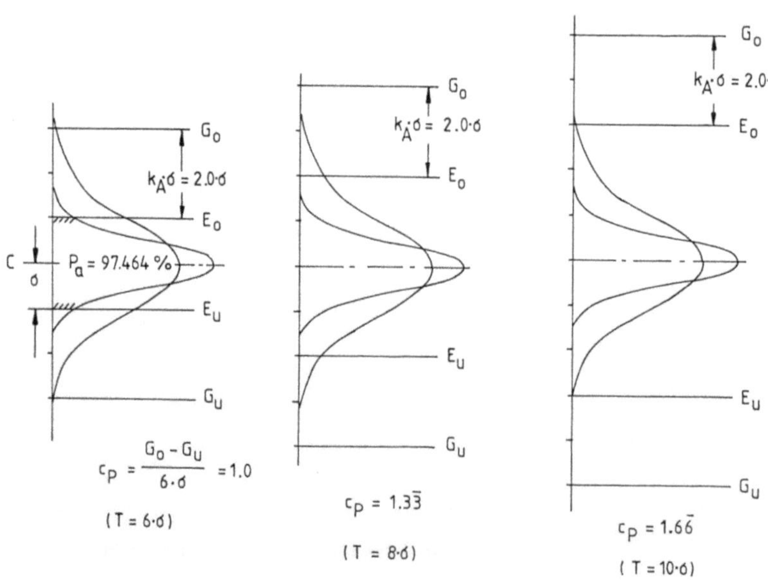

Fig. 15.3. Representation of $\bar{x}$-QCC for $n = 5$ with distributions set to the desired value (mean value) and different process capabilities; the limiting factors are in all cases the same with $k_A = 2.0$

particular with trend processes narrow intervention limits result in frequent interventions which does not matter in the presence of an automatic process control. However if process control is manual, it should in every case be checked whether it is more efficient to shift the intervention limits a little in the direction of the limit values, in order to achieve a greater clearance room for non-conformity settings and trends without fearing that non-conformities will occur in the considerable range. One possible suggestion for the examples in Fig. 15.4: at $c_P = 1.3\bar{3}$ with k_A go to $k_A \approx 2.2$ and with $c_P = 1.6\bar{6}$ to $k_A \approx 2.8$. The operational characteristics for these altered k_A values can be determined with the aid of the nomogram in Fig. 13.6 or by interpolation in Fig. 15.2. All statements up to now confine themselves to $\bar{x}$-QCC's for the case where only a weak trend exists.

With the presence of a trend of $\Delta\mu \geqq 0.1\ \sigma$/part, it is to be taken into account that the distribution of the mean values limps behind the distribution of individual values, with a sample range of $n = 5$ around the total $2\Delta\mu$. In Fig. 15.5 the operational characteristics are calculated for $\bar{x}$-QCC with $n = 5$ and the limiting factor $k_A = 2.0$ and various slopes of trend from $\Delta\mu = 0$ to $0.5\ \sigma$/part.

This calculation is further detailed for line 7 (Fig. 15.5) in connection with the left hand sketch. If as a result of the trend the centre of the distribution of individual values lies with the last of $n = 5$ consecutively taken individual values exactly on the upper intervention limit, then the interval to the upper limit

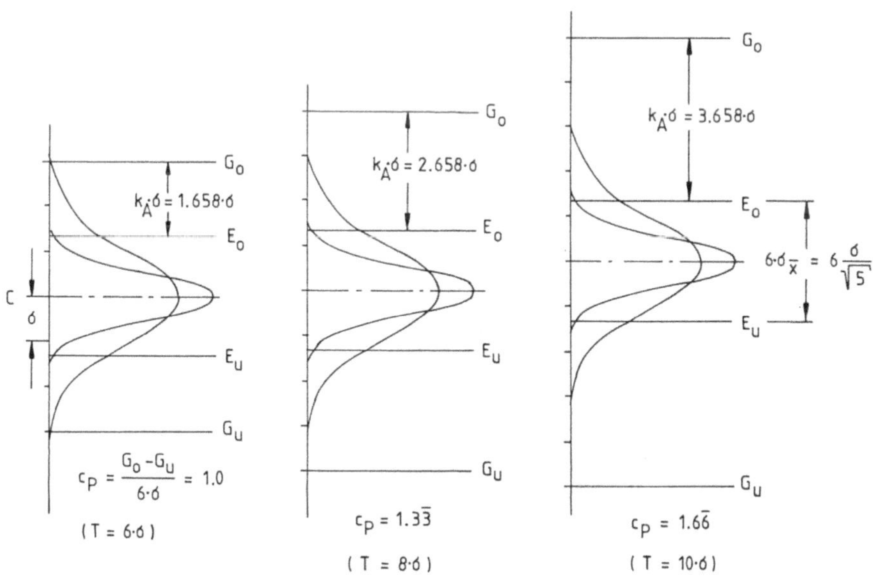

Fig. 15.4. Representation of $\bar{x}$-QCC for $n = 5$ with distributions set to the desired value (mean value) and different process capabilities; in all three cases the intervention limits are the limits of the $6\sigma_{\bar{x}}$ range $P_a = 99.73\%$

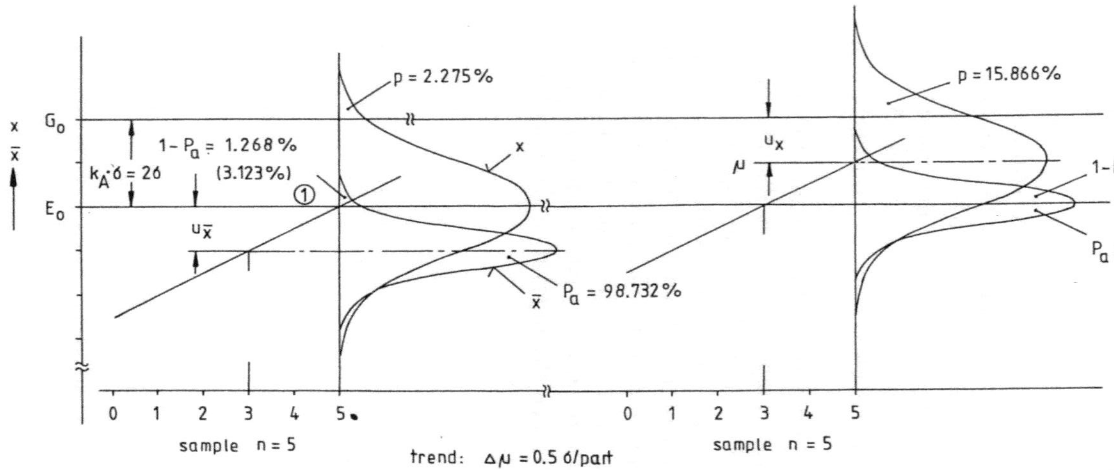

mean distance of the last of n = 5 individual values from the upper limiting value and according error proportion p			mean distance of the means of samples n = 5 from the upper intervention limit and according acceptable probability P_a $u_{\bar{x}} = (u_x - k_A + 2\Delta\mu)\sqrt{n} = (u_x - 2 + 2\Delta\mu)\sqrt{5}$								
			Δμ = 0		Δμ = 0.1 σ/part		Δμ = 0.3σ/part		Δμ = 0.5σ/part		
line no.	$u_x = \frac{G_o - \mu}{\sigma}$	p in %	$u_{\bar{x}}$	P_a	$u_{\bar{x}}$	P_a	$u_{\bar{x}}$	P_a	$u_{\bar{x}}$	P_a	note
1	3.5	0,023	3.354	0.99960	3.801	0.99993					
2	3.25	0,058	2.795	0.99741	3.242	0.99940					
3	3.0	0,135	2.236	0.98732	2.683	0.99635	3.587	0.99983			
4	2.75	0,298	1.677	0.95323	2.124	0.98316	3.019	0.99874			
5	2.5	0,621	1.118	0.86822	1.565	0.94121	2.460	0.99305	3.354	0.99960	
6	2.25	1,222	0.559	0.71192	1.006	0.84519	1.801	0.96415	2.795	0.99741	
7	2.0	2,275	0,000	0.50000	0.447	0.67256	1.342	0.91020	2.236	0.98732	cf. left hand sketch
8	1.75	4,006	-0.559	0.28808	-0.112	0.45541	0.783	0.78319	1.677	0.95323	
9	1.5	6,681	-1.118	0.13178	-0.671	0.25111	0.224	0.58862	1.118	0.86822	
10	1.25	10,565	-1.677	0.04677	-1.230	0.10935	-0.335	0.36881	0.559	0.71192	
11	1.0	15,866	-2.236	0.01268	-1.789	0.03681	-0.894	0.18566	0.000	0.50000	cf. right hand sketch
12	0.84	20,000	-2.594	0.00474	-2.147	0.01590	-1.252	0.10528	-0.358	0.36017	

Fig. 15.5. Calculation of the operational characteristics for process control from mean values of the range $n = 5$ for the limiting factor $k_A = 2.0$ and for various slopes $\Delta\mu$ of the trend with diagrammatic illustration of two cases with $\Delta\mu = 0.5\sigma$/part

① $1 - P_A = 3.123\%$ results from consideration of the fact that with a trend of 0.5σ/part, the spread of the mean values as well as the individual values is about 20% greater, cf Fig. 14.20 $\hat{\sigma}_2, \hat{\sigma}_3$ and $\hat{\sigma}_4$

② $1 - P_a = 50\%$ is also valid for the increased spread as a result of the trend value in σ units $u_x = 2.0$. Thus from the u-table (Table A1) the non-conformity probability $p = 2.275\%$. If the trend was not present (sudden shift of the process position to this position) then the distribution of the mean values would lie on the same location as that of the individual values and the probability for the intervention would be $1 - P_a = 50\%$, Fig. 15.2 with the operational characteristics for a process without trend and $k_A = 2.0$.

With a trend of $\Delta\mu = 0.5\ \sigma$/part, however, the distribution of the mean values lies at $2\Delta\mu = 1\ \sigma$ lower, with the result that the proportion of mean values above the upper intervention limit amounts to only 1.268%, the intervention probability is only $1 - P_a = 1.268\%$. Taking account of the fact that the mean values as a result of the trend are somewhat more strongly spread than with $\sigma_{\bar{x}} = \sigma/n$, $1 - P_a$ becomes equal to 3.123%.

In line 11 of Fig. 15.5 which is represented for $\Delta\mu = 0.5$/part as the right hand sketch, the error probability $p = 15.866\%$ if the distribution of the mean values lies exactly on the intervention limit. With and without account taken of the influences of the trends on the standard deviation of the mean value, the intervention probability is $1 - P_a = 50\%$.

The operational characteristics calculated in Fig. 15.5 are represented in Fig. 15.6. With an increasing trend the operational characteristics become flatter.

A comparison of the Fig. 15.6 and 15.2 leads to the recognition that the OC's without trend for k_A cover a similar range to those with trend and $k_{AT} = k_A + 2\Delta\mu$. For example, the OC for $k_A = 1.0$ in Fig. 15.2 runs the same as the OC for $k_A = 2.0$ in Fig. 15.6 and a trend with the steepness $\Delta\mu = 0.5\sigma$/part.

From this there is a conclusion to be drawn: where a trend is present the OC remains unchanged for the mean value QCC with k_A, if the limiting factor

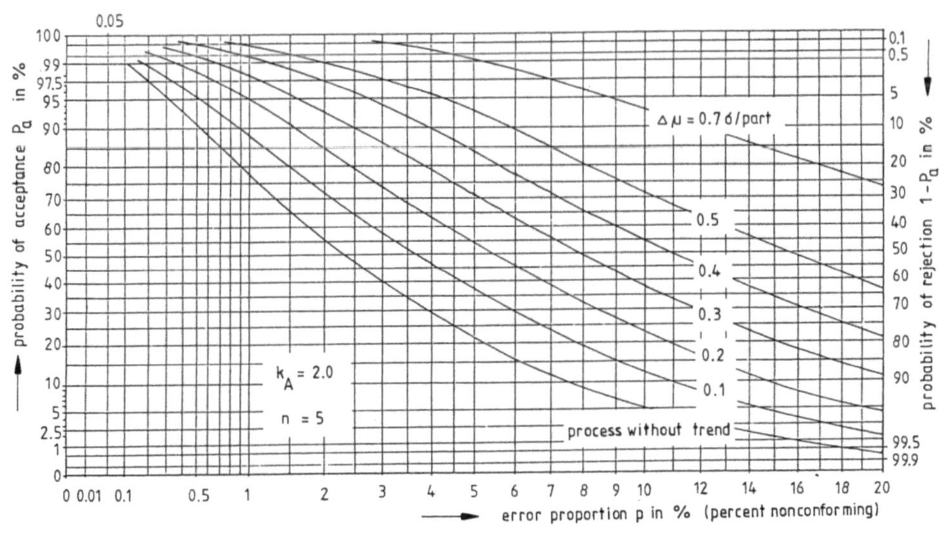

Fig. 15.6. Operational characteristics for process control from mean values of the range $n = 5$ for the limiting factor $k_A = 2.0$ and for various slopes $\Delta\mu$ of the trend

1	2	3	4	5	6	7
i	x_i	μ_i	x_i	D_i	$\bar{x}$	s_x / s_x^2
1	6	7	7			
2	11	8	13	6		
3	5	9	8	-5	10.0	2.549510
4	8	10	12	4		6.5
5	5	11	10	-2		
6	8	7	9			
7	7	8	9	0		
8	6	9	9	0	9.0	2.828427
9	1	10	5	-4		8.0
10	8	11	13	8		
11	8	7	9			
12	6	8	8	-1		
13	6	9	9	1	9.4	1.140175
14	6	10	10	1		1.3
15	6	11	11	1		
16	8	7	9			
17	6	8	8	-1		
18	9	9	12	4	10.0	1.870829
19	5	10	9	-3		3.5
20	7	11	12	3		
21	3	7	4			
22	9	8	11	7		
23	4	9	7	-4	8.2	2.774887
24	5	10	9	2		7.7
25	5	11	10	1		
26	5	7	6			
27	6	8	8	2		
28	6	9	9	1	8.4	1.816590
29	4	10	8	-1		3.3
30	6	11	11	3		
31	6	7	7			
32	7	8	9	2		
33	6	9	9	0	9.2	2.280351
34	4	10	8	-1		5.2
35	8	11	13	5		
36	5	7	6			
37	3	8	5	-1		
38	2	9	5	0	7.6	3.209361
39	6	10	10	5		10.3
40	7	11	12	2		
41	10	7	11			
42	8	8	10	-1		
43	7	9	10	0	10.4	1.816590
44	4	10	8	-2		3.3
45	8	11	13	5		
46	5	7	6			
47	6	8	8	2		
48	3	9	6	-2	8.4	2.607681
49	6	10	10	4		6.8
50	7	11	12	2		

from population with $\mu = 6$ und $\sigma = 2.01$

$\sum \bar{x} = 90.6 \quad \sum s_x^2 = 55.9$

$\bar{\bar{x}} = 9.06 \quad \overline{s_x^2} = 5.59$

$\sqrt{\overline{s_x^2}} = \hat{\sigma} = 2.364318$ in accordance with estimated value $\hat{\sigma}_2$ in figure 14.20

$\sum s_x = 22.894401$

$\frac{\bar{s}}{a_n} = \hat{\sigma} = 2.435575$ in accordance with estimated value $\hat{\sigma}_4$ in figure 14.20

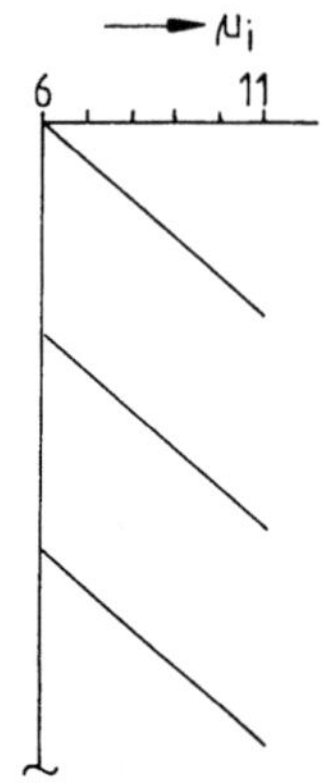

evaluation by difference method:

D_j		n_j
-5	/	1
-4	//	2
-3	/	1
-2	///	3
-1	//////	6
0	/////	5
1	/////	5
2	//////	6
3	//	2
4	///	3
5	///	3
6	/	1
7	/	1
8	/	1

$n = 40$

$\bar{D} = 1.075$

$s_D = 3.024579$

$\hat{\sigma} = \frac{s_D}{\sqrt{2}} = 2.138700$

in accordance with estimated value in figure 14.14

Fig. 15.7. Simulation of trend processes with sample tests of the range $n = 5$ with a trend of $\Delta\mu = 0.5\sigma$/part which begins at C (centre) and runs over five parts

is raised to $2\Delta\mu$ at $k_{AT} = k_A + 2\Delta\mu$, as long as the sample range for the mean value QCC is set at $n = 5$. With a sample range often chosen in practice of $n = 3$, the limiting factor must be raised to $k_{AT} = k_A + \Delta\mu$ in order for the k_A valid OC's to remain.

In order to see the drift of the mean values with a trend in another way, in Fig. 15.7 the already often used random sample test from an ND model has been divided into five sample tests, in which a trend with the slope $\Delta\mu = 0.5\sigma$/part is simulated. The mean values μ_i in column 3 climb from $\mu_0 = 6$ to $\mu_1 = 7$, $\mu_2 = 8$, $\mu_3 = 9$, $\mu_4 = 10$ and $\mu_5 = 11$. The individual values spread around this were part for part raised respectively to $0.5\sigma = 0.5 \cdot 2 = 1$, column 4. The mean values $\bar{x}$ of the five sample tests in column 6 give a total mean value of $\bar{x} = 9.06$; this lies at μ_3. In Fig. 15.7 further estimated values for σ of the total population are calculated; it should be noted that they conform with the estimated values in the other given figures.

					$\frac{T}{2\sigma} = 3$		$\frac{T}{2\sigma} = 3.5$		$\frac{T}{2\sigma} = 4.0$	
					$k_A = 1.658$		$k_A = 2.158$		$k_A = 2.658$	
sample no.	$z = \frac{\mu_3 - C}{\sigma}$	$u_{\bar{x}} = (1.342 - z)\sqrt{5}$	P_a	$\prod P_a$	$u_x = \frac{G_o - \mu_5}{\sigma}$	p in %	u_x	p in %	u_x	p in %
1	0.3	2.330	0.99010	0.99010	2.5	0.621	3	0.135	3.5	0.023
2	0.8	1.212	0.88724	0.87846	2.0	2.275	2.5	0.621	3.0	0.135
3	1.3	0.094	0.53745	0.47213	1.5	6.681	2.0	2.275	2.5	0.621
4	1.8	-1.024	0.15292	0.07220	1.0	15.866	1.5	6.681	2.0	2.275
5	2.3	-2.142	0.01610	0.00116	0.5	30.854	1.0	15.866	1.5	6.681
6	2.8	-3.260	[illegible]	[illegible]	[illegible]	50.000	0.5	30.853	1.0	15.866

explanations:

μ_3 mean position of the distributions for x, lying at the third value of samples with n = 5

μ_5 mean position of the distributions for x, lying at the last value of samples with n = 5

$u_{\bar{x}} = \frac{(E_o - \mu_3)}{\sigma}\sqrt{5}$, other writing for $u_{\bar{x}}$

$k_A = \frac{G_o - C}{\sigma} - 1.342$ is limiting factor for $\bar{x}$-chart

u_x mean distance of the last of n = 5 values of the sample to the upper limiting value in σ-units

$\frac{T}{2\sigma} = \frac{G_o - C}{\sigma}$ mean distance from the centre to the upper limiting value in σ-units

$\frac{T}{2\sigma} = 4.5$		$\frac{T}{2\sigma} = 5.0$	
$k_A = 3.158$		$k_A = 3,658$	
u_x	p in %	u_x	p in %
4.0	0.003	4.5	0.001
3.5	0.023	4.0	0.003
3.0	0.135	3.5	0.023
2.5	0.621	3.0	0.135
2.0	2.275	2.5	0.621
1.5	6.681	2.0	2.275

Fig. 15.8. Simulation of a process with a linear trend with the slope $\Delta\mu = 0.1\sigma$/part; the process is controlled with a $\bar{x}$-QCC with $n = 5$, whose intervention limits are 99.73%. The operational characteristics for different widths of tolerance zones are calculated. This simulation is represented in Fig. 15.9

There is a further special point about trend processes, which is important for the operation of original value charts, and which should be already clear from the discussion of mean value charts although this special point is less important with mean value charts. With trend processes and running sample test removals the intervention probability becomes greater the more often the production batch is tested. This is illustrated with the help of Figs. 15.8 and 15.9. Here a $\bar{x}$-QCC for $n=5$ with $6\sigma_{\bar{x}}$ limits is calculated and designed. There is a linear trend with $\Delta\mu=0.1\sigma$/part and the distribution of the individual values is set up during the course of the process exactly on the mean value.

After the production of five parts the distribution of the individual values lies at μ_5 and that of the mean values at μ_3. The acceptance probability amounts to $P_a=99.010\%$. Where the upper limit value lies at $T/2\sigma=3$ distant from the middle of the tolerance zone, the error probability for the fifth part of the first sample test is $p=0.621\%$ or relatively less with large tolerances. If no intervention is made—and the probability for this is high—the process runs further with the trend. After a further five produced parts $\mu_5=C+1\sigma$ and $P_a=0.887\,24$, Fig. 15.8. The error probability for the tenth part, the last of the second sample test is $p=2.275\%$, as long as $T=6\sigma$. The probability that the process runs further is $P_a=0.887\,24$; the total probability that there will be no intervention until the third sample test is $\prod P_a=0.990\,10\cdot 0.887\,24=0.878\,46=87.846\%$. The total probability that the process will run after the third sample test is $P_a=47.213\%$. If the process was suddenly in the mean value position for the third sample test or if the first two sample tests did not take place, then P_a would be 53.745%. The difference is indeed small; but it is larger, if the trend has a smaller slope. Also with the original value QCC, the effect of product formation is more marked.

The OC's calculated in Fig. 15.8 taking account of the lag of the mean values and for $\prod P_a$ are represented in Fig. 15.10, the above depending on the interval of the individual value distribution in σ units; μ_5 is the middle position for the last respective value of the five sample tests. This OC is completely independent of the tolerance in the example.

In the lower part of the Fig. 15.10 the OC's are drawn as $P_a=f(p)$ or $1-P_a=f(p)$ for differing sizes of tolerance zones. The statement made in connection with Fig. 15.4 is here confirmed. With good to very good process control—according to individual case—intervention limits as 99.73% limits are not necessary; there could be a smaller value for the limiting factor k_A.

As a further example a trend process is simulated which is controlled with a $\bar{x}$-QCC. Process analysis has the process capability $c_P=1.3\bar{3}$ or $T=8s$ given. Further the random sample test has been taken from the ND model with the parameters $\mu=6$ and $\sigma=2.01$, Fig. 15.11, column 2. Then the limit values lie at $G_o=\mu+4\sigma=6+4\cdot 2=14$ and $G_u=6-8=-2$, Fig. 15.12. Furthermore it is established that the process analysis has resulted in a trend with the slope $\Delta\mu=0.1\sigma$/part. This trend is given in the random sample test, at first without process control, column 3, in Fig. 15.11.

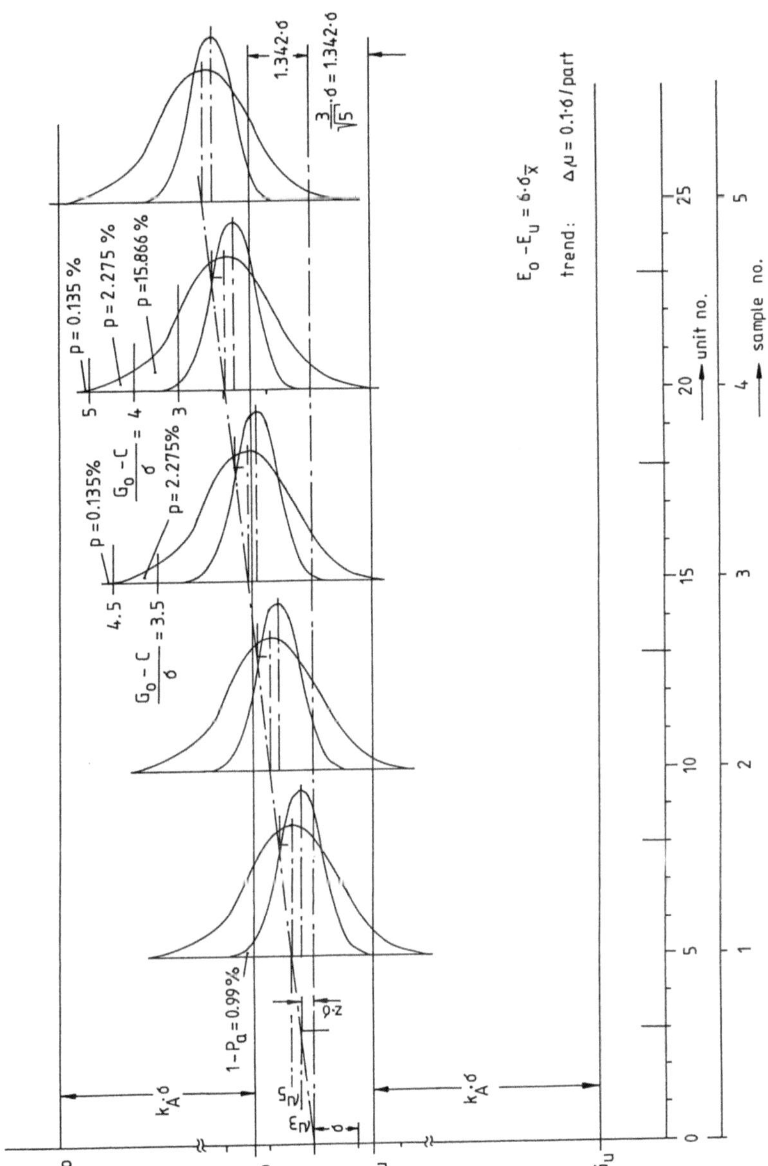

Fig. 15.9. Representation of a process with linear trend; the process is controlled by a $\bar{x}$-QCC with $n = 5$, whose intervention limits are 99.73% ($6\sigma_{\bar{x}}$ limits)

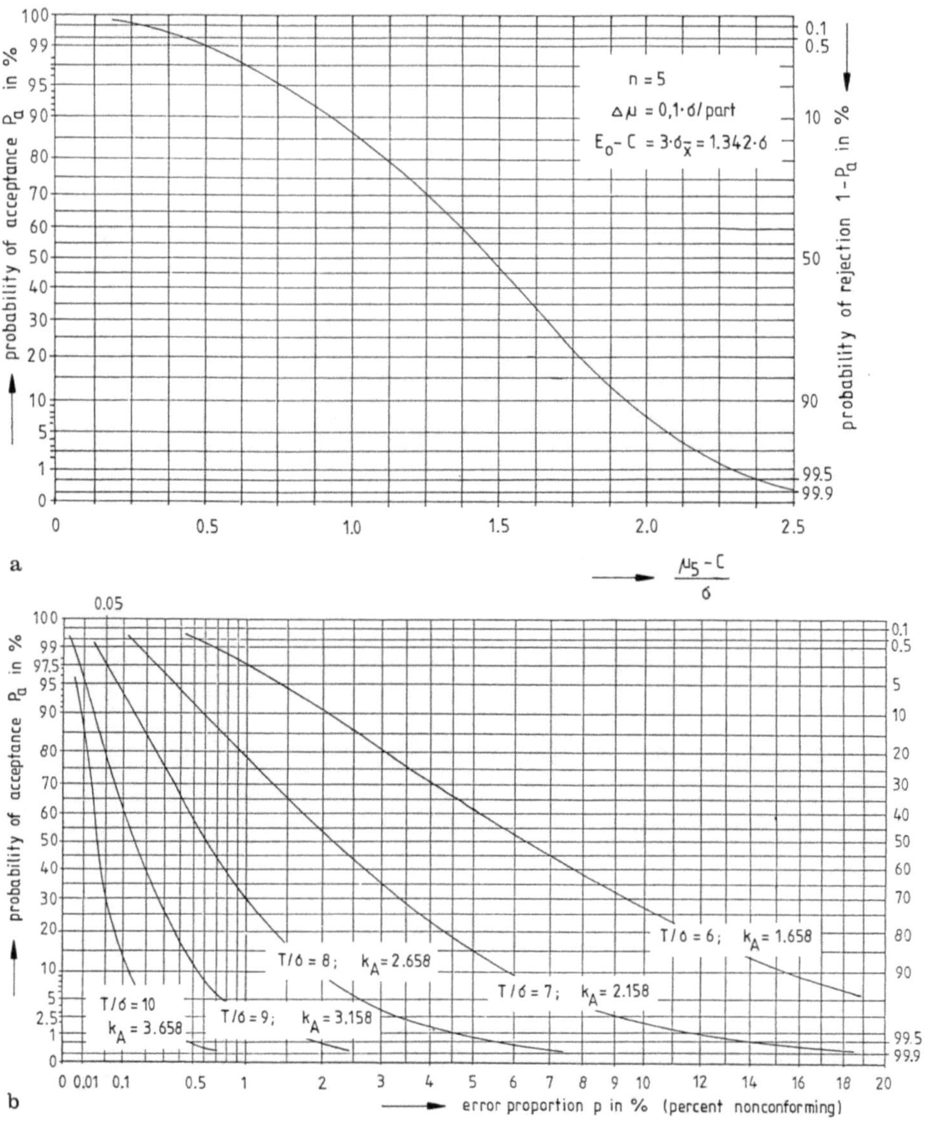

Fig. 15.10 a, b. Operations characteristics for a process with linear trend; the process is controlled with a $\bar{x}$-QCC with $n = 5$; **a** P_a dependent on the interval of the individual value distributions after the last $n = 5$ parts, **b** P_A dependent on the error proportion for different widths of tolerance zone

Fig. 15.11. Simulation of a trend process controlled with a $\bar{x}$-QCC with $n = 5$. Evaluation of the individual values of column 5: $n = 50$; $\bar{x} = 7.86$; $s_x = 2.654\,300$; $6{\cdot}s_x = 15.93$;

$$\frac{G_o - \bar{x}}{s_x} = \frac{14 - 7.86}{2.654} = 2.313$$

without trend		trend of $\Delta\mu = 0.1\,\sigma$/pt. not controlled	trend of $\Delta\mu = 0.1\,\sigma$/part and controlled by $\bar{x}$ - QCC						
1	2	3	4	5	6	7	8	9	10
i	x_i	x_i	μ_i	x_i	D_i	$\bar{x}$ / intervention	⊕ s	s^2	correction
1	6	6.2	6.2	6.2					
2	11	11.4	6.4	11.4	5.2				
3	5	5.6	6.6	5.6	-5.8	7.6	2.4698	6.1	
4	8	8.8	6.8	8.8	3.2				
5	5	6.0	7.0	6.0	-2.8				
6	8	9.2	7.2	9.2	3.2				
7	7	8.4	7.4	8.4	-0.8				
8	6	7.6	7.6	7.6	-0.8	7.6	2.8284	8.0	
9	1	2.8	7.8	2.8	-4.8				
10	8	10.0	8.0	10.0	7.2				
11	8	10.2	8.2	10.2	0.2				
12	6	8.4	8.4	8.4	-1.8				
13	6	8.6	8.6	8.6	0.2	9.0	0.7071	0.5	
14	6	8.8	8.8	8.8	0.2				
15	6	9.0	9.0	9.0	0.2				
16	8	11.2	9.2	11.2	2.2				
17	6	9.4	9.4	9.4	-1.8				
18	9	12.6	9.6	12.6	3.2	10.6	1.5166	2.3	by 4.6 to 5.4
19	5	8.8	9.8	8.8	-3.8				
20	7	11.0	10.0	11.0	2.2				
		$D_{21} = -3.8$				⊕			
21	3	7.2	5.6	2.6	(-3.8 = 2.6 - (11 - 4.6) = (2.6 + 4.6) - 11)				
22	9	13.4	5.8	8.8	6.2				
23	4	8.6	6.0	4.0	-4.8	5.2	2.3022	5.3	
24	5	9.8	6.2	5.2	1.2				
25	5	10.0	6.4	5.4	0.2				
26	5	10.2	6.6	5.6	0.2				
27	6	11.4	6.8	6.8	1.2				
28	6	11.6	7.0	7.0	0.2	6.4	0.9487	0.9	
29	4	9.8	7.2	5.2	-1.8				
30	6	12.0	7.4	7.4	2.2				
31	6	12.2	7.6	7.6	0.2				
32	7	13.4	7.8	8.8	1.2				
33	6	12.6	8.0	8.0	-0.8	8.2	1.5492	2.4	
34	4	10.8	8.2	6.2	-1.8				
35	8	15.0	8.4	10.4	4.2				
36	5	12.2	8.6	7.6	-2.8				
37	3	10.4	8.8	5.8	-1.8				
38	2	9.6	9.0	5.0	-0.8	7.6	2.2583	5.1	
39	6	13.8	9.2	9.2	4.2				
40	7	15.0	9.4	10.4	1.2				
41	10	18.2	9.6	13.6	3.2				
42	8	16.4	9.8	11.8	-1.8				
43	7	15.6	10.0	11.0	-0.8	11.4	2.0248	4.1	by 5.4 to 5.0
44	4	12.8	10.2	8.2	-2.8				
45	8	17.0	10.4	12.4	4.2				
		$D_{46} = -2.8$				⊕			
46	5	14.2	5.2	4.2	(-2.8 = 4.2 - (12.4 - 5.4) = (4.2 + 5.4) - 12.4)				
47	6	15.4	5.4	5.4	1.2				
48	3	12.6	5.6	2.6	-2.8	5.0	1.6733	2.8	
49	6	15.8	5.8	5.8	3.2				
50	7	17.0	6.0	7.0	1.2				

$\sum s = 18.2784 \quad 37.5 = \sum s^2$

$\hat{\sigma}_4 = \frac{\bar{s}}{a_n} = 1.944511$

$\hat{\sigma}_2 = \sqrt{\bar{s^2}} = 1.934692$

evaluation according the difference method:

$n = 47; \quad \bar{D} = 0.370213 = 0.180590\,\hat{\sigma}/St; \quad s_D = 2.899169$

$\hat{\sigma} = \frac{s_D}{\sqrt{2}} = 2{,}050022$

evaluation with the differences D_{21} and D_{46}

$n = 49; \quad \bar{D} = 0.220408 = 0.106267\,\hat{\sigma}/St; \quad s_D = 2.933214$

$\hat{\sigma} = \frac{s_D}{\sqrt{2}} = 2.074095$ in accordance with the estimated values in figure 14.18

For the $\bar{x}$-QCC $k_A = 2.0$ is freely chosen; taking account of the trend results in $k_{AT} = k_A + 2\Delta\mu = 2 + 2\cdot 0.1 = 2.2$. For this QCC the OC is valid for $k_A = 2$ in Fig. 15.2. Thus the intervention limits lie at $E_o = G_o - k_{AT}\sigma = 14 - 2.2\cdot 2 = 9.6$ and $E_u = G_u + k_{AT}\sigma = -2 + 4.4 = 2.4$, Fig. 15.12.

Thus the intervention limits lie at a total $E_o - E_u = 7.2$ distant from each other. For the spread of mean values without account taken of the trend is expected

$$\sigma_{\bar{x}} = \frac{\sigma_x}{\sqrt{n}} = 2/\sqrt{5} = 0.8944$$

which means, that the mean values pass

$$\frac{E_o - E_u}{\sigma_{\bar{x}}} = \frac{7.2}{0.8944} = 8.05$$

times into the intervention range; this clearance room is considered large enough.

It should be accepted that the process is set up to run exactly on the mean value $C = 6$ and then runs with a linear trend to larger values, column 4 in Fig. 15.11, until a mean value of the five sample test exceeds the upper intervention limit. This is with the 4th sample test. There is an intervention and at $\bar{x} - C = 10.6 - 6 = 4.6$ it is corrected. Since the process had run with the last

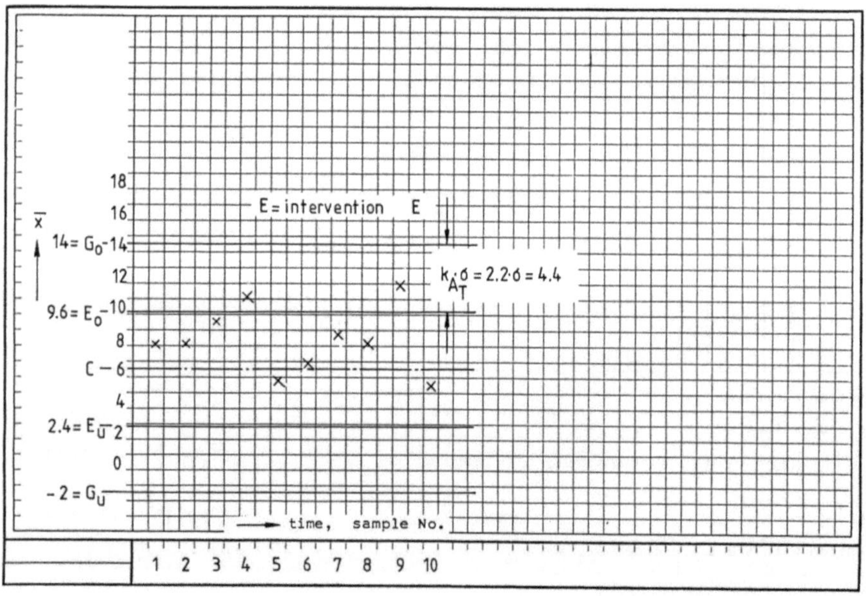

Fig. 15.12. $\bar{x}$-QCC with $n = 5$ and controlled trend process

individual value to the 4th sample test after $\mu_{20} = 10.0$ (column 4), by correction the process is set to 5.4 and then begins to drift again from there upwards, $\mu_{21} = 5.6$. After the ninth sample test a further intervention takes place and it is corrected about $\bar{x} - C = 11.4 - 6 = 5.4$ to $10.4 - 5.4 = 5.0$; following this $\mu_{46} = 5.2$.

The individual values of the simulated and controlled trend process in column 5 of the Fig. 15.11 all lie within the limit values. At the densest x_{41} lies at the upper limit value.

The further evaluation in Fig. 15.11 serves only as practice and to control whether the simulation is non-conformity-free. First 10 samples about $\sum s$ and $\sum s^2$ are evaluated to the estimated values $\hat{\sigma}_2$ and $\hat{\sigma}_4$; these conform with the parameters of the total population extremely well.

Evaluation by the difference method is first carried out without the differences D_{21} and D_{46}, since an intervention lies before these. At the result of this analysis in Fig. 15.11 it happens that the estimated value for the slope of the trend at $\bar{D} = 0.18\sigma$/part deviates considerably from the simulated slope $\Delta\mu = 0.1\sigma$/part. This suggests that the differences not included in the calculation D_{21} and D_{46} are accidentally both negative, so that the mean value $\bar{D}$ becomes too large. This establishes the requirement that painstaking process analysis should be carried out with large sample ranges ($n \geqq 200$).

A renewed evaluation including the first differences after the intervention gives $\bar{D} = 0.220408$. This value conforms from the second decimal place with all $\bar{D}$ values in Fig. 14.18 exactly; thus the simulation is non-conformity-free.

Note. In drafting these examples in this way two non-conformities were found.

From the second evaluation by the difference method it is recognisable that also after an intervention the difference in the evaluation must be included, as long as the correction total is known. In column 6 of Fig. 15.11 the differences D_{21} and D_{46} are almost identical with reference to the correction total to the differences which had been given if it had not been corrected, column 3.

This knowledge is important for process analyses in cases where the trend is steep and thus frequently involving intervention. A steep trend occurs above all when a large material volume is removed from a part. In this sort of process it happens that as well as frequent interventions, frequent tool changes are also necessary.

To analyse the good points of process control it is necessary to define a characteristic, which can be compared with process capability. It is relevant to determine the relationship of the total spread to (instant) process spread and to call it (relative) process spread. If this (relative) process spread is smaller or equal to the (relative) process capability, than process control can be called good, as long as simultaneously the total mean value of the process does not deviate or not significantly deviate from the desired value (from the mean value).

Summary

(relative) process spread (relative) process capability

$$s_P = \frac{s_x}{\hat{\sigma}} \qquad \leqq \frac{T}{6\hat{\sigma}} = c_P$$

process control is good

(relative) process spread (relative) process capability

$$s_P = \frac{s_x}{\hat{\sigma}} \qquad > \frac{T}{6\hat{\sigma}} = c_P$$

process control is bad

For the example of a process in Fig. 15.11, the individual values x_i in column 5 are evaluated according to the statistical characteristics $\bar{x}$ and s_x. The (relative) process spread is analysed as good with

$$s_P = \frac{s_x}{\hat{\sigma}} = \frac{2.654}{2.010} = 1.320 < 1.3\bar{3}$$

The interval of the total mean values to the upper limit value is

$$\frac{G_o - \bar{x}}{s_x} = \frac{14 - 7.86}{2.654} = 2.313;$$

thus the non-conformity proportion is above the upper limit value $\hat{p} = 1.036\%$.

Note: Also for analysis of the (relative) process spread—as for the analysis of process capability—a larger sample test range ($n \geqq 200$) is required as used for the simplification of the last example.

To illustrate the influence of a trend on the OC of $\bar{\bar{x}}$-QCC, in various examples in this section a trend of $\Delta\mu = 0.5\sigma$/part was set with the result that the OC are retained when the limiting factor is raised to $2\Delta\mu$, as long as the sample test range amounts to $n = 5$. This statement must be restricted to the case of a strong trend, as it does not take into account that with $n = 5$ the mean value $\bar{x}$ is always formed after five produced parts and thus only after the respective $n = 5$ parts will the test $\bar{x} \leqq E_o$ or $\bar{x} \geqq E_u$ be carried out and lead to the decision of intervention or no intervention.

In Fig. 15.13 in the upper part there is a limiting factor $k_A = 2.0$ and it is accepted that with a first sample test after 5 parts the individual value distribution lies exactly on the intervention limit, so that the error probability for the fifth part amounts to $p = 2.275\%$. The acceptance probability amounts to $P_a = 98.732\%$ and this is the probability that the process runs on with the trend without intervention to the next mean value, i.e. after five parts. So a high error probability is pre-programmed for the next five parts, for the last part of the second sample test it amounts to $p = 69.146\%$. The fact that after

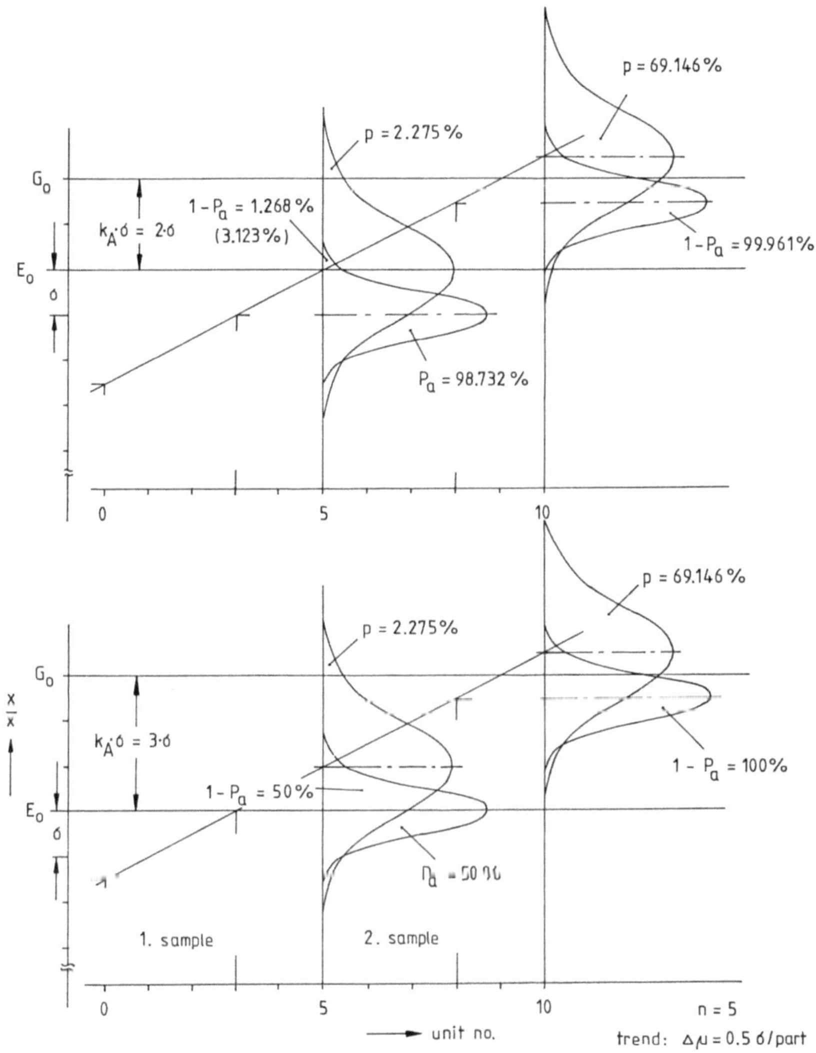

Fig. 15.13. Upper half of a $\bar{x}$-QCC for $n = 5$ with a process with steepness of trend of $\Delta\mu = 0.5\sigma$/part and two different k_A values

the second sample intervention has a very high probability, cannot make this retrogressive.

In the lower part of the Fig. 15.13 for k_A the trend is allowed $(+2\Delta\mu)$ with the result that with $p = 2.275\%$ the intervention probability amounts to $1 - P_a = 50\%$. That means that the process runs on every second time with the result that the fifth part of the second sample test with the probability of $p = 69.146\%$ that it is non-conforming and—which is even worse—can lie

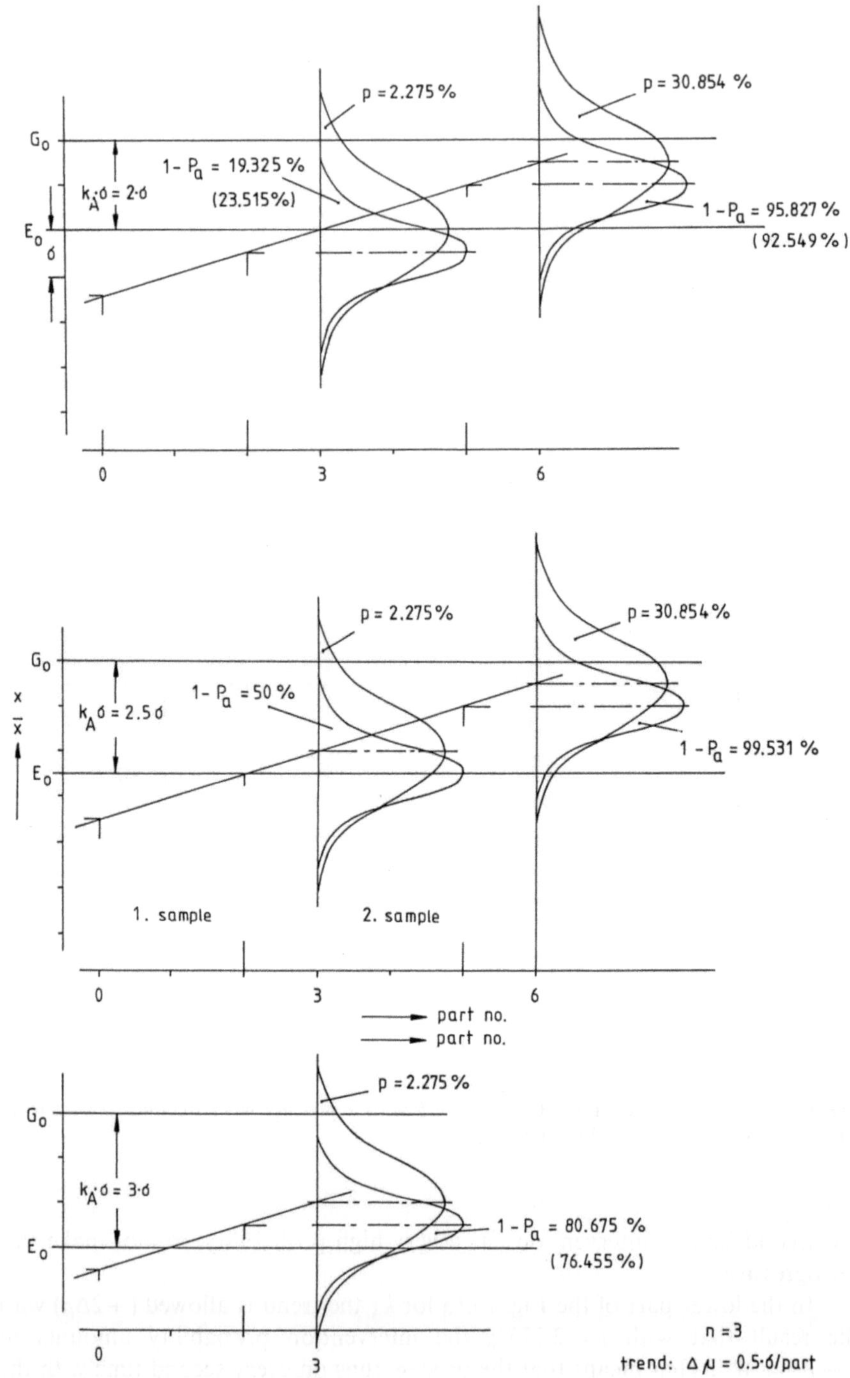

Fig. 15.14. Upper half of a $\bar{x}$-QCC for $n = 3$ with a process with a steepness of trend of $\Delta\mu = 0.5\sigma$/part and three different k_A values

significantly above G_o. With a strong trend the QCC for mean values and $n = 5$ is much too insensitive.

In Fig. 15.14 this is investigated for $n = 3$; the error probability for the last part of the second sample is at $p = 30.854\%$ much too high.

If the possibility suggested in the lower sketches of the Fig. 15.14 remains, the limiting factor should be raised further to, for example $k_A = 3.0$. This assumes that the tolerance zone must at least be $T = 2k_A\sigma + 6\sigma_{\bar{x}} = 6\sigma + \frac{6}{\sqrt{3}}\sigma = 9.46\sigma$ large as included in a process capability of $c_P = 1.58$. But also then the probability that the trend runs without intervention after the first sample test is $P_a = 20\%$.

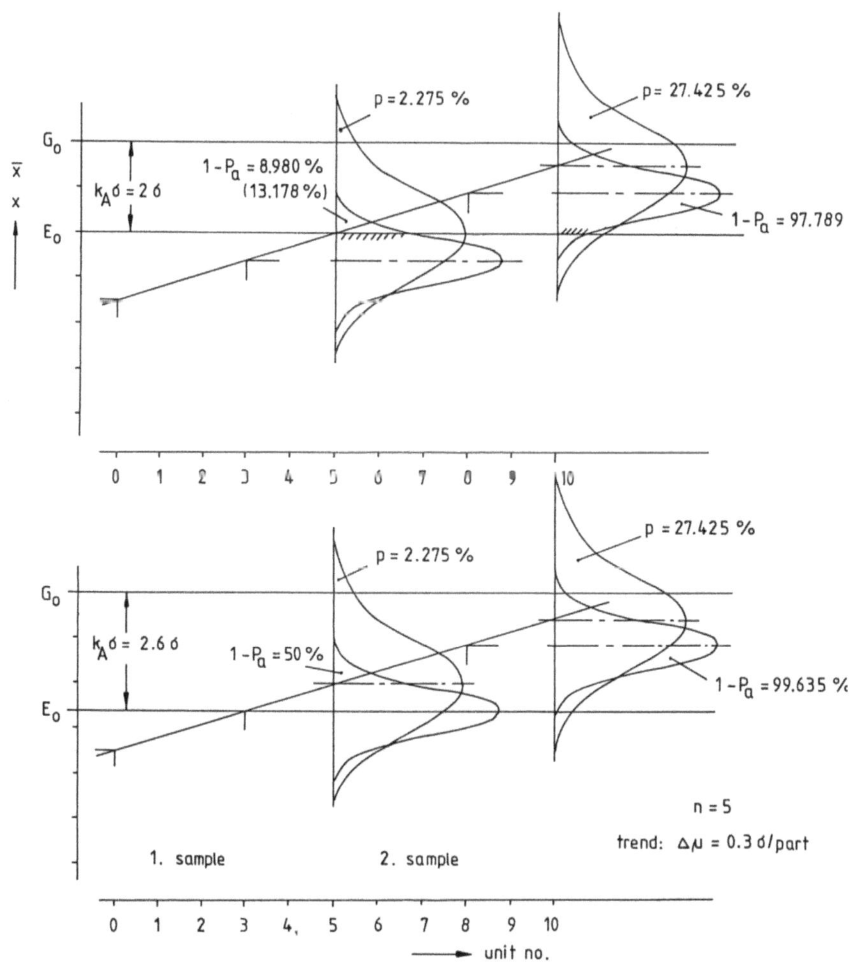

Fig. 15.15. Upper half of a $\bar{x}$-QCC for $n = 5$ with a process with a steepness of trend of $\Delta\mu = 0.3\sigma$/part and two different k_A values

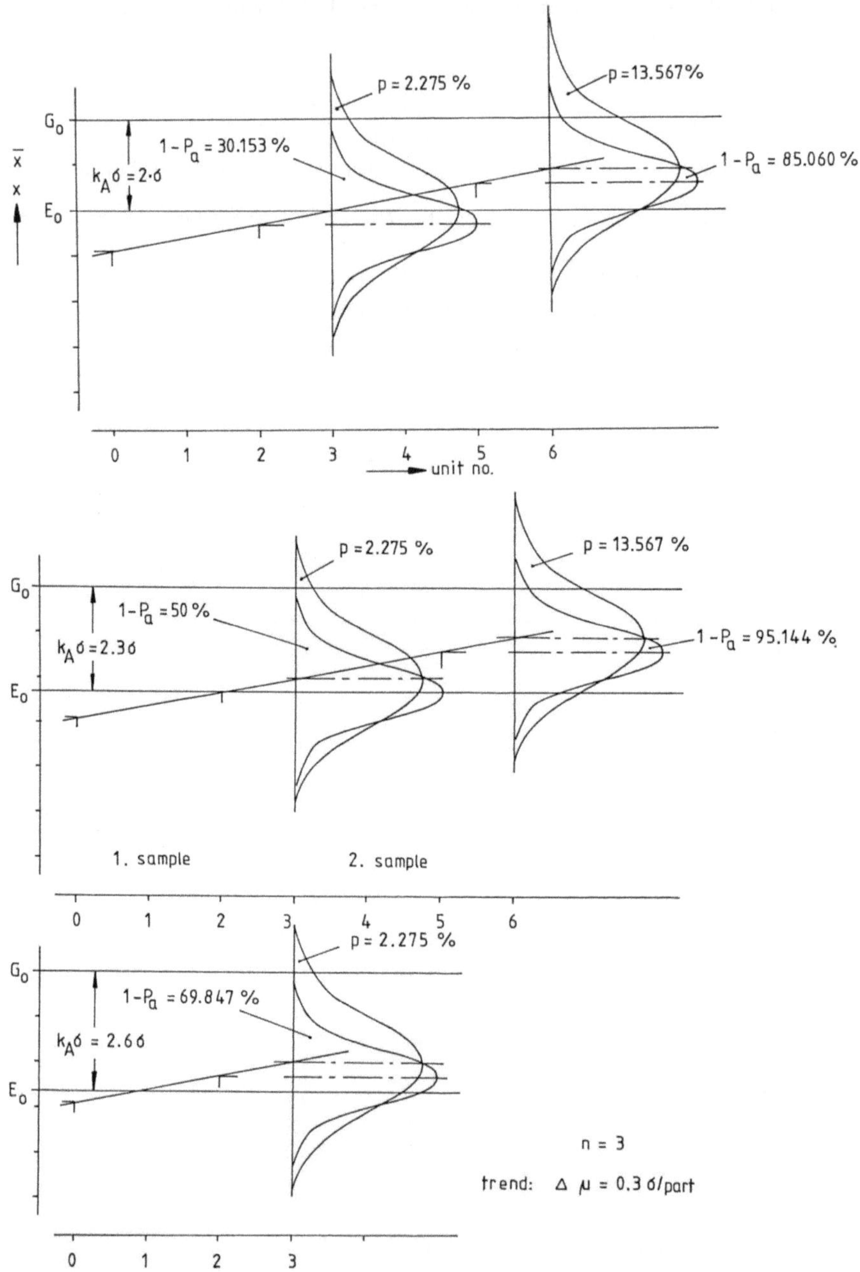

Fig. 15.16. Upper half of a $\bar{x}$-QCC for $n=3$ with a process with a steepness of trend of $\Delta\mu=0.3\sigma$/part and three different k_A values

Conclusion: with a strong trend of $\Delta\mu = 0.5\sigma$/part the $\bar{x}$-QCC cannot fulfill its proposed objective. Here the question is raised-up to which level of trend steepness should the $\bar{x}$-QCC be used?

In order to investigate this, in Fig. 15.15 a trend with a steepness of $\Delta\mu = 0.3\sigma$/part is taken for $n = 5$ with otherwise similar conditions as in Fig. 15.13. If after the first sample test the trend runs on—above with $P_a \approx 91\%$ and below with $P_a \approx 50\%$—then the error probability for the last part of the second sample test $p = 27.425\%$; that is much too high.

In Fig. 15.16 the same is done again and represented for the sample test range $n = 3$. If after the first sample test the trend runs on without intervention, the non-conformity probability for the last part of the second sample test with $p = 13.567\%$ is also too high; the problem will be somewhat lessened if k_A is

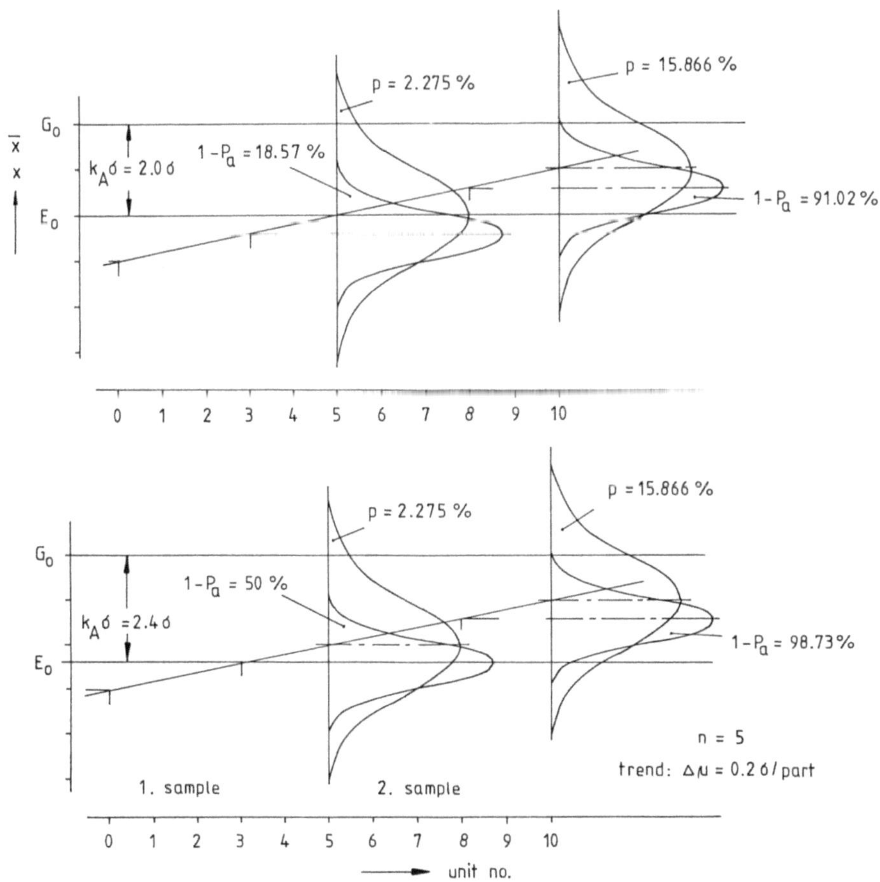

Fig. 15.17. Upper half of a $\bar{x}$-QCC for $n = 5$ values with a process with a steepness of trend of $\Delta\mu = 0.2\sigma$/part for two different k_A values

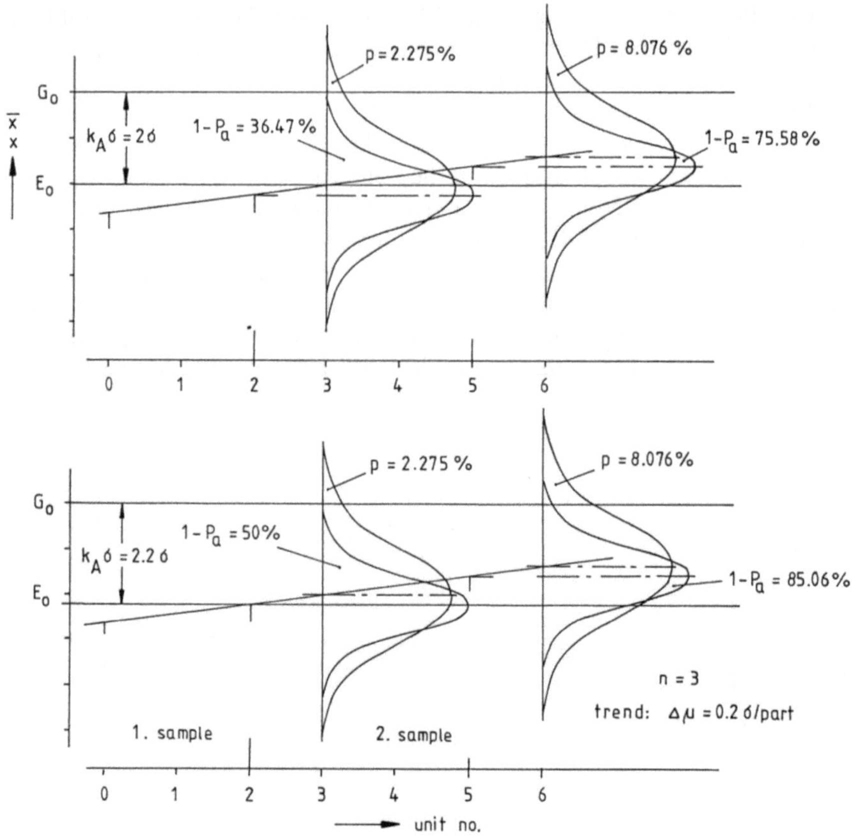

Fig. 15.18. Upper half of a $\bar{x}$-QCC for $n = 5$ with a process with a steepness of trend of $\Delta\mu = 0.2\sigma$/part for two different k_A values

increased to $k_A + 2\Delta\mu$, lowest sketch in Fig. 15.16, because then the probability for the running on of the trend without intervention is only $P_a \approx 30\%$.

In Figs 15.17 and 15.18 the case is investigated that amounts to a steepness of trend of $\Delta\mu = 0.2\sigma$/part.

If the instant non-conformity proportion is $p = 2.275\%$, than the trend runs further at $k_A = 2.0$ with a probability of $P_a \approx 80\%$ with the result that for the fifth part of the following sample test this is $p = 15.866\%$, Fig. 15.17.

With $n = 3$ under similar conditions the error probability for the last part of the second sample test is $p = 8.076\%$, Fig. 15.18.

The x-QCC is not problematic if the trend amounts to only $\Delta\mu = 0.1\sigma$/part, Fig. 15.19. In the case $n = 5$, $k_A = 2.4$ and for the example with $n = 3$, $k_A = 2.2$ have been taken.

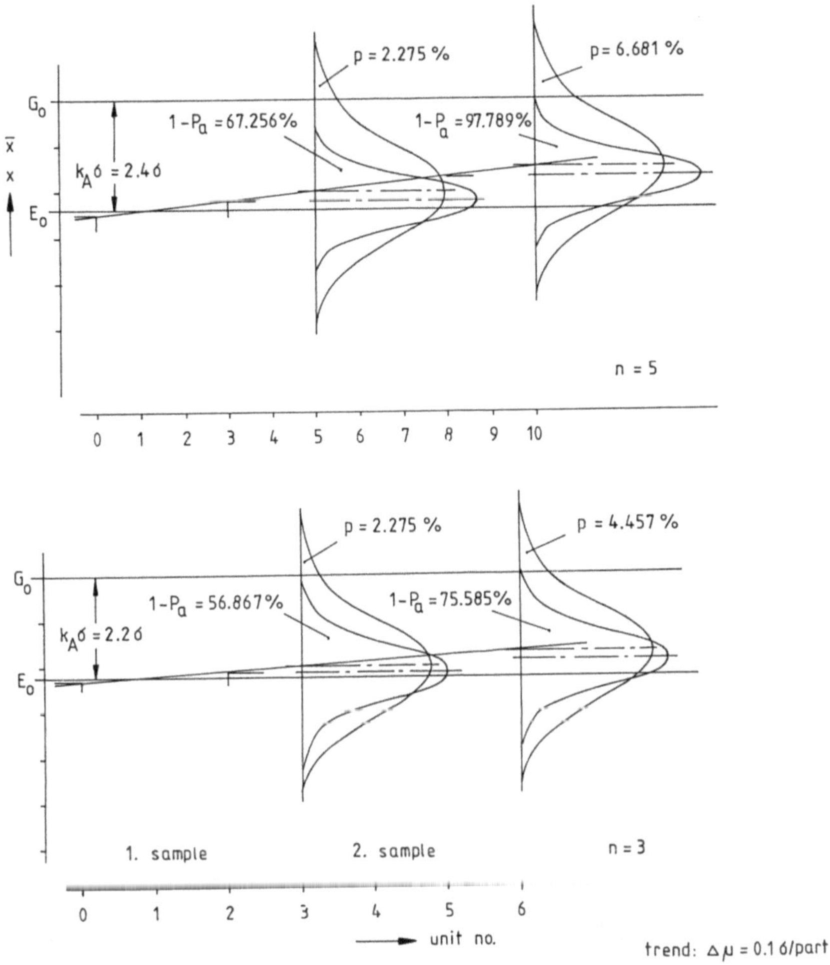

Fig. 15.19. Upper half of $\bar{x}$-QCC for $n = 3$ and $n = 5$ with processes with a steepness of trend of $\Delta\mu = 0.1\sigma$/part

In summary with the presence of trend processes the following analysis of the $\bar{x}$-QCC is recommended:

Trend in σ/part	Analysis
$\leqq 0.1$	$\bar{x}$-QCC unproblematic and usable;
0.1 to 0.3	$\bar{x}$-QCC problematic, especially with $n = 5$, with $n = 3$ only usable for very good process capability;
$\geqq 0.3$	$\bar{x}$-QCC not usuable, in all cases only x-QCC with $n = 1$ is recommended, Sect. 15.3.

15.3 Process Control by Original Value QCC's

For the repetition and further explanation of the original value charts mentioned in Chap. 13, to begin with—as for mean value charts—the case is taken where a process shows no trend or a small trend which can be ignored. Then errors occur mainly because the process position is set up wrongly or because of a random cause the process position is shifted.

The operational characteristics for original value charts can be graphically determined with the nomogram in Fig. 13.29; for arithmetical determination an example is given in Fig. 15.20.

In the Figs. 15.21 to 15.24, the operational characteristics for the original value QCC with limiting factors $k_E = 0.5$ to 2.0 and different sample test ranges

$\frac{G_o - \mu}{\sigma}$	p in %	$\frac{E_o - \mu}{\sigma}$	n = 1 P_{a_1}	n = 5 $P_{a_5} = P_{a_1}^5$	n = 9 $P_{a_9} = P_{a_1}^9$
4	0.003	3	0.99865	0.99327	0.98791
3.5	0.023	2.5	0.99379	0.96933	0.94548
3	0.135	2	0.97725	0.89131	0.81293
2.5	0.621	1.5	0.93319	0.70770	0.53670
2	2.275	1	0.84134	0.42156	0.21122
1.5	6.681	0.5	0.69146	0.15806	0.03613
1	15.866	0	0.50000	0.03125	0.00195

Fig. 15.20. Calculation of the operational characteristics of original value QCC with $k_E = 1.0$ and for different *n*

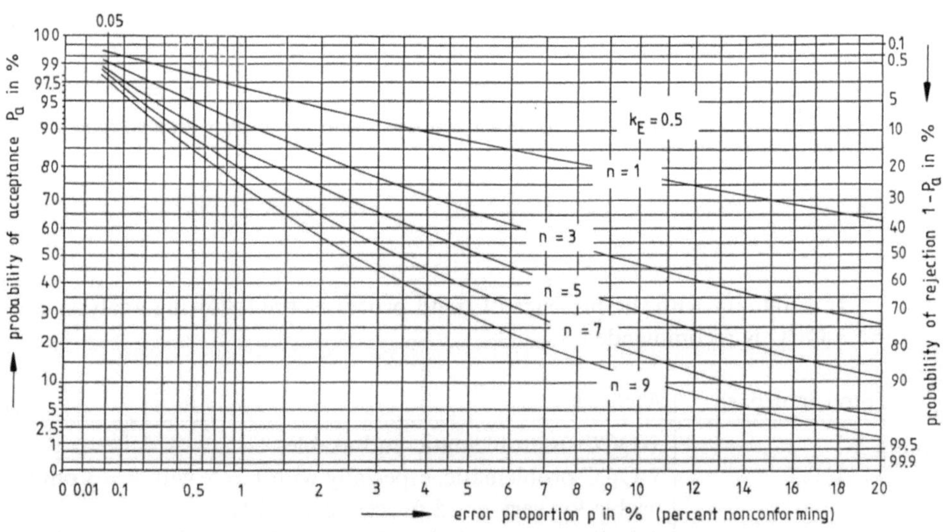

Fig. 15.21. Operational characteristics of original value QCC with $k_E = 0.5$ and for different *n*

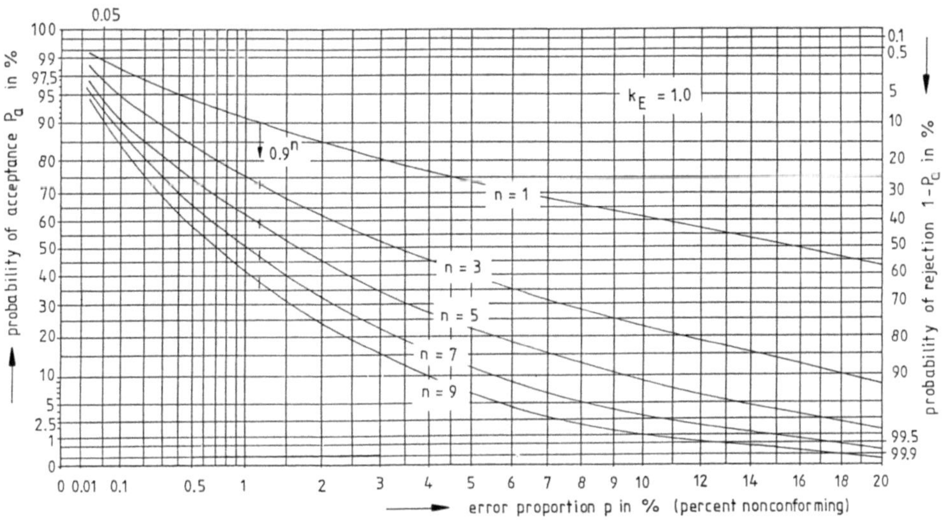

Fig. 15.22. Operational characteristics of original value QCC with $k_E = 1.0$ and for different n

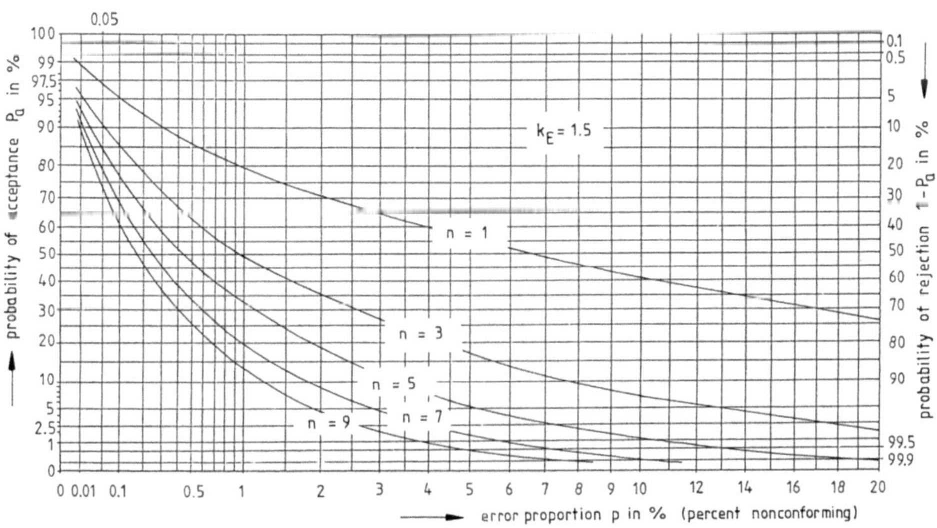

Fig. 15.23. Operational characteristics of original value QCC with $k_E = 1.5$ and for different n

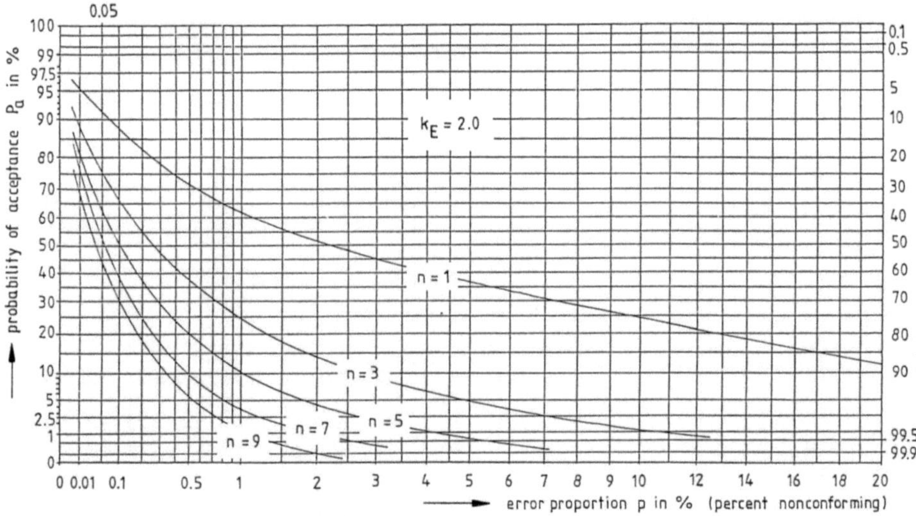

Fig. 15.24. Operational characteristics of original value QCC with $k_E = 2.0$ and for different n

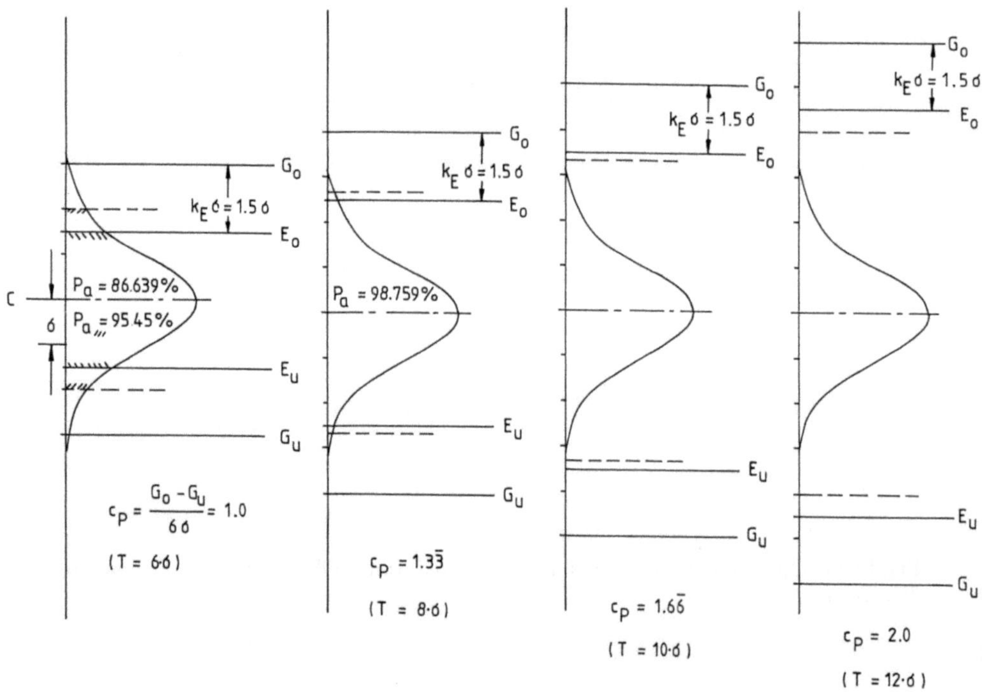

Fig. 15.25. Representation of original value QCC with $n = 1$ and with $k_E = 1.5$ for different c_P values

n are represented. The curves for $n > 1$ hold equally for the execution of an original value chart with the sample test range n as also for the n-th continuous application of an original value chart with $n = 1$.

From the rule of thumb, according to which above a short time period, on average with every non-conformity proportion it should be calculated that an acceptance probability of 50% is expected, a limiting factor—taking $n = 5$—of $k_E \approx 1.0$ to 1.5 is measured.

The influence of (relative) process capability is—for simplicity half for $n = 1$—represented in Fig. 15.25 for the optional limiting factor $k_E = 1.5$. From the diagram it is clear that original value charts with $n = 1$—and first right with $n > 1$—can only be used if $c_P \geqq 1.3\bar{3}$. With $c_P > 1.6\bar{6}$ the choice is to be made of a greater k_E value, to improve the control to the mean value.

The schematic use of the rule that independent of the process capability as acceptance range the range $E_o - E_u = 6\sigma$ (or 99%-range) is established, is represented in Fig. 15.26.

For $c_P = 1.0$ a process control is impossible; it must be calculated with (in percentage) two position non-conformity proportions. This statement is valid exclusively for this particular case, where no trend is present and the QCC is periodically applied. With very good to excellent process capability there result—by the use of the 6σ-rule—k_E values which can be estimated as too large.

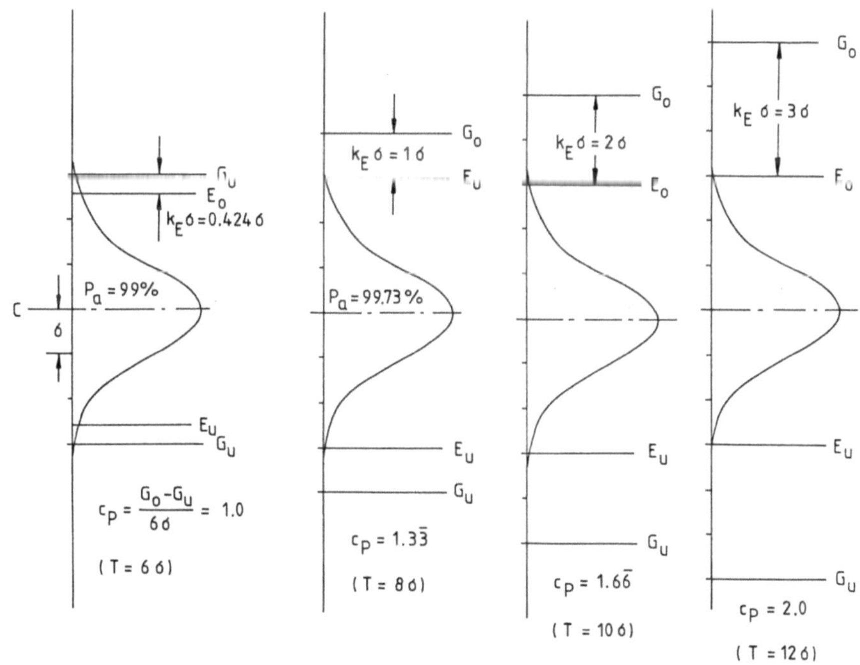

Fig. 15.26. Representation of original value QCC with $n = 1$ and with $P_A = 99.73\%$ (99%) with distribution position on the mean value for different c_P-values

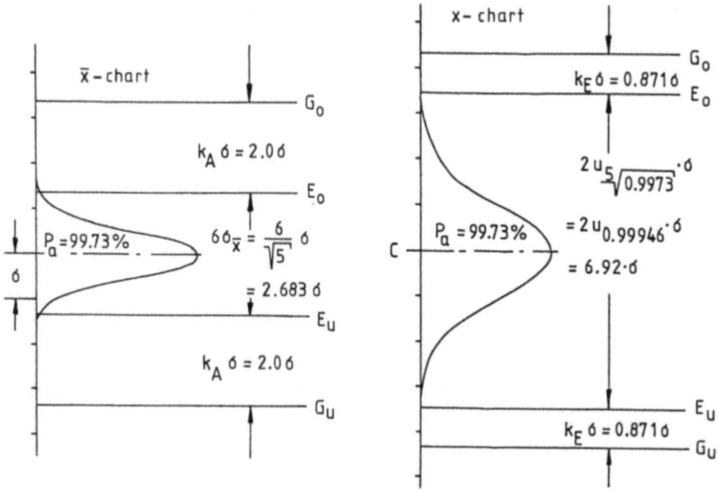

necessary $T = (2.683 + 2\cdot 2.0)\,\sigma$ $\quad T = (6.92 + 2\cdot 0.871)\,\sigma$

$= 6.683\sigma$ $(6.304\sigma$ für $P_a = 99\%)$ $\quad = 8.662\,\sigma$ $(7.922\,\sigma$ für $P_a = 99\,\%)$

$c_p = 1.114$ (1.051) $\quad c_p = 1.444$ (1.320)

Fig. 15.27. Comparison of a $\bar{x}$-QCC with an x-QCC on the basis: $P_A = 99.73\%$, if process distribution is set exactly on C; both OC's go through $P_{50} = 2.275\%$; $n = 5$

A comparison between mean value QCC and original value QCC follows in the Figs. 15.27 to 15.30. In order to have a basis for comparison, it was established that with both charts the acceptance probability $P_a = 99.73\%$, if the individual values lie with their mean value μ at the mean value C. Furthermore for both charts with $p_{50} = 2.275\%$ a common point of OC was established. The calculation of the operational characteristics follows in Fig. 15.28 and its

line No.	$\frac{G_o - \mu}{\sigma}$	p in %	$\bar{x}$- chart with $k_A = 2.0$: $\frac{E_o - \mu}{\sigma} = u_x$	$u_{\bar{x}} = u_x \cdot \sqrt{5}$	P_a	x- chart with $k_E = 0.871$: $\frac{E_o - \mu}{\sigma}$	P	$P_a = P^5$
1	4	0.003	2	4.472	1.0000	3.129	0.99913	0.99566
2	3.5	0.023	1.5	3.354	0.99960	2.629	0.99572	0.97878
3	3	0.135	1	2.236	0.98732	2.129	0.98337	0.91957
4	2.5	0.621	0.5	1.118	0.86822	1.629	0.94834	0.76704
5	2	2.275	0	0	0.50000	1.129	0.87055	0.50000
6	1.5	6.681	-0.5	-1.118	0.13178	0.629	0.73533	0.21499
7	1	15.866	-1	-2.236	0.01268	0.129	0.55132	0.05093
8	0.842	20.000	-1.158	-2.589	0.00481	-0.029	0.48843	0.02780

Fig. 15.28. Calculation of the operational characteristics for the $\bar{x}$-charts with $k_A = 2.0$ and for the x-charts with $k_E = 0.871$; sample test range $n = 5$; both OC's go through the point $p_{50} = 2.275\%$

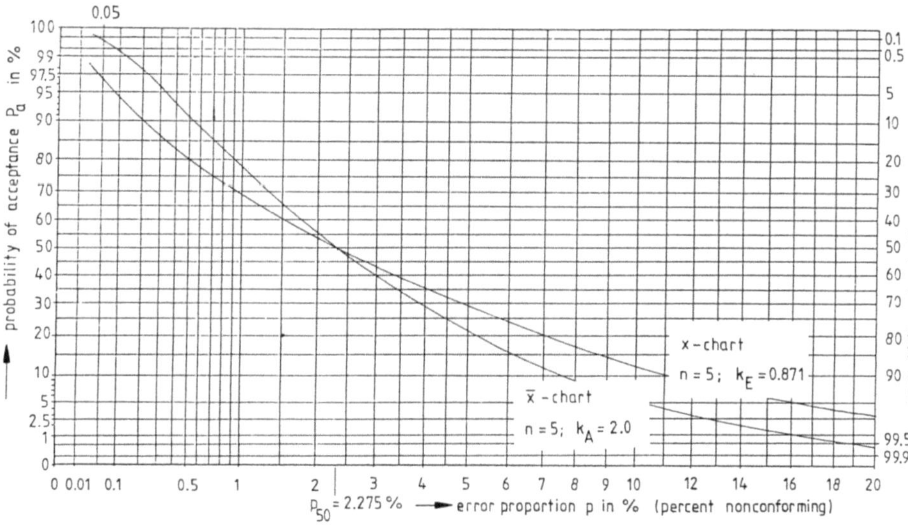

Fig. 15.29. Representation of the operation characteristics for the $\bar{x}$-QCC with $k_A = 2.0$ and for the x-QCC with $k_E = 0.871$; sample test range $n = 5$; these OC's are independent of the size of the tolerance zone

graphical representation is given in Fig. 15.29. Apart from the slightly higher dividing capacity of the $\bar{x}$-QCC both OC's run approximately the same.

As a result of the comparison between the $\bar{x}$-QCC and the x-QCC this qualitative statement is produced, that with the original value charts the limiting factor k_E is smaller than the comparable k_A value with the mean value charts; for this reason they need considerably more space for the acceptance range than the $\bar{x}$-QCC. The original value charts demand a greater c_P-value.

For a better understanding of the function of both comparable charts Fig. 15.27 is given again in Fig. 15.30 with the additional representation of the situation $p_{50} = 2.275\%$.

The illustration up to now of process control by original value QCC has restricted itself to the case where no trend was present.

For the case where a trend is present with the steepness $\Delta\mu \geqq 0.1\sigma$/part a periodically executed QCC will not suffice, it must be continuously applied. In Sect. 15.2 it was shown that then the application of a mean value QCC becomes problematical; this is why there is a need for the original value QCC.

Since an original value QCC with $n > 1$ has the same OC as the process control by an n-th applied $n = 1$-QCC, the conclusion is drawn from the latter that it will react faster to non-conformities.

In Fig. 15.31 the total operational characteristics of an $n = 1$-QCC with $k_E = 1.0$ and a trend with the slope $\Delta\mu = 0.3\sigma$/part are calculated. The products calculated in the last but one column assume that the trend of $G_o - 4.3\sigma$ runs and that its slope is constant. The OC calculated on these assumptions and

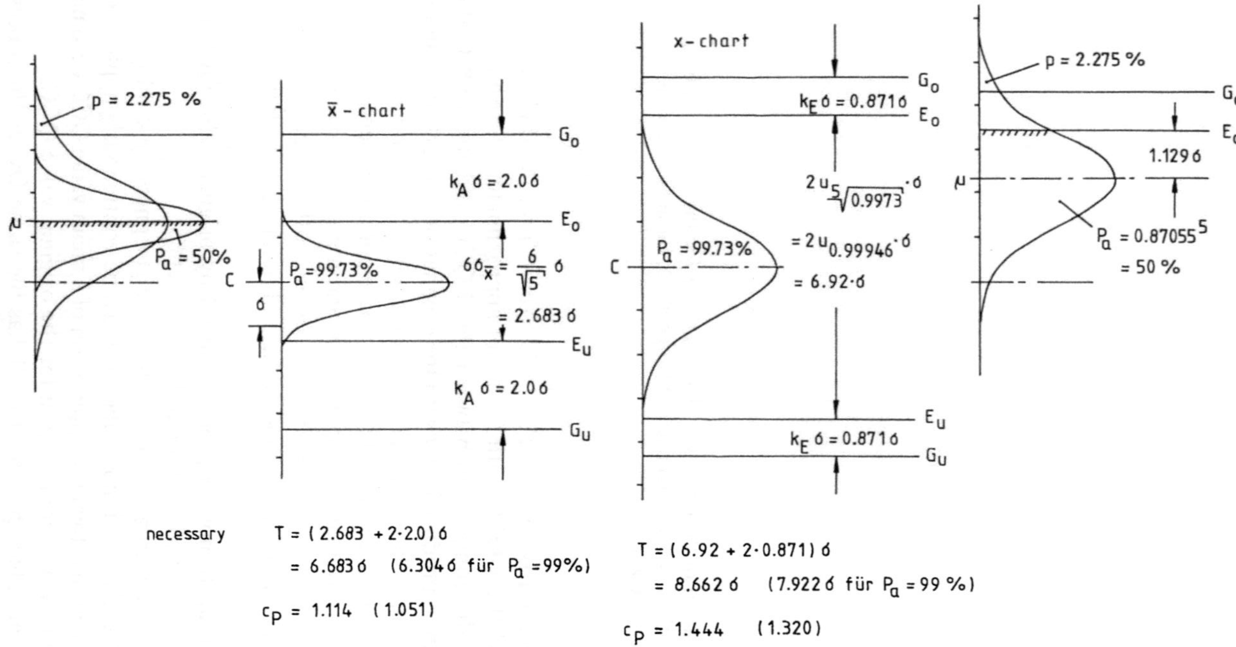

Fig. 15.30. Comparison of a $\bar{x}$-QCC with an x-QCC as Fig. 15.27; in addition there are inserted the situations for $p_{50} = 2.275\%$

line No.	$\frac{G_o - \mu}{\sigma}$	p in %	$\frac{E_c - \mu}{\sigma}$	P_a	$\prod P_a$	$\prod P_a$ start at line 8
1	4	0.003	3.0	0.99865	0.99865	
2	3.7	0.011	2.7	0.99653	0.99518	
3	3.4	0.034	2.4	0.99180	0.98702	
4	3.1	0.097	2.1	0.98214	0.96940	
5	2.8	0.256	1.8	0.96407	0.93456	
6	2.5	0.621	1.5	0.93319	0.87213	
7	2.2	1.390	1.2	0.88493	0.77177	
8	1.9	2.872	0.9	0.81594	0.62972	0.81594
9	1.6	5.480	0.6	0.72575	0.45702	0.69217
10	1.3	9.680	0.3	0.61791	0.28240	0.36591
11	1.0	15.866	0.0	0.50000	0.14119	0.18295
12	0.7	24.196	-0.3	0.38209	0.05395	0.06990

Fig. 15.31. Calculation of the acceptance curve of a QCC with $n = 1$ carried out continually on a process with a constant trend slope of $\Delta\mu = 0.3\sigma$/part and limiting factor $k_E = 1.0$

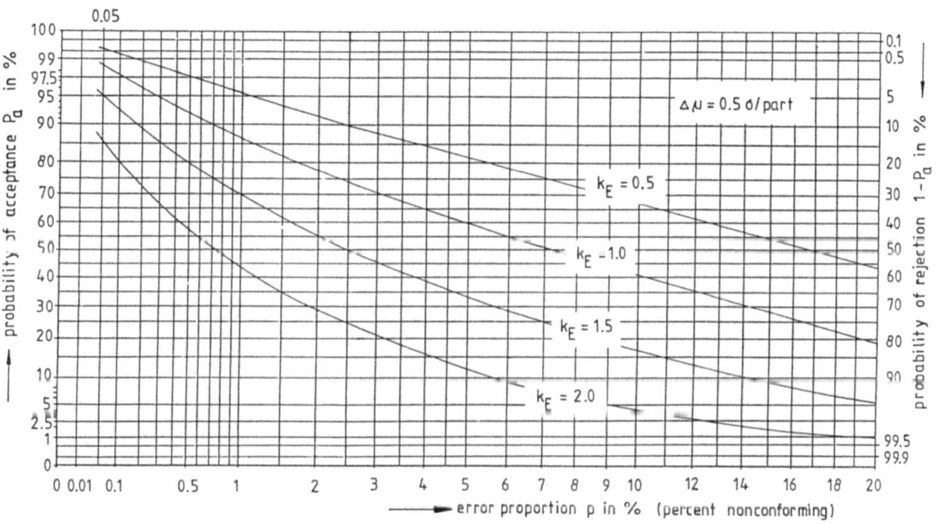

Fig. 15.32. Operational characteristics for QCC with $n = 1$ carried out continually on a trend process with constant slope $\Delta\mu = 0.5\sigma$/part and different limiting factors k_E

more are represented in Fig. 15.33; the Figs. 15.32 and 15.34 contain operational characteristics for other slopes $\Delta\mu$.

A comparison of the Fig. 15.32 to 15.34 shows that the OC's become steeper if the trend shows a small slope. This becomes clearer from Figs. 15.35 to 15.38 from which can be read that with a trend of a slope $\Delta\mu = 0.1\sigma$/part the limiting factor $k_E = 1.0$ is useable ($p_{50} = 1.6\%$) while with a trend to slope $\Delta\mu = 0.5\sigma$/part a limiting factor $k_E \geqq 1.5$ is necessary.

Against this calculation and representation the objection can be raised that even with trend processes incorrect setting is possible or random influences can suddenly change the set position so that then the operational characteristics

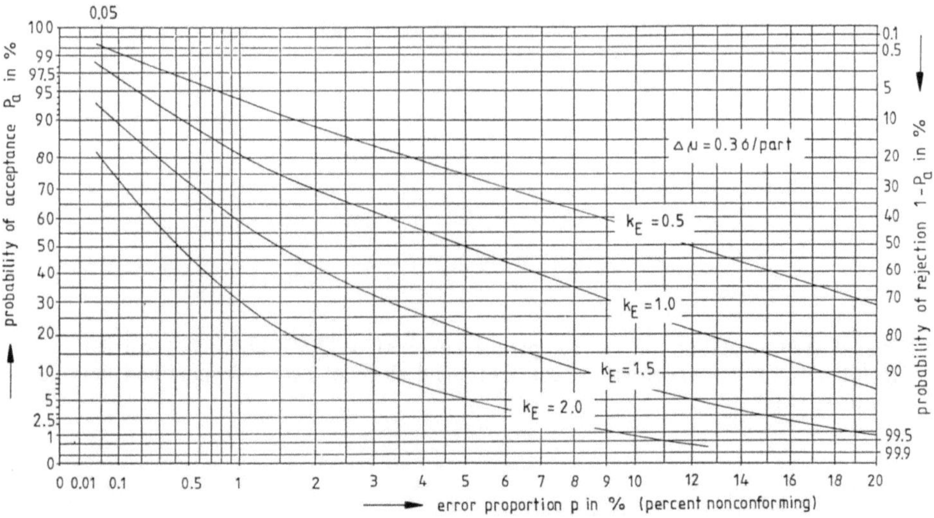

Fig. 15.33. Operational characteristics for continually executed QCC with $n = 1$ of a trend process with constant slope $\Delta\mu = 0.3\sigma$/part and different limiting factors k_E

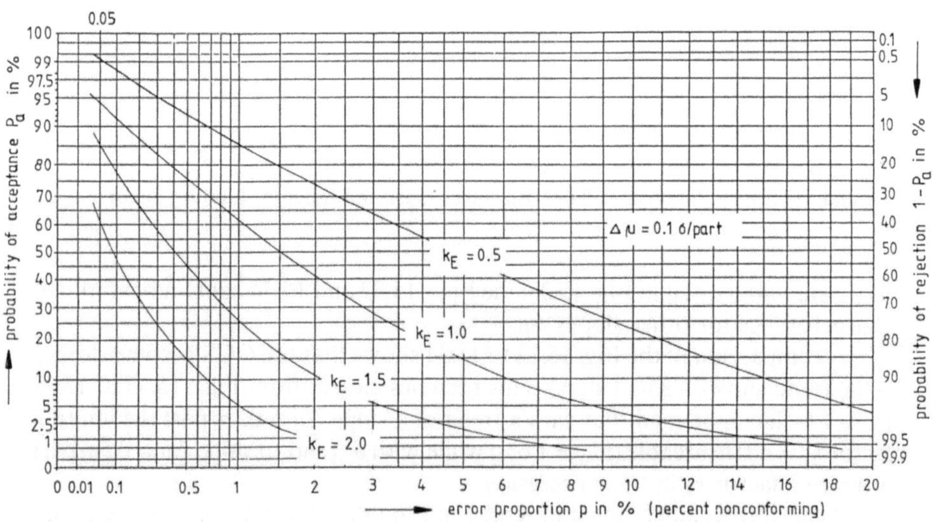

Fig. 15.34. Operational characteristics for continually executed QCC with $n = 1$ of a trend process with the constant slope $\Delta\mu = 0.1\sigma$/part and different limiting factors k_E

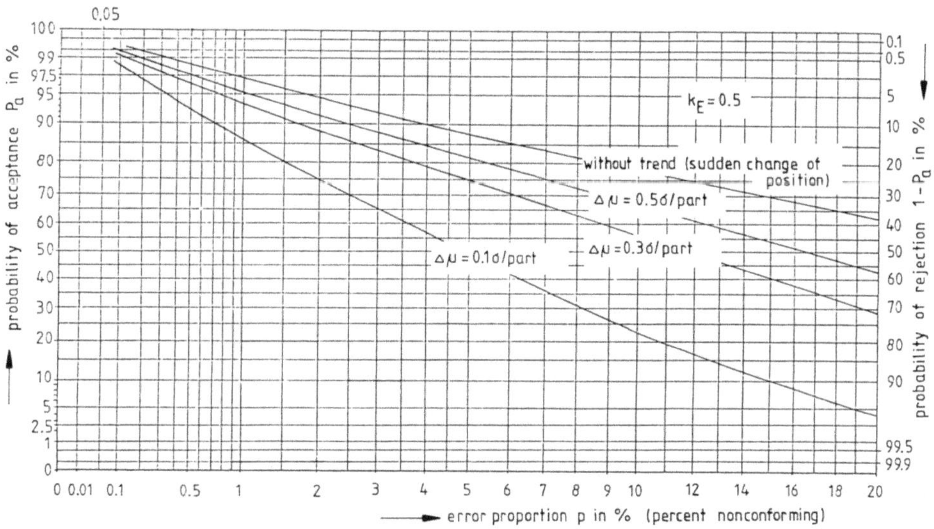

Fig. 15.35. Operational characteristics for continually executed QCC with $n = 1$ with the limiting factor $k_E = 0.5$ and different trend slopes $\Delta\mu$

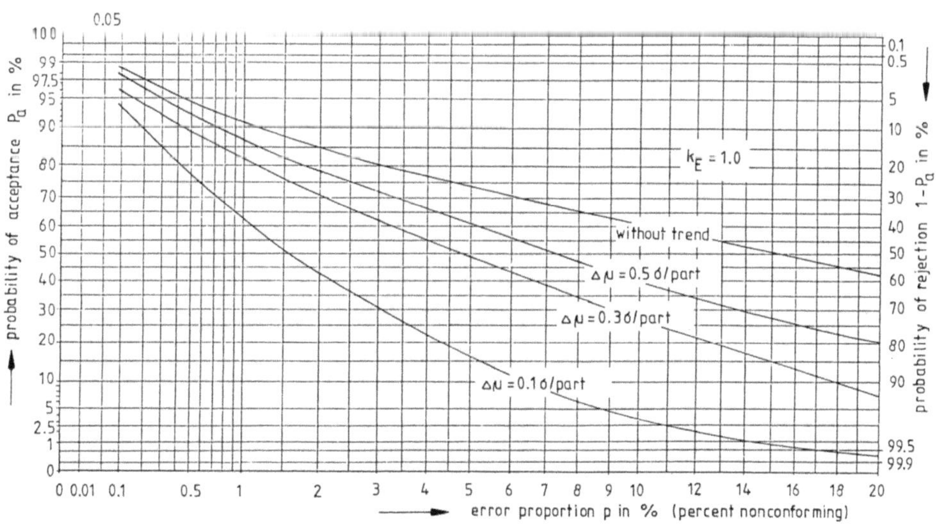

Fig. 15.36. Operational characteristics for a continually executed QCC with $n = 1$ with the limiting factor $k_E = 1.0$ and different trend slopes $\Delta\mu$

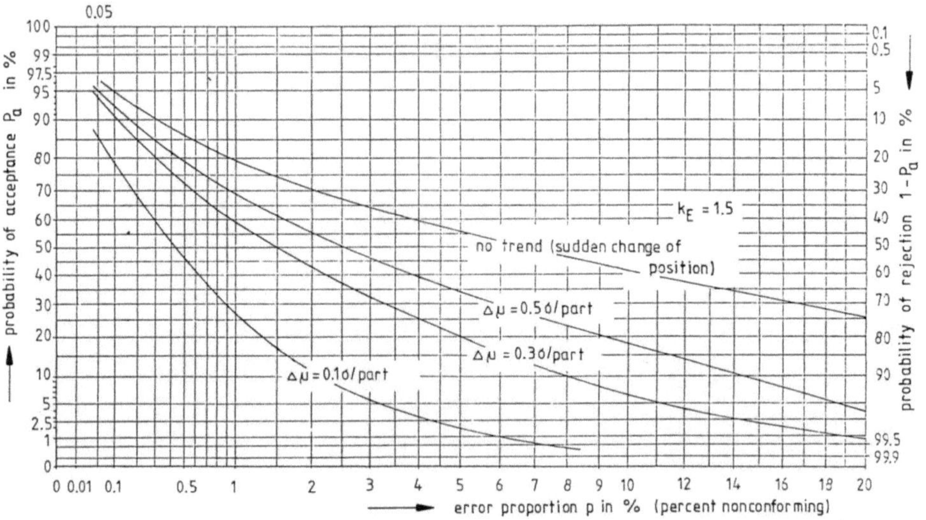

Fig. 15.37. Operational characteristics for continually executed QCC with $n = 1$ and the limiting factor $k_E = 1.5$ with different trend slopes $\Delta\mu$

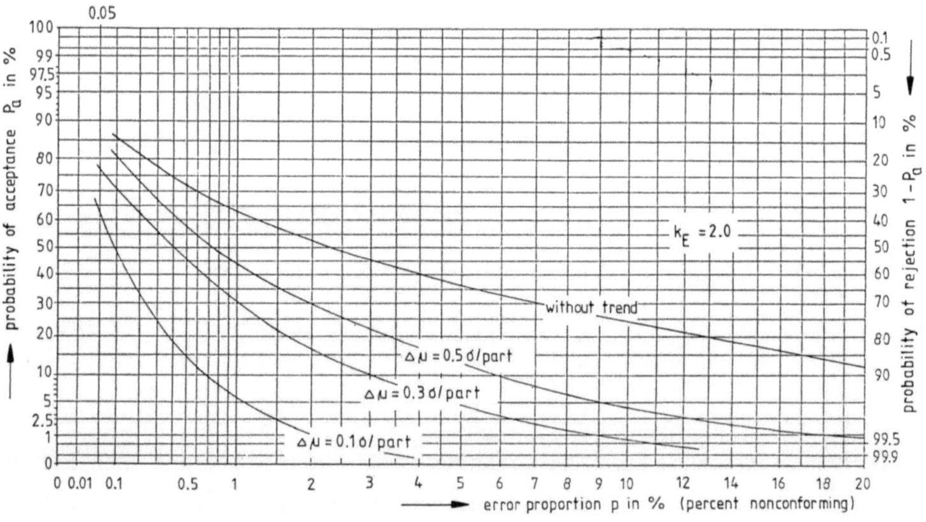

Fig. 15.38. Operational characteristics for continually executed QCC with $n = 1$ and the limiting factor $k_E = 2.0$ with different trend slopes $\Delta\mu$

represented in Figs. 15.32 to 15.38 will not hold. This objection is correct but not aggravating.

In order to show this, in Fig. 15.31 in the last column, in addition the assumption is met that the trend of $G_o - 2.2\sigma$ runs with the result that for the first produced part the non-conformity probability $p = 2.872\%$. With further produced parts $\prod P_A$ converges very quickly on nil.

For a better understanding, in Fig. 15.39 another type of representation is chosen by which the calculation through diagrammatic representation of the

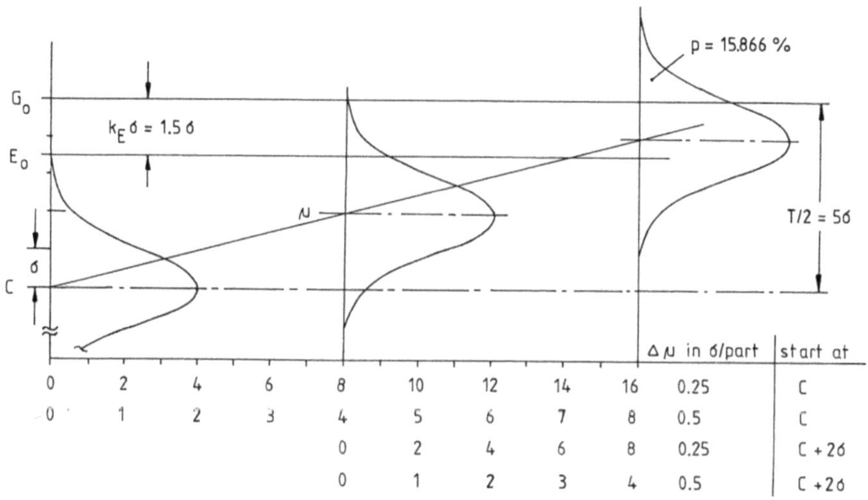

value	note	part i 1	2	3	4	5	6
distance $u = \frac{G_o - \mu}{\sigma}$		4.75	4.5	4.25	4.0	3.75	3.5
error proport. p in %							0.023
distance $u = \frac{E_o - \mu}{\sigma}$		3.25	3.0	2.75	2.5	2.25	2.0
probability of acceptance P_a		0.99942	0.99865	0.99702	0.99397	0.98778	0.97725
product $\prod P_a$	$\Delta\mu$ = 0.25 σ/ part starting at C	0.99942	0.99807	0.99510	0.98892	0.97683	0.95461
product $\prod P_a$	$\Delta\mu$ = 0.25 σ/ part starting at C + 2σ						
product $\prod P_a$	$\Delta\mu$ = 0.5 σ/ part starting at C		0.99865		0.99242		0.96987
product $\prod P_a$	$\Delta\mu$ = 0.5 σ/ part starting at C + 2σ						

Fig. 15.39. Calculation of the acceptance product of linear trends with the diagramatic representation of the trends. Process control results through continually applied $n = 1$ QCC with the limiting factor $k_E = 1.5$

7	8	9	10	11	12	13	14	15	16
3.25	3.0	2.75	2.5	2.25	2.0	1.75	1.5	1.25	1.0
0.058	0.135	0.298	0.621	1.222	2.275	4.006	6.681	10.565	15.866
1.75	1.5	1.25	1.0	0.75	0.5	0.25	0.0	-0.25	-0.5
0.95994	0.93319	0.89435	0.84134	0.77337	0.69146	0.59871	0.5	0.40129	0.30854
0.91637	0.85514	0.76480	0.64346	0.49763	0.34409	0.20601	0.10301	0.04133	0.01275
		0.89435	0.75245	0.50192	0.40238	0.24091	0.12045	0.04834	0.01491
	0.90507		0.70147		0.52653		0.26326		0.08123
			0.84134		0.58175		0.29087		0.08974

Fig. 15.39. (Continued)

trend is completed. If the process is set up at the beginning of the linear trend at $C = G_o - 5\sigma$ and the trend which has the (constant) slope of $\Delta\mu = 0.25\sigma$/part, then the non-conformity probability for the 16th produced part $p = 15.866\%$. The probability that the process runs on without correction is however only $\prod P_A = 1.275\%$.

Drifting however the process from $G_o - 3\sigma$ batch (error setting) is the probability that after a total of eight consecutive produced parts, no intervention is made $\prod P_A = 1.491\%$. The same goes for a trend with $\Delta\mu = 0.5\sigma$/part Fig. 15.39.

15.4 Process Correction

With process control by periodic or continuous quality control charts for the process position, at each point where an intervention limit is exceeded, the process setting is corrected. Ideally the correction would be about the total of the deviation of the instant mean value of the distribution with respect to the mean value $|\mu_i - C|$; the instant mean value μ_i is however not exactly known.

If no trend is present or the slope of the trend is slight, then—in cases of mean value charts—the $\bar{x}$-values are the best estimated values for the instant mean value μ; from these the $|\bar{x} - C|$-values can be used as correction totals. It should however be considered that the mean values which lead to intervention come predominantly from the outer lying half of the mean value distribution. In Fig. 15.40 the case is represented where 50% of the mean values lie above

without mean deviation		with mean deviation			
i	x_i	μ_i	x_i	mean value $\bar{x}$ with	without intervention
1	6	8.7	8.7		
2	11	8.7	13.7		
3	5	8.7	7.7	9.7	
4	8	8.7	10.7		
5	5	8.7	7.7		
6	8	8.7	10.7		
7	7	8.7	9.7		
8	6	8.7	8.7		8.7
9	1	8.7	3.7		
10	8	8.7	10.7		
11	8	8.7	10.7		
12	6	8.7	8.7		
13	6	8.7	8.7	9.1	
14	6	8.7	8.7		
15	6	8.7	8.7		
16	8	8.7	10.7		
17	6	8.7	8.7		
18	9	8.7	11.7	9.7	
19	5	8.7	7.7		
20	7	8.7	9.7		
21	3	8.7	5.7		
22	9	8.7	11.7		
23	4	8.7	6.7		7.9
24	5	8.7	7.7		
25	5	8.7	7.7		
26	5	8.7	7.7		
27	6	8.7	8.7		
28	6	8.7	8.7		8.1
29	4	8.7	6.7		
30	6	8.7	8.7		
31	8	8.7	8.7		
32	7	8.7	9.7		
33	6	8.7	8.7	8.9	
34	4	8.7	6.7		
35	8	8.7	10.7		
36	5	8.7	7.7		
37	3	8.7	5.7		
38	2	8.7	4.7		7.3
39	6	8.7	8.7		
40	7	8.7	9.7		
41	10	8.7	12.7		
42	8	8.7	10.7		
43	7	8.7	9.7	10.1	
44	4	8.7	6.7		
45	8	8.7	10.7		
46	5	8.7	7.7		
47	6	8.7	8.7		
48	3	8.7	5.7		8.1
49	6	8.7	8.7		
50	7	8.7	9.7		

$\bar{\bar{x}} = 9.5 \qquad \dfrac{\mu_i - C}{\bar{\bar{x}} - C} = \dfrac{2.7}{3.5} = 0.7714$

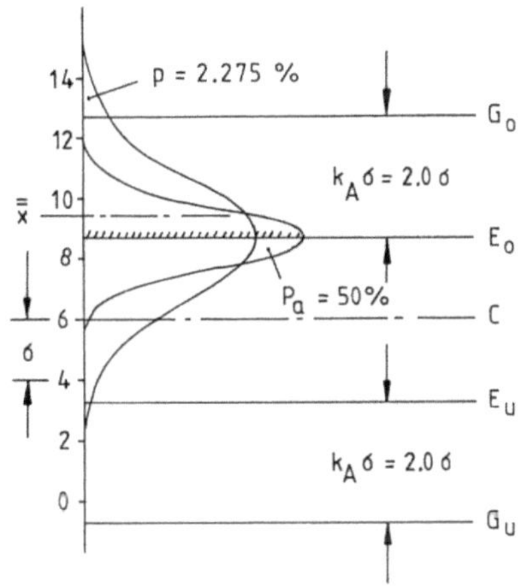

Fig. 15.40. Simulation of a normally distributed process which lies with its mean value on the upper intervention limit of $n = 5 - \bar{x}$-QCC

the upper intervention limit. The simulated, normally distributed process is shifted upwards by

$$\frac{3}{\sqrt{5}}\sigma = 2.68 \approx 2.7.$$

From the ten mean values five lie above E_o. The total mean value of the mean values, which led to intervention, is $\bar{x} = 9.5$. Thus the ratio of mean deviation to mean value deviation is

$$\frac{\mu_i - C}{\bar{\bar{x}} - C} = \frac{2.7}{3.5} = 0.77.$$

Thus the optimal correction with $n = 5 - \bar{x}$-QCC is the total $K \approx 0.75|\bar{x} - C|$.

This can just be a reference value; if a more exact correction total is required, then the simulation range must be enlarged.

From a discussion of the optimal correction with the original value QCC again this leads to the frequently mentioned view that the random process spread can be reduced by process control. So far as is possible, the temperature of a room or its oven is very precisely controlled or the rpm of a turbine, so it must be possible to control a production process exactly to the desired value.

In Fig. 15.41 the extreme case is represented which will correct about the total deviation from mean value determined for a part. Process capability is good with $c_p = 1.3\bar{3}$ ($T = 8\sigma$). After the setting up of the process the setter

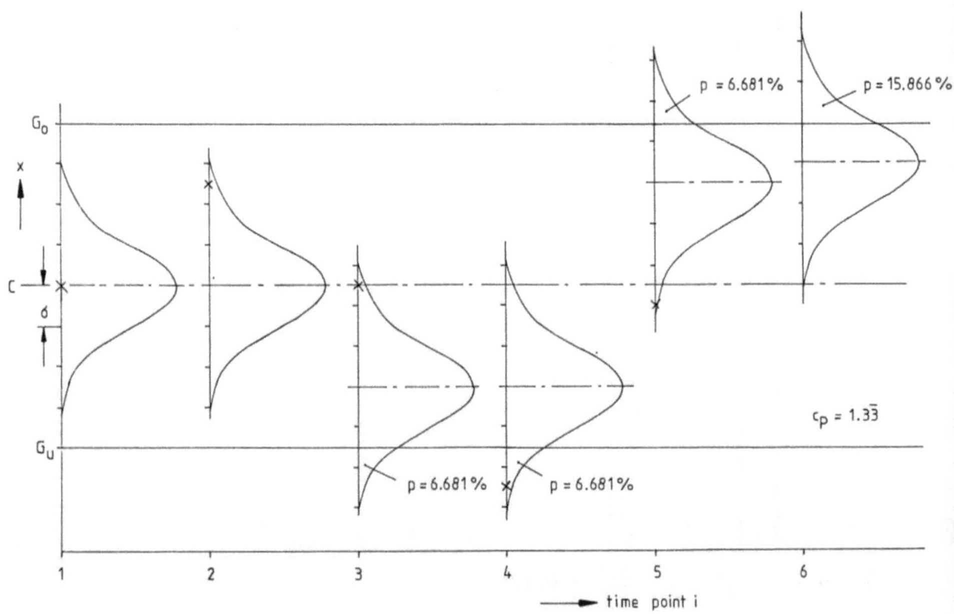

Fig. 15.41. Manual process control by correction of current deviation to $K_i = -(x_i - C)$

without correction		with correction of current deviation				
			intervention		deviation	correction
i	x_i	μ_i	x_i	C	$x_i - C$	$-(x_i - C)$
1	6	6	6			
2	11	6	11	x	5	-5
3	5	1	0	x	-6	6
4	8	7	9	x	3	-3
5	5	4	3	x	-3	3
6	8	7	9	x	3	-3
7	7	4	5	x	-1	1
8	6	5	5	x	-1	1
9	1	6	1	x	-5	5
10	9	11	13	x	7	-7
11	6	4	6			
12	6	4	1	x	-2	2
13	6	6	6			
14	6	6	6			
15	6	6	6			
16	8	6	8	x	2	-2
17	6	4	4	x	-2	2
18	9	6	9	x	3	-3
19	5	3	2	x	-4	4
20	7	7	8	x	2	-2
21	3	5	2	x	-4	4
22	9	9	12	x	6	-6
23	4	3	1	x	-5	5
24	5	8	7	x	1	-1
25	5	7	6			
26	5	7	6			
27	6	7	7	x	1	-1
28	6	6	6			
29	4	6	1	x	-2	2
30	5	8	8	x	2	-2
31	6	6	6			
32	7	6	7	x	1	-1
33	6	5	5	x	-1	1
34	4	6	4	x	-2	2
35	8	9	10	x	4	-4
36	5	4	3	x	-3	3
37	3	7	4	x	-2	2
38	8	8	5	x	-1	1
39	6	10	10	x	4	-4
40	7	6	7	x	1	-1
41	10	5	9	x	3	-3
42	8	2	4	x	-2	2
43	7	4	5	x	-1	1
44	4	5	3	x	-3	3
45	8	8	10	x	4	-4
46	5	4	3	x	-3	3
47	6	7	7	x	1	-1
48	3	6	3	x	-3	3
49	6	9	9	x	3	-3
50	7	6	7	x	1	-1
51		5				$\sum K = -1 = \mu_{51} - \mu_1$

evaluation of correct values:

-7	/
-6	/
-5	/
-4	///
-3	~~////~~
-2	///
-1	~~////~~ /
0	~~////~~ ////
1	~~////~~
2	~~////~~ /
3	~~////~~
4	//
5	//
6	/

$n = 50$

$\bar{K} = -0.02$

$\sum K = -1$

evaluation of controlled x_i:

$n = 50$

$\bar{x} = 6.0$

$s = 2.899683$

Fig. 15.42. Simulation of a process where every deviation from the desired value is corrected to 100%; the 50 x_i-values without correction are a random sample test from a ND model with the parameters $\mu = 6$ and $\sigma \approx 2$

measures $x_1 = C$ and assumes that the process is set exactly on the mean value, which is correct. After some time, at time point 2, nothing has changed about the process position, which however the setter cannot know. He removes a part and determines the size value $x_2 = C + 2.5\sigma$ and at exactly this deviation he corrects the setting of the process. To be safe he removes the part produced immediately after the correction, whose size value lies randomly at $x_3 = C$. From this the setter decides that the process now is set at the centre, however in fact the non-conformity proportion $p = 6.68\%$. The continuation of this example finally leads to time point 6 where the non-conformity proportion $p = 15.866\%$.

A further drastic example with number values is given in Fig. 15.42. The regular correction of each deviation from the desired value $C = 6$ leads to the total spread of the process with respect to the remaining spread of about 50% unnecessarily increased.

It should be accepted that the process needs "a certain clearance room"; however the process should be held "in narrow limits", Figs. 15.43 and 15.44. The intervention limits are established at $E_u = 3.5$ and $E_o = 8.5$. Every intervention, which lies outside this range, is fully corrected. However—or just because—the total spread lies again at about 50% over the remaining spread. Here the question arises, whether this is to be traced to narrow limits or to a false correction. To clarify the latter, in Fig. 15.43 the totals of mean deviations and the sums of the deviations of all individual values which lead to intervention are formed and determined. The ratio is

$$\frac{\text{Mean value deviation}}{\text{Individual value deviation}} \approx 0.5.$$

From this it can be concluded that the total process can be improved by narrow limits, in that respectively only up to half of the respective individual value deviations are corrected, Figs. 15.45 and 15.46. In fact despite the narrow limits, the occurring spread with s_x is considerably smaller than in Figs. 15.43 and 15.44. However "narrow limits" are not to be unreservedly recommended. More appropriate are measured wide limits which do not lead to unnecessary interventions but nervertheless are sufficient to protect against errors, Fig. 15.46. None of the 50 individual values leads to an intervention; the total spread is thus not larger than the remaining spread.

In Fig. 15.45 in addition the differences of controlled x_i-values are formed and evaluated; this is an effective method of controlling the accuracy of the simulation of a process.

Finally the question will be tested as to which is the optimal correction total, if the intervention limits measured by the original value QCC lie far apart. In Fig. 15.47 an original value chart is sketched, in which the intervention limits lie at $E_u = 0.5$ and $E_o = 11.5$; the ND model passes through the acceptance range, if their mean value conforms with C. Furthermore different deviations from the middle position are simulated and the mean value of the deviation totals of the individual values is calculated above E_o. In the position $\mu = C + 3$

	without correction	with correction						
1	2	3		4		5	6	7
i	x_i	μ_i	$\lvert\mu_i - C\rvert$	x_i	$\lvert x_i - C\rvert$	intervention	deviation $x_i - C$	correction $-(x_i - C)$
1	6	6		6				
2	11	6	0	11	5	x	5	-5
3	5	1	5	0	6	x	-6	6
4	8	7	1	9	3	x	3	-3
5	5	4	2	3	3	x	-3	3
6	8	7	1	9	3	x	3	-3
7	7	4		5				
8	6	4		4				
9	1	4	2	-1	7	x	-7	7
10	8	11	5	13	7	x	7	-7
11	8	4		6				
12	6	4		4				
13	6	4		4				
14	6	4		4				
15	6	4		4				
16	8	4		6				
17	6	4		4				
18	9	4		7				
19	5	4	2	3	3	x	-3	3
20	7	7		8				
21	3	7		4				
22	9	7	1	10	4	x	4	-4
23	4	3	3	1	5	x	-5	5
24	5	8		7				
25	5	8		7				
26	5	8		7				
27	6	8		8				
28	6	8		8				
29	4	8		6				
30	6	8		8				
31	6	8		8				
32	7	8	2	9	3	x	3	-3
33	6	5		5				
34	4	5	1	3	3	x	-3	3
35	8	8	2	10	4	x	4	-4
36	5	4	2	3	3	x	-3	3
37	3	7		4				
38	2	7	1	3	3	x	-3	3
39	6	10	4	10	4	x	4	-4
40	7	6		7				
41	10	6	0	10	4	x	4	-4
42	8	2		4				
43	7	2	4	3	3	x	-3	3
44	4	5	1	3	3	x	-3	3
45	8	8	2	10	4	x	4	-4
46	5	4	2	3	3	x	-3	3
47	6	7		7				
48	3	7		4				
49	6	7		7				
50	7	7		8				

sample

$n = 50$

$\bar{x} = 6.06$

$s_x = 1.963026$

$\overline{\lvert\mu_i - C\rvert} = 2.0476$

$\overline{\lvert x_i - C\rvert} = 3.9524$

$$\frac{\overline{\lvert\mu_i - C\rvert}}{\overline{\lvert x_i - C\rvert}} = 0.5181 \approx 0.5$$

Fig. 15.43. Simulation of a process which is continually controlled with a $n = 1$-QCC in "narrow limits". At intervention the full deviation of the individual value is corrected. Intervention limits $E_u = 3.5$; $E_0 = 8.5$

evaluation of the controlled x_i-values

x_j		n_j	$\sum n_j$	G_j in %
-1	/	1	1	1
0	/	1	2	3
1	/	1	3	5
2		0	3	5
3	//// ///	8	11	21
4	//// ////	10	21	41
5	//	2	23	45
6	////	4	27	53
7	//// //	7	34	67
8	//// /	6	40	79
9	///	3	43	85
10	////	5	48	95
11	/	1	49	97
12		0	49	97
13	/	1	50	99

$n = 50$

$\bar{x} = 5.92$

$s_x = 2.981850$

(relative) process dispersion

$$\frac{s_x}{\hat{\sigma}} = \frac{2.981850}{1.963026} = 1.519007$$

Note: for $\hat{\sigma}$ the standard deviation of the uncontrolled process was choisen

Fig. 15.43. (Continued)

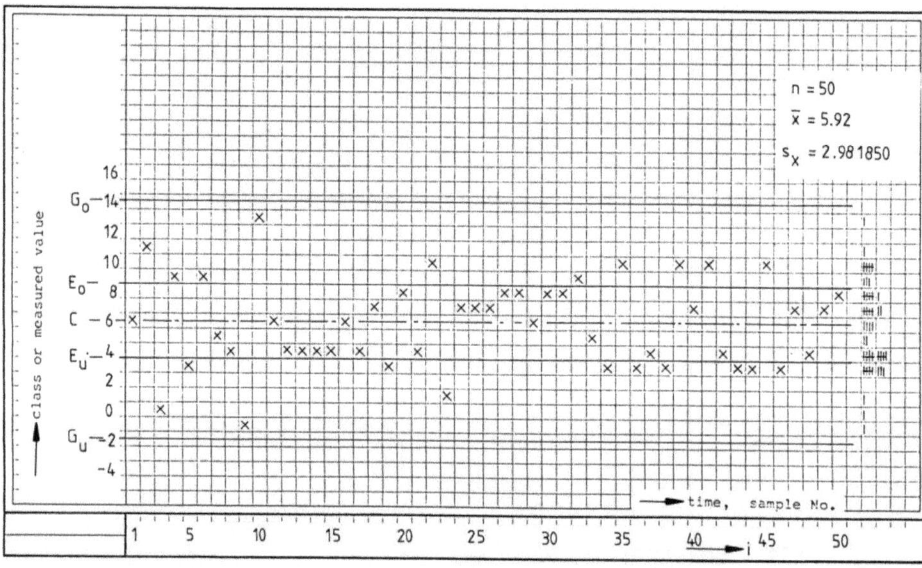

Fig. 15.44. Representation of a $n = 1$-QCC with "narrow limits" and the process simulated in Fig. 15.43

without correction		with correction			deviation	correction (roundet)	difference
i	x_i	μ_i	x_i	inter-vention	$A_i = x_i - C$	$K_i = -A_i/2$	$D_i = x_i - x_{i-1}$
1	6	6	5				
2	11	6	11	x	5	-2.5	5
3	5	3.5	2.5	x	-3.5	1.8	-6 = 2.5 - (11 - 2.5)
4	8	5.3	7.3				3 = 7.3 - (2.5 + 1.8)
5	5	5.3	4.3				-3
6	8	5.3	7.3				3
7	7	5.3	6.3				-1
8	6	5.3	5.3				-1
9	1	5.3	0.3	x	-5.7	2.8	-5
10	8	8.1	10.1	x	4.1	-2.0	7
11	8	6.1	8.1				0
12	6	6.1	6.1				-2
13	6	6.1	6.1				0
14	6	6.1	6.1				0
15	6	6.1	6.1				0
16	8	6.1	8.1				2
17	6	6.1	6.1				-2
18	9	6.1	9.1	x	3.1	-1.6	3
19	5	4.5	3.5				-4
20	7	4.5	5.5				2
21	3	4.5	1.5	x	-4.5	2.2	-4
22	9	6.7	9.7	x	3.7	-1.9	6
23	4	4.8	2.8	x	-3.2	1.6	-5
24	5	6.4	5.4				1
25	5	6.4	5.4				0
26	5	6.4	5.4				0
27	6	6.4	6.4				1
28	6	6.4	6.4				0
29	4	6.4	4.4				-2
30	6	6.4	6.4				2
31	6	6.4	6.4				0
32	7	6.4	7.4				1
33	6	6.4	6.4				-1
34	4	6.4	4.4				-2
35	8	6.4	8.4				4
36	5	6.4	5.4				-3
37	3	6.4	3.4	x	-2.6	1.3	-2
38	2	7.7	3.7				-1
39	6	7.7	7.7				4
40	7	7.7	8.7	x	2.7	-1.4	1
41	10	6.3	10.3	x	4.3	-2.2	3
42	8	4.1	6.1				-2
43	7	4.1	5.1				-1
44	4	4.1	2.1	x	-3.9	2.0	-3
45	8	6.1	8.1				4
46	5	6.1	5.1				-3
47	6	6.1	6.1				1
48	3	6.1	3.1	x	-2.9	1.4	-3
49	6	7.5	7.5				3
50	7	7.5	8.5				1

evaluation of differences:

x_j		n_j
-6	/	1
-5	//	2
-4	//	2
-3	/////	5
-2	//////	6
-1	/////	5
0	////////	8
1	//////	6
2	///	3
3	/////	5
4	///	3
5	/	1
6	/	1
7	/	1

$n = 49$

$\bar{D} = 0.020408$

$s_D = 2.933214$

$\hat{\sigma} = \frac{s_D}{\sqrt{2}} = 2.074095$

in accordance with figure 14.18

evaluation of the controlled x_i

x_i- values:

$\sum K_i = 1.5 = \mu_{50} - \mu_0 = 7.5 - 6.0$

$n = 50$

$\bar{x} = 6.058$

$s_x = 2.287329$

(relative) process deviation $\frac{s_x}{\hat{\sigma}} = \frac{2.287329}{2.074095} = 1.102808$

Fig. 15.45. Simulation of a process which is controlled with a $n = 1$-QCC with "narrow limits"; at intervention the correction total $K_i = -0.5A_i$ is corrected

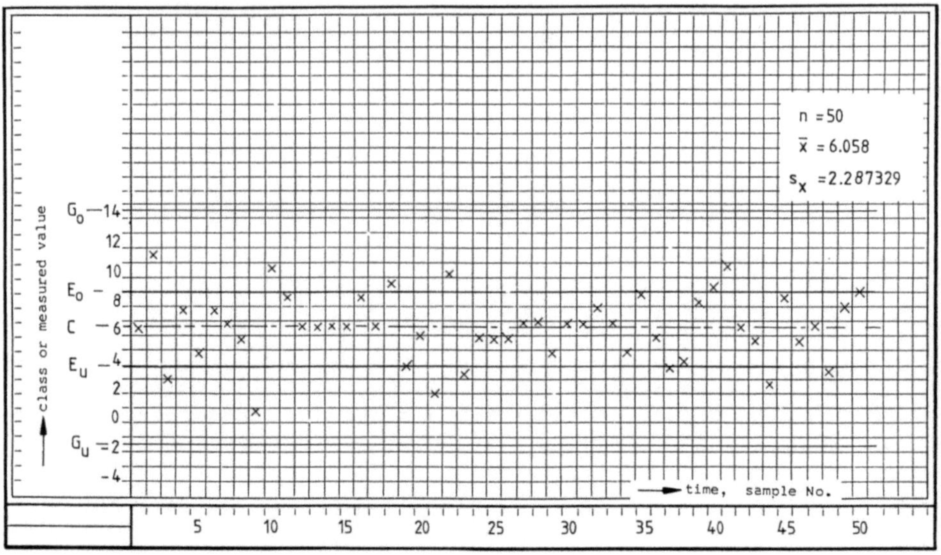

a

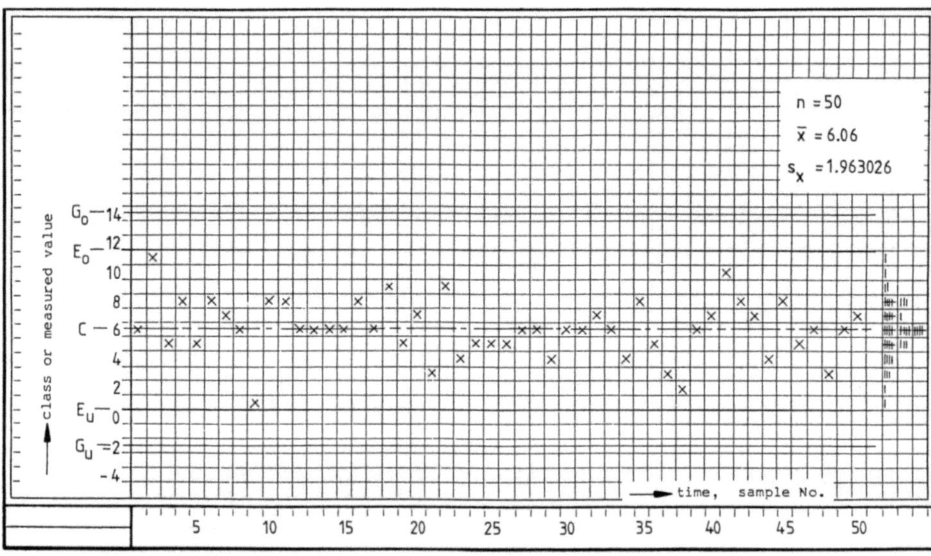

b

Fig. 15.46 a, b. Representation of an $n = 1$-QCC with narrow limits and the process simulated in Fig. 15.45 ($K = -0.5A$) (**a**) and wide limits, so that no intervention is used in the process which is centre set and running without disruption (**b**)

mean position of the distribution of the individuel values	$\mu = C+1$	$\mu = C+2$	$\mu = C+3$	$\mu = C+4$	$\mu = C+5$	$\mu = C+6$
mean of the deviations of the x values above E_0	$\bar{A} = 6$	$\bar{A} = 6.5$	$\bar{A} = 6.6$	$\bar{A} = 6.72$	$\bar{A} = 6.95$	$\bar{A} = 7,3$
optimal correction total	$K = -0.17\bar{A}$	$K = -0.31\bar{A}$	$K = -0.45\bar{A}$	$K = -0.59\bar{A}$	$K = -0.72\bar{A}$	$K = -0.82\bar{A}$

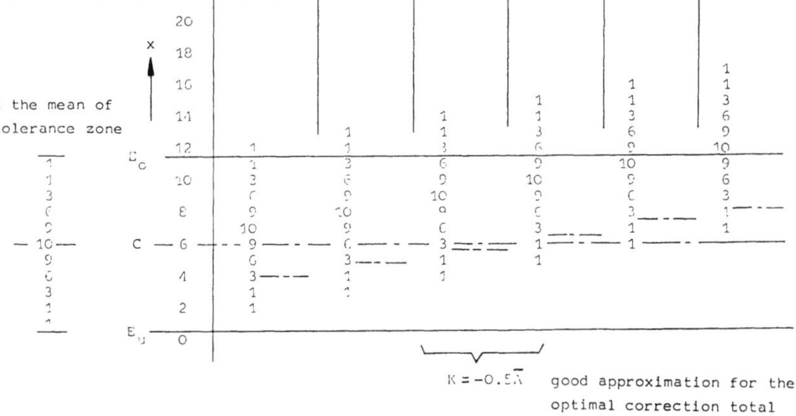

Fig. 15.47. Empirical determination of the optimal correction total with different deviations of the distribution of individual values with respect to the middle of the tolerance zone C; the limit values are not drawn

for example, this is $\bar{A} = \frac{1}{5}(3 \cdot 6 + 7 + 8) = 6.6$. Since the distribution lies at the total 3 above C, the optimal correction would in this situation be $K = 0.45\bar{A}$; than from this situation C on average would be corrected. Since the true process position is never known, $K = 0.5A$ is a useable approximation. The short, additional middle line in Fig. 15.47 shows where the control is taken through the approximate correction. With small deviations "over-steer" can happen, with large deviations there may be "under-steer".

All discussions up to now on process correction have confined themselves to the case where there is no trend present or where it slope is slight. Then—independently of the position of the intervention limits—$K = -0.5A$ is the correction total recommended.

While with a process without trend and a $n = 1-$QCC appropriate corrections should be made to half of the last deviation total, $K_i = -0.5A_i$, with a trend process a larger correction total is necessary. In Fig. 15.48 a process with a trend slope of $\Delta\mu = 0.5\sigma$/part is simulated. With each intervention half the total of the last deviation is corrected. The process "climbs" upwards and stays in the upper tolerance half "hanging". The mean value is at $\bar{x} = 9.692$ much too high, Fig. 15.49. The (relative) process spread is at $s_P = 1.2167$ under the (relative) process capability.

A control of the accuracy of the simulation follows by the difference method, among others over the sum of the correction values, Fig. 15.48. This control also follows in different further examples, without expressing this again.

without trend		with trend		intervention	deviation	correction	difference
i	x_i	μ_i	x_i	E	A_i	$K_i = -0.5\,A_i$	$D_i = x_i - x_{i-1}$
1	6	7	7				
2	11	8	13	x	7	−3.5 to 4.5	6
3	5	5.5	4.5				−5 = 4.5 − (13 − 3.5)
4	8	6.5	8.5				4
5	5	7.5	6.5				−2
6	8	8.5	10.5				4
7	7	9.5	10.5				0
8	6	10.5	10.5				0
9	1	11.5	6.5				−4
10	8	12.5	14.5	x	8.5	−4.2 to 8.3	8
11	8	9.3	11.3				1 = 10.3 − (14.5 − 4.2)
12	6	10.3	10.3				−1
13	6	11.3	11.3				1
14	6	12.3	12.3	x	6.3	−3.2 to 9.1	1
15	6	10.1	10.1				1 = 10.1 − (12.3 − 3.2)
16	8	11.1	13.1	x	7.1	−3.6 to 7.5	3
17	6	8.5	8.5				−1
18	9	9.5	12.5	x	6.5	−3.2 to 6.3	4
19	5	7.3	6.3				−3
20	7	8.3	9.3				3
21	3	9.3	6.3				−3
22	9	10.3	13.3	x	7.3	−3.6 to 6.7	7
23	4	7.7	5.7				−4
24	5	8.7	7.7				2
25	5	9.7	8.7				1
26	5	10.7	9.7				1
27	6	11.7	11.7	x	5.7	−2.8 to 8.9	2
28	6	9.9	9.9				1
29	4	10.9	8.9				−1
30	6	11.9	11.9	x	5.9	−3.0 to 8.9	3
31	6	9.9	9.9	x			1
32	7	10.9	11.9	x	5.9	−3.0 to 7.9	2
33	6	8.9	8.9				0
34	4	9.9	7.9				−1
35	8	10.9	12.9	x	6.9	−3.4 to 7.5	5
36	5	8.5	7.5				−2
37	3	9.5	6.5				−1
38	2	10.5	6.5				0
39	6	11.5	11.5				5
40	7	12.5	13.5	x	7.5	−3.8 to 8.7	2
41	10	9.7	13.7	x	7.7	−3.8 to 5.9	4
42	8	6.9	8.9				−1
43	7	7.9	8.9				0
44	4	8.9	6.9				−2
45	8	9.9	11.9	x	5.9	−3.0 to 6.9	5
46	5	7.9	6.9				−2
47	6	8.9	8.9				2
48	3	9.9	6.9				−2
49	6	10.9	10.9				4
50	7	11.9	12.9	x	6.9	−3.4 to 8.5	2

evaluation of the differences:

D_j		n_j
−5	/	1
−4	//	2
−3	//	2
−2	/////	5
−1	//////	6
0	/////	5
1	////////	8
2	//////	6
3	///	3
4	/////	5
5	///	3
6	/	1
7	/	1
8	/	1

$n = 49$

$\bar{D} = 1.020408$

$s_D = 2.933214$

$$\hat{\sigma} = \frac{s_D}{\sqrt{2}} = 2.074095$$

in accordance with figure 14.18, column 3

$$\sum K_i = -47.5$$

evaluation of the controlled x_i: $\mu_{50} - \mu_o = 8.5 - 6.0 = 2.5$

$$= 50\Delta\mu + \sum K_j = 50 - 47.5 = 2.5$$

$n = 50$

$\bar{x} = 9.692$

$s_x = 2.523590$

(relative) process spread $s_P = \dfrac{s_x}{\hat{\sigma}} = 1.2167$

Fig. 15.48. Simulation of a process with a trend slope of $\Delta\mu = 0.5\sigma$/part. The process is continually controlled with an $n = 1$-QCC. Tolerance $T = 8\sigma$; process capability $c_P = 1.3\bar{3}$; limiting factor $k_E = 1.25$

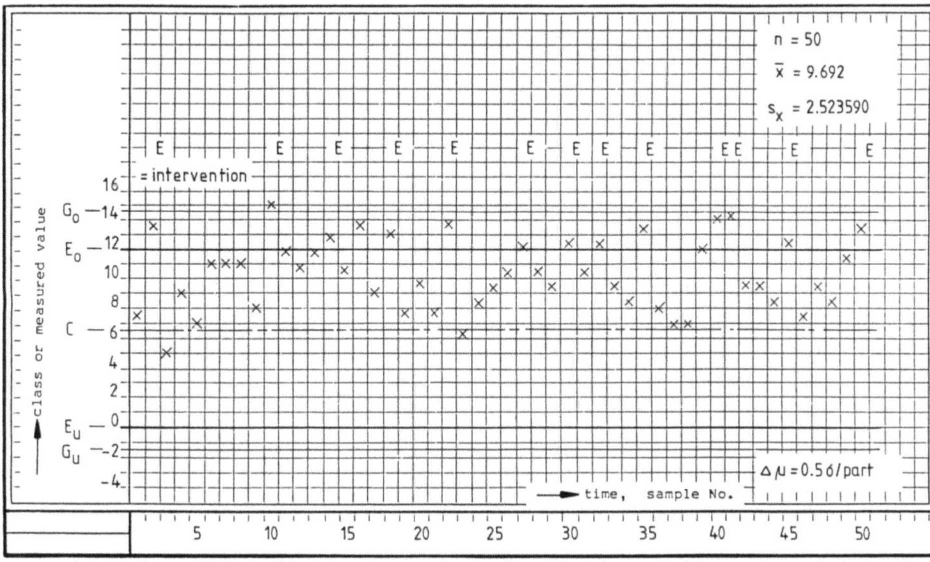

Fig. 15.49. Representation of the process simulated in Fig. 15.48

In Fig. 15.50 a process is renewed with a simulated trend slope of $\Delta\mu = 0.5\sigma$/part, however at every intervention the whole total of the last deviation is corrected. The total mean value lies at $\bar{x} = 7.72$, somewhat over the mean value, also the (relative) process spread is at $s_P = 1.4645$ rather too high. In total, however, with this spread a production batch lying at the centre with few non-conformities can be expected, Fig. 15.51.

In Fig. 15.52 there is again simulated a process with a trend slope of $\Delta\mu = 0.5\sigma$/part, however here the process capability is $c_P = 1.6\bar{6}$ or $T = 10\sigma$. Since control is with the same limiting factors $k_E = 1.25$ as in Fig. 15.50, the mean value at $\bar{x} = 9.14$ is much too high and the spread is at $s_x = 3.58$ also larger. Thus the (relative) process spread $s_x/\sigma = 1.73$ is slightly larger than c_P. This would be acceptable if the mean value $\bar{x}$ lay closer to the mean size, Fig. 15.53.

In order to achieve this, there are two possibilities, which can be found for simultaneous use. On the one hand, the intervention limits towards which the trend is drifting could be shifted to the middle; on the other hand by the choice of a greater correction total a better centring of the total mean value on the mean size can be achieved.

In Fig. 15.54 a process is simulated with a trend slope of $\Delta\mu = 0.5\sigma$/part and which is continually controlled by an $n = 1 -$ QCC, by which the intervention limits are set asymmetrically. The upper intervention limit E_o was shifted with respect to the last example towards the middle, so that the process of drifting will be stopped earlier. Also at every intervention $K_i = -(A + 2)$ is corrected, i.e. it will be deliberately "over-steered". Thus at the lower intervention

without trend		with trend		intervention	deviation	correction	difference
i	x_i	μ_i	x_i	E	$A_i = x_i - C$	$K_i = -A_i$	$D_i = x_i - x_{i-1}$
1	6	7	7				
2	11	8	13	x	7	−7 to 1	6
3	5	2	1				−5 = 1 − (13 − 7)
4	8	3	5				4
5	5	4	3				−2
6	8	5	7				4
7	7	6	7				0
8	6	7	7				0
9	1	8	3				−4
10	8	9	11				8
11	8	10	12	x	6	−6 to 4	1
12	6	5	5				−1 = 5 − (12 − 6)
13	6	6	6				1
14	6	7	7				1
15	6	8	8				1
16	8	9	11				3
17	6	10	10				−1
18	9	11	14	x	8	−8 to 3	4
19	5	4	3				−3
20	7	5	6				3
21	3	6	3				−3
22	9	7	10				7
23	4	8	6				−4
24	5	9	8				2
25	5	10	9				1
26	5	11	10				1
27	6	12	12	x	6	−6 to 6	2
28	6	7	7				1
29	4	8	6				−1
30	6	9	9				3
31	6	10	10				1
32	7	11	12	x	6	−6 to 5	2
33	6	6	6				0
34	4	7	5				−1
35	8	8	10				5
36	5	9	8				−2
37	3	10	7				−1
38	2	11	7				0
39	6	12	12	x	6	−6 to 6	5
40	7	7	8				2
41	10	8	12	x	6	−6 to 2	4
42	8	3	5				−1
43	7	4	5				0
44	4	5	3				−2
45	8	6	8				5
46	5	7	6				−2
47	6	8	8				2
48	3	9	6				−2
49	6	10	10				4
50	7	11	12	x	6	−6 to 5	2

evaluation of the differences:

D_j		n_j
−5	/	1
−4	//	2
−3	//	2
−2	/////	5
−1	//////	6
0	/////	5
1	////////	8
2	//////	6
3	///	3
4	/////	5
5	///	3
6	/	1
7	/	1
8	/	1

$n = 49$

$\bar{D} = 1.020408$

$s_D = 2.933214$

$\hat{\sigma} = \frac{s_D}{\sqrt{2}} = 2.074095$

in accordance with figure 14.18, column 3

evaluation of the controlled x_i

$n = 50$

$\bar{x} = 7.72$

$s_x = 3.037453$

(relative) process spread $s_p = 1.4645$

$\sum K_i = -51$

$\mu_{50,new} - \mu_0 = 5 - 6 = -1$

$= 50\Delta\mu + \sum K_i = 50 - 51 = -1$

Fig. 15.50. Simulation of a process with the trend slope $\Delta\mu = 0.5\sigma$/part. The process is continually controlled with an $n = 1$-QCC. Tolerance $T = 8\sigma$; process capability $c_p = 1.3\overline{3}$; limiting factor $k_E = 1.25$

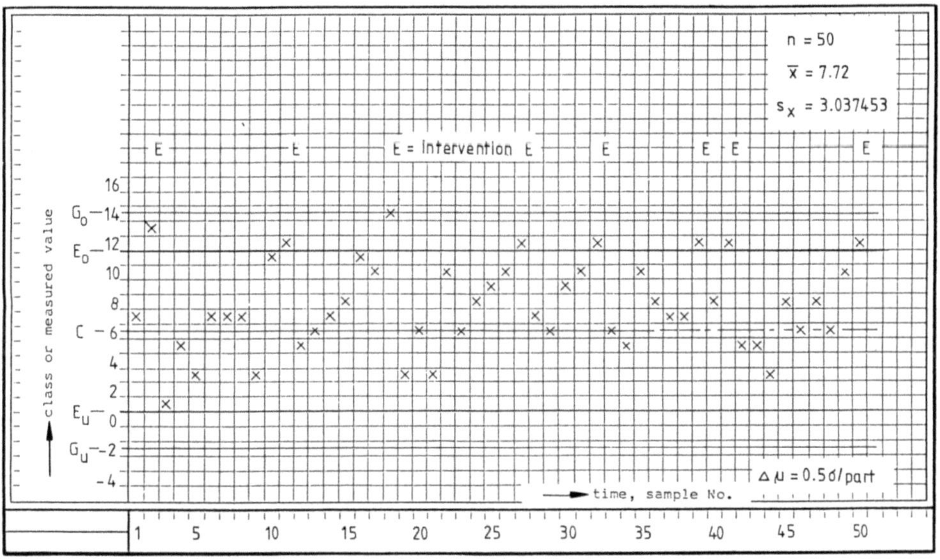

Fig. 15.51. Representation of the process simulated in Fig. 15.50

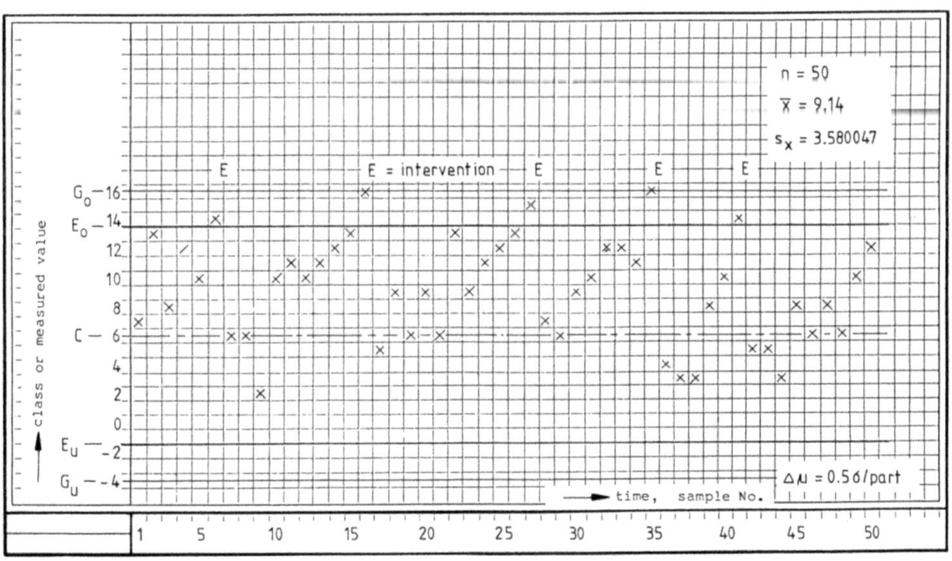

Fig. 15.53. Representation of the process simulated in Fig. 15.52

without trend		with trend				
				intervention	deviation	correction
i	x_i	μ_i	x_i	E	$A_i = x_i - C$	$K_i = -A_i$
1	6	7	7			
2	11	8	13			
3	5	9	8			
4	8	10	12			
5	5	11	10			
6	8	12	14	x	8	-8 to 4
7	7	5	6			
8	6	6	6			
9	1	7	2			
10	8	8	10			
11	8	9	11			
12	6	10	10			
13	6	11	11			
14	6	12	12			
15	6	13	13			
16	8	14	16	x	10	-10 to 4
17	6	5	5			
18	9	6	9			
19	5	7	6			
20	7	8	9			
21	3	9	6			
22	9	10	13			
23	4	11	9			
24	5	12	11			
25	5	13	12			
26	5	14	13			
27	6	15	15	x	9	-9 to 6
28	6	7	7			
29	4	8	6			
30	6	9	9			
31	6	10	10			
32	7	11	12			
33	6	12	12			
34	4	13	11			
35	8	14	16	x	10	-10 to 4
36	5	5	4			
37	3	6	3			
38	2	7	3			
39	6	8	8			
40	7	9	10			
41	10	10	14	x	8	-8 to 2
42	8	3	5			
43	7	4	5			
44	4	5	3			
45	8	6	8			
46	5	7	6			
47	6	8	8			
48	3	9	6			
49	6	10	10			
50	7	11	12			

$\sum K_i = -45$

$\mu_{50} - \mu_0 = 11 - 6 = 5$

$= 50\Delta\mu + \sum K_i = 50 - 45 = 5$

evaluation of controlled x_i:

$n = 50$

$\bar{x} = 9.14$

$s_x = 3.580047$

(relative) process dispersion $s_p = 1.7261$

Fig. 15.52. Simulation of a process with a trend slope $\Delta\mu = 0.5\sigma$/part. The process is continually controlled with an $n=1$-QCC. Tolerance $T = 10\sigma$; process capability $c_p = 1.6\bar{6}$; limiting factor $k_E = 1.25$ on both sides

without trend		with trend				
				intervention	deviation	correction
i	x_i	μ_i	x_i	E	$A_i = x_i - C$	$K_i = -(A_i + 2)$
1	6	7	7			
2	11	8	13	x	7	-9 to -1
3	5	0	-1			
4	8	1	3			
5	5	2	1			
6	8	3	5			
7	7	4	5			
8	6	5	5			
9	1	6	1			
10	8	7	9			
11	8	8	10			
12	6	9	9			
13	6	10	10			
14	6	11	11	x	5	-7 to 4
15	6	5	5			
16	8	6	8			
17	6	7	7			
18	9	8	11	x	7	-7 to 1
19	5	2	1			
20	7	3	4			
21	3	4	1			
22	9	5	8			
23	4	6	4			
24	5	7	6			
25	5	8	7			
26	5	9	8			
27	6	10	10			
28	6	11	11	x	5	-7 to 4
29	4	5	3			
30	6	6	6			
31	6	7	7			
32	7	8	9			
33	6	9	9			
34	4	10	8			
35	8	11	13	x	7	-9 to 2
36	5	3	2			
37	3	4	1			
38	2	5	1			
39	6	6	6			
40	7	7	8			
41	10	8	12	x	6	-8 to 0
42	8	1	3			
43	7	2	3			
44	4	3	1			$\sum K_i = -47$
45	8	4	6			
46	5	5	4			
47	6	6	6			$\mu_{50} - \mu_0 = 9 - 6 = 3$
48	3	7	4			
49	6	8	8			$= 50\,\Delta\mu + \sum K_i = 50 - 47 = 3$
50	7	9	10			

evaluation of controlled x_i:

$n = 50$

$\bar{x} = 6.18$

$s_x = 3.589611$

(relative) process dispersion $s_p = 1.7307$

Fig. 15.54. Simulation of a process with a trend slope $\Delta\mu = 0.5\sigma$/part. The process is continually controlled with an $n = 1$-QCC. Tolerance $T = 10\sigma$; process capability $c_p = 1.6\bar{6}$; limiting factor above $k_{E,o} = 2.75$, limiting factor below $k_{E,u} = 0.75$

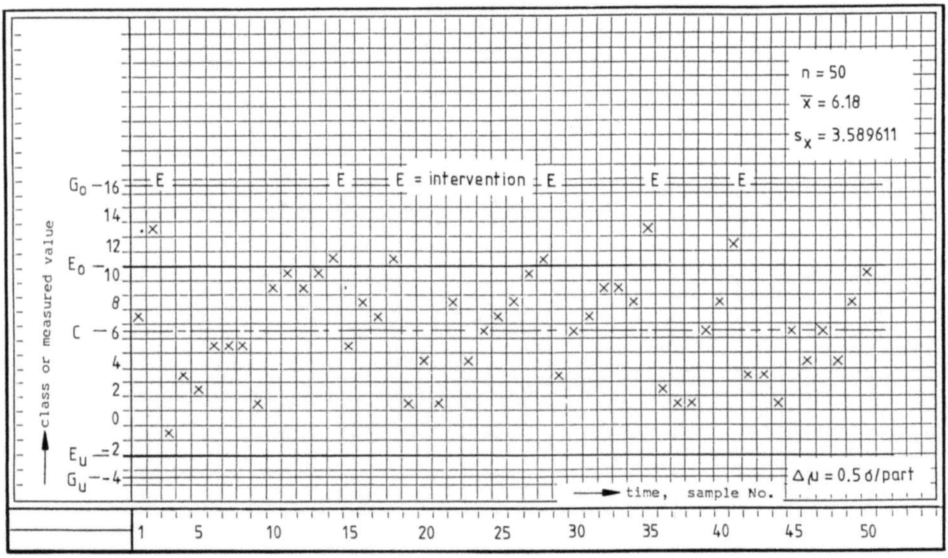

Fig. 15.55. Representation of the process simulated in Fig. 15.54

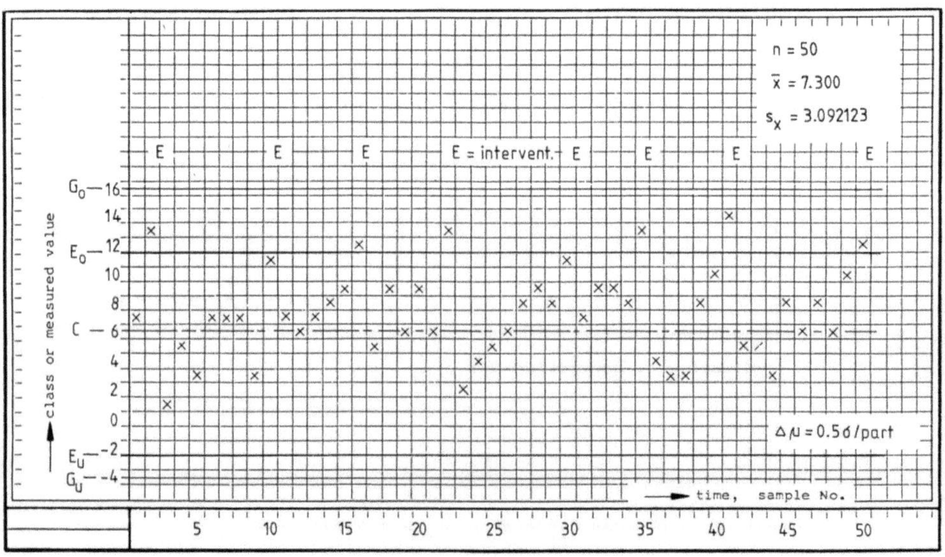

Fig. 15.57. Representation of the process simulated in Fig. 15.56

without trend		with trend		intervention	deviation	correction	difference	
i	x_i	μ	x_i	E	Δ_i	$K_i = -\Delta_i$	$D_i = x_i - x_{i-1}$	
1	6	7	7					
2	11	8	13	x	7	-7 to 1	6	
3	5	2	1				-5	= 1 - (13 - 7)
4	8	3	5				4	
5	5	4	3				-2	
6	8	5	7				4	
7	7	6	7				0	
8	6	7	7				0	
9	1	8	3				-4	
10	8	9	11	x	5	-5 to 4	8	
11	8	5	7				1	= 7 -(11 - 5)
12	6	6	6				-1	
13	6	7	7				1	
14	6	8	8				1	
15	6	9	9				1	
16	8	10	12	x	6	-6 to 4	3	
17	6	5	5				-1	= 5 - (12 - 6)
18	9	6	9				4	
19	5	7	6				-3	
20	7	8	9				3	
21	3	9	6				-3	
22	9	10	13	x	7	-7 to 3	7	
23	4	4	2				-4	
24	5	5	4				2	
25	5	6	5				1	
26	5	7	6				1	
27	6	8	8				2	
28	6	9	9				1	
29	4	10	8				-1	
30	6	11	11	x	5	-5 to 6	3	
31	6	7	7				1	
32	7	8	9				2	
33	6	9	9				0	
34	4	10	8				-1	
35	8	11	13	x	7	-7 to 4	5	
36	5	5	4				-2	
37	3	6	3				-1	
38	2	7	3				0	
39	6	8	8				5	
40	7	9	10				2	
41	10	10	14	x	8	-8 to 2	4	
42	8	3	5				-1	
43	7	4	5				0	
44	4	5	3				-2	
45	8	6	8				5	
46	5	7	6				-2	
47	6	8	8				2	
48	3	9	6				-2	
49	6	10	10				4	
50	7	11	12	x	6	-6 to 5	2	

evaluation of the differences:

D_j		n_j
-5	/	1
-4	//	2
-3	//	2
-2	/////	5
-1	//////	6
0	/////	5
1	////////	8
2	//////	6
3	///	3
4	/////	5
5	///	3
6	/	1
7	/	1
8	/	1

$n = 49$

$\bar{D} = 1.020408$

$s_D = 2.933214$

$\hat{\sigma} = \frac{s_D}{\sqrt{2}} = 2.074095$

in accordance with figure 14.18, column 3

evaluation of the controlled x_i:

$n = 50$

$\bar{x} = 7.300$

$s_x = 3.092123$

(relative) process spread $s_p = \frac{s_x}{\hat{\sigma}} = 1.490830$

$\sum K_i = -51$

$\mu_{50,new} - \mu_0 = 5 - 6 = -1$

$= 50\,\Delta\mu + \sum K_i = 50 - 51 = -1$

Fig. 15.56. Simulation of a process with the same end conditions as that simulated in Fig. 15.54; only the correction total is smaller, so that interventions are twice as often

without trend		with trend and with correction					
				intervention	deviation	correction	difference
i	x_i	μ_i	x_i	E	$A_i = x_i - C$	$K_i = -(A_i + 2)$	$D_i = x_i - x_{i-1}$
1	6	6.5	6.5				
2	11	7	12	x	6	-8 auf -1	5.5
3	5	-0.5	-1.5				-5.5
4	8	0	2				3.5
5	5	0.5	-0.5				-2.5
6	8	1	3				3.5
7	7	1.5	2.5				-0.5
8	6	2	2				-0.5
9	1	2.5	-2.5				-4.5
10	8	3	5				7.5
11	8	3.5	5.5				0.5
12	6	4	4				-1.5
13	6	4.5	4.5				0.5
14	6	5	5				0.5
15	6	5.5	5.5				0.5
16	8	6	8				2.5
17	6	6.5	6.5				-1.5
18	9	7	10				3.5
19	5	7.5	6.5				-3.5
20	7	8	9				2.5
21	3	8.5	5.5				-3.5
22	9	9	12	x	6	-8 auf 1	6.5
23	4	1.5	-0.5				-4.5
24	5	2	1				1.5
25	5	2.5	1.5				0.5
26	5	3	2				0.5
27	6	3.5	3.5				1.5
28	6	4	4				0.5
29	4	4.5	2.5				-1.5
30	6	5	5				2.5
31	6	5.5	5.5				0.5
32	7	6	7				1.5
33	6	6.5	6.5				-0.5
34	4	7	5				-1.5
35	8	7.5	9.5				4.5
36	5	8	7				-2.5
37	3	8.5	5.5				-1.5
38	2	9	5				-0.5
39	6	9.5	9,5				4.5
40	7	10	11	x	5	-7 auf 3	1.5
41	10	3.5	7.5				3.5
42	8	4	6				-1.5
43	7	4.5	5.5				-0.5
44	4	5	3				-2.5
45	8	5.5	7.5				4.5
46	5	6	5				-2.5
47	6	6.5	6.5				1.5
48	3	7	4				-2.5
49	6	7.5	7.5				3.5
50	7	8	9				1.5

evaluation of differences:

D_j		n_j
-5.5	/	1
-4.5	//	2
-3.5	//	2
-2.5	/////	5
-1.5	//////	6
-0.5	/////	5
0.5	////////	8
1.5	//////	6
2.5	///	3
3.5	/////	5
4.5	///	3
5.5	/	1
6.5	/	1
7.5	/	1

$n = 49$

$\bar{D} = 0.520408$

$s_D = 2.933214$

$\hat{\sigma} = \frac{s_D}{\sqrt{2}} = 2.074095$

in accordance with figure 14.18

evaluation of controlled x_i-values:

$n = 50$

$\bar{x} = 5.25$

$s_x = 3.251766$

(relative) process dispersion $s_P = \frac{s_x}{\hat{\sigma}} = 1.567800$

Fig. 15.58. Simulation of a process with a trend slope $\Delta\mu = 0.25\sigma$/part. The process is continually controlled with an $n = 1$-QCC. Tolerance $T = 10\sigma$; process capability $c_P = 1.6\bar{6}$; limiting factor above $k_{E,o} = 2.75$; limiting factor below $k_{E,u} = 0.75$

limit from which the process is drifting, and does not lead often to unnecessary intervention was $k_{E,u}$ somewhat lessened. The process simulated in Fig. 15.54 is represented in Fig. 15.55.

The result of this measure is almost optimal. The mean value lies with $\bar{x} = 6.18$ close by the mean size and the (relative) process spread has become at $s_P = 1.73$ no bigger, but it still lies above the process capability $c_P = 1.6\bar{6}$. This can be altered by a repeat of the over-steering. In Fig. 15.56 this happens, in that correction is made not at $K_i = (A_i + 2)$ but only at $K_i = -A_i$. Thus the trend has less clearance room, the total spread of the process is reduced and the (relative) process spread slides to $s_P = 1.49$ under the value for the process capability. Thus the process control is optimised. The process is represented in Fig. 15.57.

For completion further processes are simulated. In Figs. 15.58 and 15.59 the trend slope is $\Delta\mu = 0.25\sigma$/part. The intervention limits are again asymmetrical. The total mean value is at $\bar{x} = 5.25$ a little low. Here it could be suggested that it might be better corrected with $K_i = -(A_i - 1)$.

In the Figs. 15.60 to 15.62 a process is simulated with a trend slope of $\Delta\mu = 0.1\sigma$/part. Here a correction of $K_i = -A_i$ is fully sufficient. The representation of the process without limits should show, in Fig. 15.61 how difficult it is to recognise a trend of size and direction clearly. This will only be seen clearly in Fig. 15.62 by the use of intervention.

With the examples for the control and correction of trend process discussed up to now, it is assumed that the process capability is good at $c_P = 1.3\bar{3}$ and very good at $c_P = 1.6\bar{6}$.

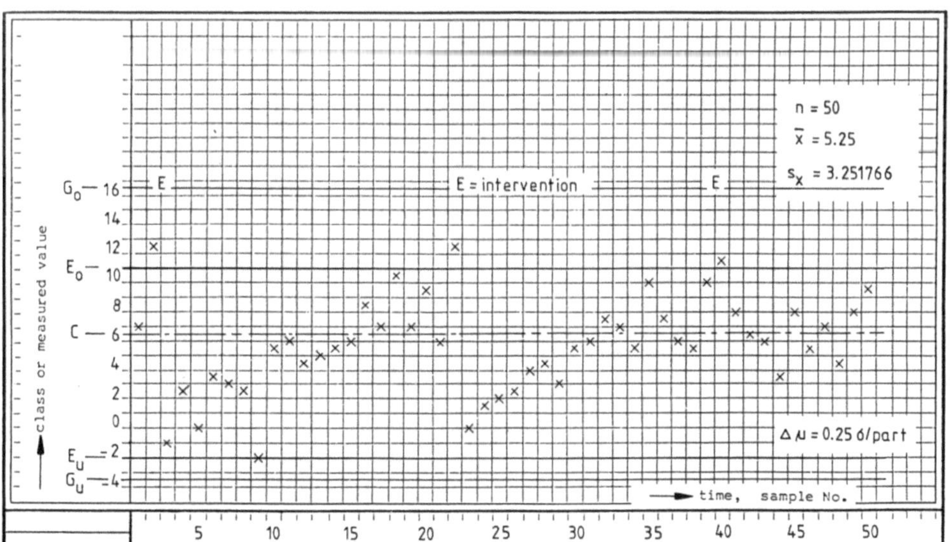

Fig. 15.59. Representation of the process simulated in Fig. 15.58

without trend		with trend and with correction					
				intervention	deviation	correction	difference
i	x_i	μ_i	x_i	E	$A_i = x_i - C$	$K_i = -A_i$	$D_i = x_i - x_{i-1}$
1	6	6.2	6.2				
2	11	6.4	11.4	x	5.4	-5.4 auf 1	5.2
3	5	1.2	0.2				-5.8
4	3	1.4	3.4				3.2
5	5	1.6	0.6				-2.8
6	8	1.8	3.8				3.2
7	7	2	3				-0.8
8	6	2.2	2.2				-0.8
9	1	2.4	-2.6	x	-8.6	8.6 auf 11	-4.8
10	8	11.2	13.2	x	7.2	-7.2 auf 4	7.2
11	8	4.2	6.2				0.2
12	6	4.4	4.4				-1.8
13	6	4.6	4.6				0.2
14	6	4.8	4.8				0.2
15	6	5	5				0.2
16	8	5.2	7.2				2.2
17	6	5.4	5.4				-1.8
18	9	5.6	8.6				3.2
19	5	5.8	4.8				-3.8
20	7	6	7				2.2
21	3	6.2	3.2				-3.8
22	9	6.4	9.4				6.2
23	4	6.6	4.6				-4.8
24	5	6.8	5.8				1.2
25	5	7	6				0.2
26	5	7.2	6.2				0.2
27	6	7.4	7.4				1.2
28	6	7.6	7.6				0.2
29	4	7.8	5.8				-1.8
30	6	8	8				2.2
31	6	8.2	8.2				0.2
32	7	8.4	9.4				1.2
33	6	8.6	8.6				-0.8
34	4	8.8	6.8				-0.8
35	8	9	11	x	5	-5 auf 4.0	4.2
36	5	4.2	3.2				-2.8
37	3	4.4	1.4				-1.8
38	2	4.6	0.6				-0.8
39	6	4.8	4.8				4.2
40	7	5	6				1.2
41	10	5.2	9.2				3.2
42	8	5.4	7.4				-1.8
43	7	5.6	6.6				-0.8
44	4	5.8	3.8				-2.8
45	8	6	8				4.2
46	5	6.2	5.2				-2.8
47	6	6.4	6.4				1.2
48	3	6.6	3.6				-2.8
49	6	6.8	6.8				3.2
50	7	7	8				1.2

evaluation of corrected x_i-values:

$n = 50$

$\bar{x} = 5.78$

$s_x = 2.995575$

(relative) process dispersion:

$$s_P = \frac{s_x}{\hat{\sigma}} = 1.444281$$

evaluation of differences:

$n = 49$

$\bar{D} = 0.220408$

$s_D = 2.933214$

$$\hat{\sigma} = \frac{s_D}{\sqrt{2}} = 2.074095$$

in accordance with figure 14.18

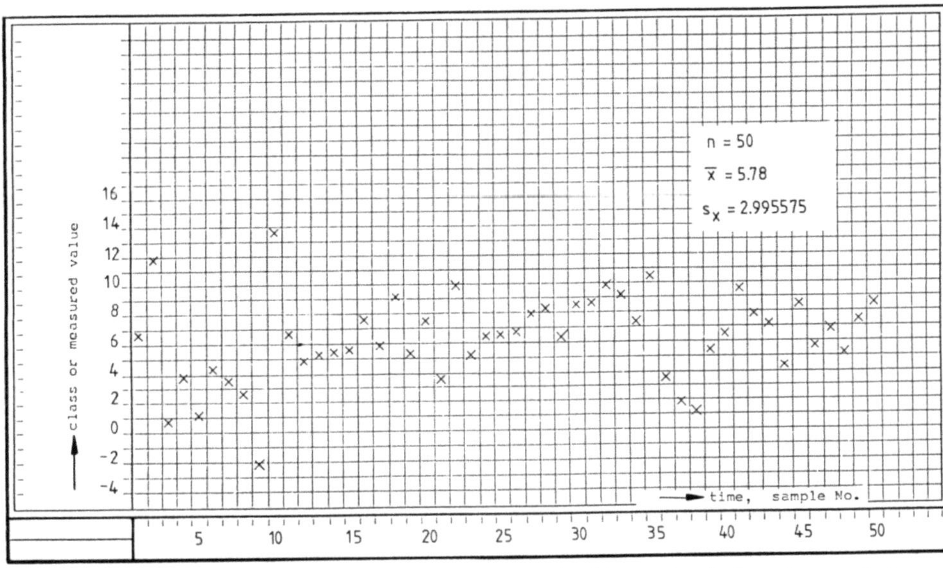

Fig. 15.61. Representation of the process simulated without the limit values and the intervention limits

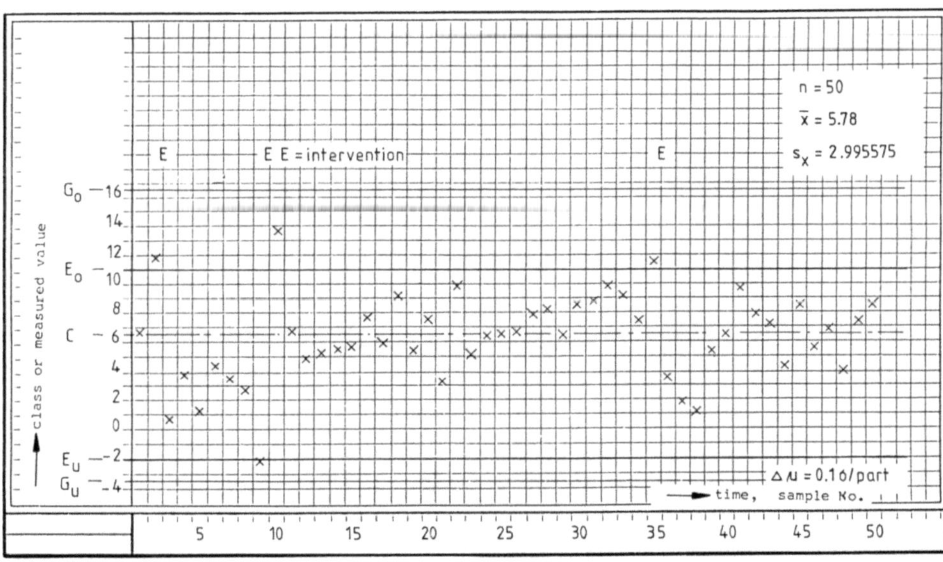

Fig. 15.62. Representation of the process simulated in Fig. 15.60 with limit values and intervention limits

Fig. 15.60. Simulation of a process with a trend slope $\Delta\mu = 0.1\sigma$/part. The process is continually controlled with an $n = 1$-QCC. Tolerance $T = 10\sigma$; process capability $c_P = 1.6\bar{6}$; limiting factor above $k_{E,o} = 2.75$, limiting factor below $k_{E,u} = 0.75$

	without trend	with trend						
				intervention	deviation	correction	difference	
i	x_i	μ_i	x_i	E	A_i	$K_i = -0.5\,A_i$	$D_i = x_i - x_{i-1}$	
1	6	6.5	6.5					
2	11	7	12	x	6	-3 auf 4	5.5	
3	5	4.5	3.5				-5.5	= 3.5 - (12 - 3)
4	8	5	7				3.5	
5	5	5.5	4.5				-2.5	
6	8	6	8				3.5	
7	7	6.5	7.5				-0.5	
8	6	7	7				-0.5	
9	1	7.5	2.5				-4.5	
10	8	8	10	x	4	-2 auf 6	7.5	
11	3	6.5	8.5				0.5	= 8,5 - (10 - 2)
12	6	7	7				-1.5	
13	6	7.5	7.5				0.5	
14	6	8	8				0.5	
15	6	8.5	8.5				0.5	
16	8	9	11	x	5	-2.5 auf 6.5	2.5	
17	6	7	7				-1.5	= 7 - (11 - 2.5)
18	9	7.5	10.5	x	4.5	-2.2 auf 5.3	3.5	
19	5	5.8	4.8				-3.5	
20	7	6.3	7.3				2.5	
21	3	6.8	3.8				-3.5	
22	9	7.3	10.3	x	4.3	-2.2 auf 5.1	6.5	
23	4	5.6	3.6				-4.5	
24	5	6.1	5.1				1.5	
25	5	6.6	5.6				0.5	
26	5	7.1	6.1				0.5	
27	6	7.6	7.6				1.5	
28	6	8.1	8.1				0.5	
29	4	8.6	6.6				-1.5	
30	6	9.1	9.1	x	3.1	-1.6 auf 7.5	2.5	
31	6	8	8				0.5	
32	7	8.5	9.5	x		-1.8 auf 6.7	1.5	
33	6	7.2	7.2				-0.5	
34	4	7.7	5.7				-1.5	
35	8	8.2	10.2	x	4.3	-2.0 auf 6.2	4.5	
36	5	6.7	5.7				-2.5	
37	3	7.2	4.2				-1.5	
38	2	7.7	3.7				-0.5	
39	6	8.2	8.2				4.5	
40	7	8.7	9.7	x	3.7	-1.8 auf 6.9	1.5	
41	10	7.4	11.4	x	5.4	-2.7 auf 4.7	3.5	
42	8	5.2	7.2				-1.5	
43	7	5.7	6.7				-0.5	
44	4	6.2	4.2				-2.5	
45	8	6.7	8.7				4.5	
46	5	7.2	6.2				-2.5	
47	6	7.7	7.7				1.5	
48	3	8.2	5.2				-2.5	
49	6	8.7	8.7				3.5	
50	7	9.2	10.2	x	4.2	-2.1 auf 7.1	1.5	

$\sum K_i = -23.9$

evaluation of differences:

D_j		n_j
-5.5	/	1
-4.5	//	2
-3.5	//	2
-2.5	/////	5
-1.5	//////	6
-0.5	/////	5
0.5	////////	8
1.5	//////	6
2.5	///	3
3.5	/////	5
4.5	///	3
5.5	/	1
6.5	/	1
7.5	/	1

$n = 49$

$\bar{D} = 0.520408$

$s_D = 2.933214$

$\hat{\sigma} = 2.074095$

in accordance with figure 14.18

evaluation of controlled x_i-values:

$n = 50$

$\bar{x} = 7.256$

$s_x = 2.264199$

$\mu_{50,neu} - \mu_0 = 7.1 - 6 = 1.1$

$= 50\,\Delta\mu + \sum K_i = 50\ 0.5 - 23.9 = 1.1$

(relative) process dispersion $s_p = \dfrac{s_x}{\hat{\sigma}} = 1.091656$

Fig. 15.63. Simulation of a process with a trend slope $\Delta\mu = 0.25\sigma$/part. The process is continually controlled with an $n = 1$-QCC. Tolerance $T = 7\sigma$; process capability $c_p = 1.6\bar{6}$; limiting factor above $k_{E,o} = 2$; limiting factor below $k_{E,u} = 1.25$

Finally the question should be asked as to which control and correction must be undertaken if the process capability lies at the lower limit i.e. at $c_P \gtrsim 1$.

It is assumed that the tolerance amounts to $T = 7\sigma$, then the (relative) process capability $c_P = 1.1\bar{6}$. It is further assumed that the process has a trend slope of $\Delta\mu = 0.25\sigma/\text{part}$.

After multiple testing and repetition the simulation given in Fig. 15.63 was found as a solution. Because of the narrow intervention limits—$E_o - E_u = 3.75\sigma$—correction will be made at half the total of the last deviation.

The result is bewildering. The total spread $s_x = 2.2642$ lies in spite of the trend at less than 10% above the remaining spread; the (relative) process spread amounts to $s_P = s_x/\hat{\sigma} = 1.09$. In any case on average every fifth time an intervention must be made, which can only be handled by automatic process control, but it does work.

From the representation of the controlled process in Fig. 15.64 it can be seen that even when the limit values lie at $G_u = 0$ and $G_o = 12$—this would mean $T = 6\sigma$—production with only scarce non-conformities would be possible. The interval to the upper limit value in s_x-units then amounts to

$$u_x = \frac{G_o - \bar{x}}{s_x} = \frac{12 - 7.256}{2.2642} = 2.095;$$

thus an error proportion of $p \approx 2\%$ above G_o can be expected.

A c_P value of $c_P \gtrsim 1.0$ is factual—even with a trend slope of $\Delta\mu = 0.25\sigma/\text{part}$—the lower limit of the process capability, assuming that process control and process correction are optimised.

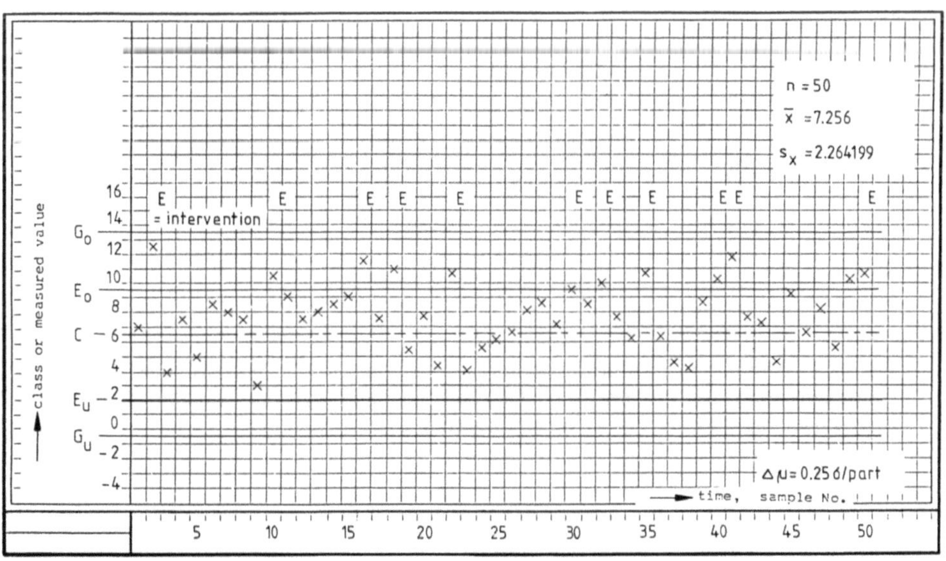

Fig. 15.64. Representation of the process simulated in Fig. 15.63

	without trend	with trend						
				intervention	deviation	correction	difference	
i	x_i	μ_i	x_i	E	A_i	$K_i = -2.5$	$D_i = x_i - x_{i-1}$	
1	6	6.5	6.5					
2	11	7	12	x	6	-2.5 to 4.5	5.5	
3	5	5	4				-5.5	= 4-(12-2.5)
4	8	5.5	7.5				3.5	
5	5	6	5				-2.5	
6	8	6.5	8.5				3.5	
7	7	7	8				-0.5	
8	6	7.5	7.5				-0.5	
9	1	8	3				-4.5	
10	8	8.5	10.5	x	4.5	-2.5 to 6	7.5	
11	8	6.5	8.5				0.5	= 8.5-(10.5-2.5)
12	6	7	7				-1.5	
13	6	7.5	7.5				0.5	
14	6	8	8				0.5	
15	6	8.5	8.5				0.5	
16	8	9	11	x	5	-2.5 to 6.5	2.5	
17	6	7	7				-1.5	= 7-(11-2.5)
18	9	7.5	10.5	x	4.5	-2.5 to 5	3.5	
19	5	5.5	4.5				-3.5	= 4.5-(10.5-2.5)
20	7	6	7				2.5	
21	3	6.5	3.5				-3.5	
22	9	7	10	x	4	-2.5 to 4.5	6.5	
23	4	5	3				-4.5	= 3-(10-2.5) ; and so forth at the following interventions
24	5	5.5	4.5				1.5	
25	5	6	5				0.5	
26	5	6.5	5.5				0.5	
27	6	7	7				1.5	
28	6	7.5	7.5				0.5	
29	4	8	6				-1.5	
30	6	8.5	8.5				2.5	
31	6	9	9				0.5	
32	7	9.5	10.5	x	1.5	-2.5 to 7	1.5	
33	6	7.5	7.5				-0.5	
34	4	8	6				-1.5	
35	8	8.5	10.5	x	4.5	-2.5 to 6	4.5	
36	5	6.5	5.5				-2.5	
37	3	7	4				-1.5	
38	2	7.5	3.5				-0.5	
39	6	8	8				4.5	
40	7	8.5	9.5	x	3.5	-2.5 to 6	1.5	
41	10	6.5	10.5	x	4.5	-2.5 to 4	3.5	
42	8	4.5	6.5				-1.5	
43	7	5	6				-0.5	
44	4	5.5	3.5				-2.5	
45	8	6	8				4.5	
46	5	6.5	5.5				-2.5	
47	6	7	7				1.5	
48	3	7.5	4.5				-2.5	
49	6	8	8				3.5	
50	7	8.5	9.5	x	4.5	-2.5 to 6	1.5	

D_j		n_j
-5.5	/	1
-4.5	//	2
-3.5	//	2
-2.5	/////	5
-1.5	//////	6
-0.5	/////	5
0.5	////////	8
1.5	//////	6
2.5	///	3
3.5	/////	5
4.5	///	3
5.5	/	1
6.5	/	1
7.5	/	1

$n = 49$

$\bar{D} = 0.520408$

$s_D = 2.933214$

$\hat{\sigma} = s_D/\sqrt{2} = 2.074095$

in accordance with figure 14.18

evaluation of controlled x_i-values:

$n = 50$

$\bar{x} = 7.11$

$s_x = 2.330564$

(relative) process dispersion

$$s_p = \frac{s_x}{\hat{\sigma}} = \frac{2.330564}{2.074095} = 1.123654$$

$$\sum K_i = 10\cdot(-2.5) = -25$$

checking:

$$\mu_{50,new} - \mu_0 = 6 - 6 = 0 = 50\Delta\mu + \sum K_i = 25 - 25 = 0$$

Fig. 15.65. Simulation of a process with a trend slope $\Delta\mu = 0.25\sigma$/part. The process is continually controlled with an $n = 1$-QCC. Tolerance $T = 7\sigma$; process capability $c_p = 1.1\bar{6}$; limit values: $G_o = 13$; $G_u = -1$; intervention limits: $E_o = 9$; $E_u = 1.5$

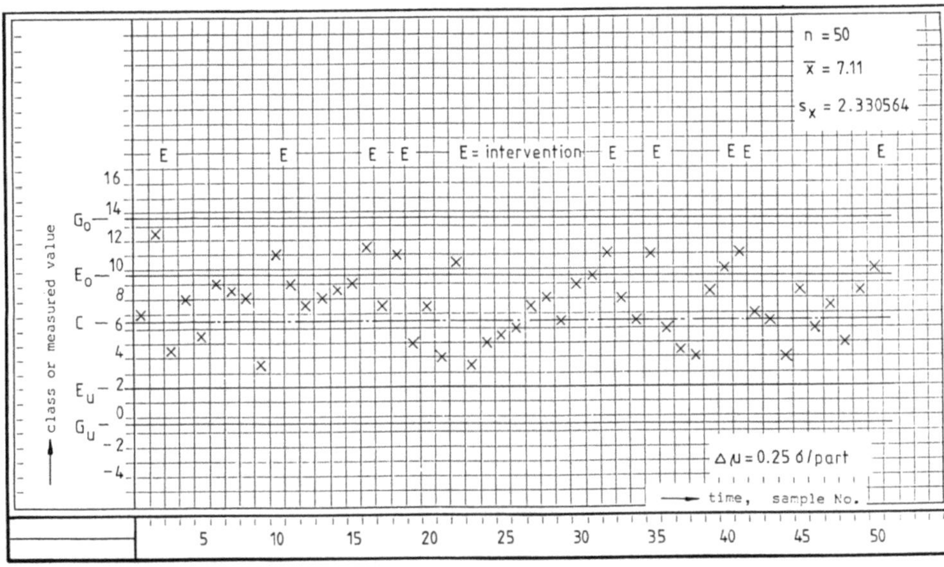

Fig. 15.66. Representation of the process simulated in Fig. 15.65

In summary, according to the size of the process capability and the slope of the trend, process control can be optimised by the appropriate establishment of intervention limits and correction amounts. For the case where automatic process control is available, so as not to disrupt by frequent setting corrections, with trend processes it is recommended that operation is by narrow limits, at least on the process side to which the trend tends. The last example, Figs. 15.63 and 15.64, show that even with a strong trend ($\Delta\mu = 0.25\sigma$/part) production can be controlled so that the total spread lies only about 10% over the (instant) remaining spread.

Frequently it is suggested that every intervention corrects the setting position to a constant correction amount, independent of the size of the "current" deviation. This case is simulated in Fig. 15.65 with representation in Fig. 15.66. The example shows that this can be used with success. The total spread is with $s_x = 2.33$ relatively slight and the relative process spread is at $s_P = 1.124$ only about 12% larger than the remaining spread.

16 Bibliography

Books

1 Böttger, F.: Erzielung von Fertigungsvorteilen durch Anwendung statistischer Gesetze auf die Toleranzberechnung. Diss. TH Aachen 1961
2 DGQ 11–04: Begriffe und Formelzeichen im Bereich der Qualitätssicherung. Berlin: Beuth 1979
3 DGQ 16–30: Qualitätsregelkarten. Berlin: Beuth 1979
4 DGQ 16–43: Stichprobenpläne für quantitative Merkmale. Berlin: Beuth 1982
5 DGQ-SAQ-ÖVQ 16–01: Stichprobenprüfung anhand qualitativer Merkmale. Berlin: Beuth 1982
6 Goubeaud, F.: Berechnung von Toleranzketten über Datenfernverarbeitung. Kernforschungszentrum Karlsruhe, KfK—CAD 88, April 1979
7 Graf, U.; Henning, H.-J.; Stange, K.: Formeln und Tabellen der mathematischen Statistik. Berlin: Springer 1966
8 Kreyszig, E.: Statistische Methoden und ihre Anwendungen. Göttingen: Vandenhoeck & Ruprecht 1968
9 Sachs, L.: Statistische Methoden. 5. Aufl. Berlin: Springer 1982
10 Smirnow, N.W.; Dunin-Barkowski, I.W.: Mathematische Statistik in der Technik. Berlin: VEB Deutscher Verlag der Wissenschaften 1969

Periodical Extracts

11 Deixler, A.: Vorschlag zur Reform der bestehenden Toleranzauffassung statistischer Methoden. Qualitätskontrolle 7 (1962) 45–49
12 Geiger, W.: Einige Gesichtspunkte zur statistischen Tolerierung. Schweiz. Maschinenmarkt 19 (1975), CH 9403 Goldach
13 Geiger, W.: "Grenzwerte" oder die fatale "± Toleranz". DIN-Mitt. 55 (1976) 596–599
14 Göldner, R.: Zur Auswertung kleiner Stichproben im Wahrscheinlichkeitsnetz, QZ 16 (1971) 156–160
15 Goubeaud, F.: Toleranzkopplung in Theorie und Praxis. Eine Möglichkeit zur Senkung der Fertigungskosten. Feinwerktechnik 63 (1959) 133–142
16 Henning, H.J.; Wartmann, R.: Stichproben kleinen Umfangs im Wahrscheinlichkeitsnetz. Mitteilungsbl. math. Statistik 9 (1957) 168–181
17 Schlötel, E.: Toleranzfestlegung unter Berücksichtigung statistischer Gesichtspunkte. Qualitätskontrolle 13 (1968) 111–119
18 Stuhlmann, W.; Schmidt, P.-W.: Statistische Tolerierung als Problem der Fertigung und der Prüfung. Werkstattechnik 55 (1965) 585–589
19 Trumphold, H.; Beck, Ch.: Optimale Toleranzfestlegung unter Berücksichtigung der Maßkettentheorie und der statistischen Eigenschaften des Fertigungsprozesses. Fertigungstech. Betr. 21 (1971) 242–246

20 Wilrich, P.-Th.: Nomogramme zur Ermittlung von Stichprobenplänen für messende Prüfung bei einer einseitig vorgeschriebenen Toleranzgrenze. QZ 15 (1970) 61–65 und 181–187
21 Wilrich, P.-Th.: Qualitätsregelkarten bei vorgegebenen Grenzwerten. QZ 24 (1979) 260–271
22 Zollikofer, O.: Industrie Organisation 20 (1951) 1–8

DIN-Standards and Other Rules and Principles

23 DIN 406 T. 1: Maßeintragung in Zeichnungen; Arten
24 DIN 406 T. 2: Maßeintragung in Zeichnungen; Regeln
25 DIN 1319 T. 3: Grundbegriffe der Meßtechnik; Begriffe für die Meßunsicherheit und für die Beurteilung von Meßgeräten und Meßeinrichtungen
26 DIN 1683 T. 1: Grundrohgußteile aus Stahlguß; Allgemeintoleranzen, Bearbeitungszugaben
27 DIN 6930 T. 2: Stanzteile aus Stahl; Allgemeintoleranzen
28 DIN 7150: ISO-Toleranzen und ISO-Passungen für Längenmaße von 1 bis 500 mm; Einführung
29 DIN 7151: ISO-Grundtoleranzen für Längenmaße von 1 bis 500 mm Nennmaß
30 DIN 7152: Bildung von Toleranzfeldern aus den ISO-Grundabmaßen für Nenmaße von 1 bis 500 mm
31 DIN 7154 T. 1: ISO-Passungen für Einheitsbohrung; Toleranzfelder, Abmaße in µm
32 DIN 7154 T. 2: ISO-Passungen für Einheitswelle; Toleranzfelder, Abmaße in µm
33 DIN 7157: Passungsauswahl; Toleranzfelder, Abmaße, Paßtoleranzen
34 DIN 7168 T. 1: Allgemeintoleranzen; Längen und Winkelmaße
35 DIN 7172 T. 1: ISO-Toleranzen und ISO-Abmaße für Längenmaße über 500 bis 3150 mm; Grundtoleranzen
36 DIN 7182 T. 1: Maße, Abmaße, Toleranzen und Passungen; Grundbegriffe
37 DIN 7186 T. 1: Statistische Tolerierung; Begriffe, Anwendungsrichtlinien und Zeichnungsangaben
38 DIN 7186 T. 2: Statistische Tolerierung; Grundlagen für Rechenverfahren
39 DIN 53804: Statistische Auswertungen (4 Teile)
40 DIN 55302: Statistische Auswertungsverfahren (2 Teile)
41 DIN 55303: Statistische Auswertung von Daten (2 Teile)
42 DIN 55350: Begriffe der Qualitätssicherung und Statistik (12 Teile)
43 DIN ISO 1101: Technische Zeichnungen; Maß-, Form- und Lagetolerierung, Form-, Richtungs-, Orts- und Lauftoleranzen, Allgemeines, Definitionen, Symbole, Zeichnungseintragungen
44 ISO/R 286: ISO-System für Toleranzen und Passungen; Teil 1: Grundlagen, Toleranzen und Abmaße
45 ISO/DIS 3534/2: Statistics—Vocabulary and symbols—Part 2: Statistical quality control
46 VDI/DGQ 3441: Statistische Prüfung der Arbeits- und Positionsgenauigkeit von Werkzeugmaschinen, Grundlagen

Further Reading

Books

1 Anthony, D.M.: Engineering Metrology, Oxford: Pergamon Press, 1986
2 Fuller, W.A.: Measurement Error Models. New York: John Wiley, 1987

3 Oakland, J.S.: Statistical Process Control, proceeding of International Conference. Bedford: IFS Publ. Ltd., June 1987
4 Smith, D.J.: Statistics Workshop, Technics Publication, Introduction to the Tolerancing of functional Length Dimensions, BSI Guidance Document 7309, ISSN 0-580-14541-7, 1985

Periodical Extracts

1 Morillo, G.Z.: Dynamic Metrology: A review on in-situ techniques for controlling workpiece accuracy, "Quality Assurance", The Institute of Quality Assurance (IQA), London, June 1980
2 Schuhmacher, R.B.F.: New Metrology Concepts through Product QC Methods, The American Society for Quality Control, 36 Congress Publication, May 1982
3 Trumphold, H.: The Application of Metrology for Quality Assurance in Automated Production Process, The European Organisation for Quality, Conference Proceedings, Bern, June 1985
4 Pilbrow, A.M.: Independency in Tolerancing, The British Standard Institution News, December 1985
5 Various Authors: Modern organizational and technical problems of the impact of metrology on product quality, 1 st EOQC symposium on Metrology Abstracts, October 1984

17 Appendix

Tables

Table A1. Standardised normal distribution, variable

$$u = \frac{x - \mu}{\sigma}$$

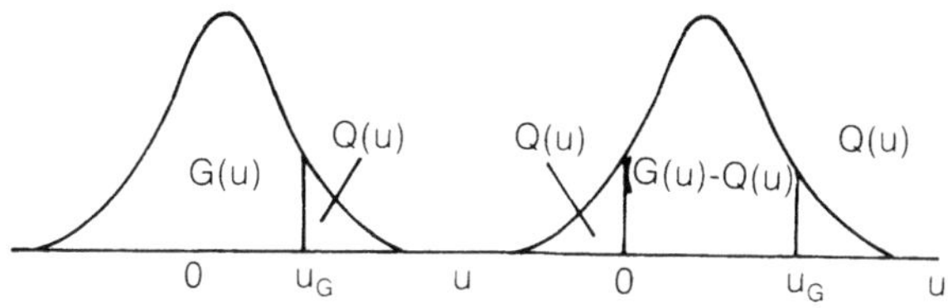

u	$G(u)$	$Q(u)$	$G(u)–Q(u)$	$g(u)$
0.00	0.500 00	0.500 00	0.000 00	0.398 94
0.01	0.503 99	0.496 01	0.007 98	0.398 92
0.02	0.507 98	0.492 02	0.015 96	0.398 86
0.03	0.511 97	0.488 03	0.023 93	0.398 76
0.04	0.515 95	0.484 05	0.031 91	0.398 62
0.05	0.519 94	0.480 06	0.039 88	0.398 44
0.06	0.523 92	0.476 08	0.047 84	0.398 22
0.07	0.527 90	0.472 10	0.055 81	0.397 97
0.08	0.531 88	0.468 12	0.063 76	0.397 67
0.09	0.535 86	0.464 14	0.071 71	0.397 33
0.10	0.539 83	0.460 17	0.079 66	0.396 95
0.11	0.543 80	0.456 20	0.087 59	0.396 54
0.12	0.547 76	0.452 24	0.095 52	0.396 08
0.13	0.551 72	0.448 28	0.103 43	0.395 59
0.14	0.555 67	0.444 33	0.111 34	0.395 05
0.15	0.559 62	0.440 38	0.119 24	0.394 48
0.16	0.563 56	0.436 44	0.127 12	0.393 87
0.17	0.567 49	0.432 51	0.134 99	0.393 22
0.18	0.571 42	0.428 58	0.142 85	0.392 53
0.19	0.575 35	0.424 65	0.150 69	0.391 81
0.20	0.579 26	0.420 74	0.158 52	0.391 04
0.21	0.583 17	0.416 83	0.166 33	0.390 24
0.22	0.587 06	0.412 94	0.174 13	0.389 40
0.23	0.590 95	0.409 05	0.181 91	0.388 53
0.24	0.594 83	0.405 17	0.189 67	0.387 62
0.25	0.598 71	0.401 29	0.197 41	0.386 67
0.26	0.602 57	0.397 43	0.205 14	0.385 68
0.27	0.606 42	0.393 58	0.212 84	0.384 66
0.28	0.610 26	0.389 74	0.220 52	0.383 61
0.29	0.614 09	0.385 91	0.228 18	0.382 51
0.30	0.617 91	0.382 09	0.235 82	0.381 39

u	$G(u)$	$Q(u)$	$G(u)–Q(u)$	$g(u)$
0.31	0.621 72	0.378 28	0.243 44	0.380 23
0.31	0.621 72	0.378 28	0.243 44	0.380 23
0.32	0.625 52	0.374 48	0.251 03	0.379 03
0.33	0.629 30	0.370 70	0.258 60	0.377 80
0.34	0.633 07	0.366 93	0.266 14	0.376 54
0.35	0.636 83	0.363 17	0.273 66	0.375 24
0.36	0.640 58	0.359 42	0.281 15	0.373 91
0.37	0.644 31	0.355 69	0.288 62	0.372 55
0.38	0.648 03	0.351 97	0.296 05	0.371 15
0.39	0.651 73	0.348 27	0.303 46	0.369 73
0.40	0.655 42	0.344 58	0.310 84	0.368 27
0.41	0.659 10	0.340 90	0.318 19	0.366 78
0.42	0.662 76	0.337 24	0.325 51	0.365 26
0.43	0.666 40	0.333 60	0.332 80	0.363 71
0.44	0.670 03	0.329 97	0.340 06	0.362 13
0.45	0.673 64	0.326 36	0.347 29	0.360 53
0.46	0.677 24	0.322 76	0.354 48	0.358 89
0.47	0.680 82	0.319 18	0.361 65	0.357 23
0.48	0.684 39	0.315 61	0.368 77	0.355 53
0.49	0.687 93	0.312 07	0.375 87	0.353 81
0.50	0.691 46	0.308 54	0.382 93	0.352 07
0.51	0.694 97	0.305 03	0.389 95	0.350 29
0.52	0.698 47	0.301 53	0.396 94	0.348 49
0.53	0.701 94	0.298 06	0.403 89	0.346 67
0.54	0.705 40	0.294 60	0.410 80	0.344 82
0.55	0.708 84	0.291 16	0.417 68	0.342 94
0.56	0.712 26	0.287 74	0.424 52	0.341 05
0.57	0.715 66	0.284 34	0.431 32	0.339 12
0.58	0.719 04	0.280 96	0.438 09	0.337 18
0.59	0.722 40	0.277 60	0.444 81	0.335 21
0.60	0.725 75	0.274 25	0.451 49	0.333 22

Table A1. (Continued)

u	$G(u)$	$Q(u)$	$G(u)-Q(u)$	$g(u)$
0.61	0.729 07	0.270 93	0.458 14	0.331 21
0.62	0.732 37	0.267 63	0.464 74	0.329 18
0.63	0.735 65	0.264 35	0.471 31	0.327 13
0.64	0.738 91	0.261 09	0.477 83	0.325 06
0.65	0.742 15	0.257 85	0.484 31	0.322 97
0.66	0.745 37	0.254 63	0.490 75	0.320 86
0.67	0.748 57	0.251 43	0.497 14	0.318 74
0.68	0.751 75	0.248 25	0.503 50	0.316 59
0.69	0.754 90	0.245 10	0.509 81	0.314 43
0.70	0.758 04	0.241 96	0.516 07	0.312 25
0.71	0.761 15	0.238 85	0.522 30	0.310 06
0.72	0.764 24	0.235 76	0.528 48	0.307 85
0.73	0.767 30	0.232 70	0.534 61	0.305 63
0.74	0.770 35	0.229 65	0.540 70	0.303 39
0.75	0.773 37	0.226 63	0.546 75	0.301 14
0.76	0.776 37	0.223 63	0.552 75	0.298 87
0.77	0.779 35	0.220 65	0.558 70	0.296 59
0.78	0.782 30	0.217 70	0.564 61	0.294 31
0.79	0.785 24	0.214 76	0.570 47	0.292 00
0.80	0.788 14	0.211 85	0.576 29	0.289 69
0.81	0.791 03	0.208 97	0.582 06	0.287 37
0.82	0.793 89	0.206 11	0.587 78	0.285 04
0.83	0.796 73	0.203 27	0.593 46	0.282 69
0.84	0.799 55	0.200 45	0.599 09	0.280 34
0.85	0.802 34	0.197 66	0.604 68	0.277 98
0.86	0.805 11	0.194 89	0.610 21	0.275 62
0.87	0.807 85	0.192 15	0.615 70	0.273 24
0.88	0.810 57	0.189 43	0.621 14	0.270 86
0.89	0.813 27	0.186 73	0.626 53	0.268 48
0.90	0.815 94	0.184 06	0.631 88	0.266 09
0.91	0.818 59	0.181 41	0.637 18	0.263 69
0.92	0.821 21	0.178 79	0.642 43	0.261 29
0.93	0.823 81	0.176 19	0.647 63	0.258 88
0.94	0.826 39	0.173 61	0.652 78	0.256 47
0.95	0.828 94	0.171 06	0.657 89	0.254 06
0.96	0.831 47	0.168 53	0.662 94	0.251 64
0.97	0.833 98	0.166 02	0.667 95	0.249 23
0.98	0.836 46	0.163 54	0.672 91	0.246 81
0.99	0.838 91	0.161 09	0.677 83	0.244 39
1.00	0.841 34	0.158 66	0.682 69	0.241 97
1.01	0.843 75	0.156 25	0.687 50	0.239 55
1.02	0.846 14	0.153 86	0.692 27	0.237 13
1.03	0.848 50	0.151 51	0.696 99	0.234 71
1.04	0.850 83	0.149 17	0.701 66	0.232 30
1.05	0.853 14	0.146 86	0.706 28	0.229 88
1.06	0.855 43	0.144 57	0.710 86	0.227 47
1.07	0.857 69	0.142 38	0.715 38	0.225 06
1.08	0.859 93	0.140 07	0.719 86	0.222 65
1.09	0.862 14	0.137 86	0.724 29	0.220 25
1.10	0.864 33	0.135 67	0.728 67	0.217 85
1.11	0.866 50	0.133 50	0.733 00	0.215 46
1.12	0.868 64	0.131 36	0.737 29	0.213 07
1.13	0.870 76	0.129 24	0.741 52	0.210 69
1.14	0.872 86	0.127 14	0.745 71	0.208 31
1.15	0.874 93	0.125 07	0.749 86	0.205 94
1.16	0.876 98	0.123 02	0.753 95	0.203 57
1.17	0.879 00	0.121 00	0.758 00	0.201 21
1.18	0.881 00	0.119 00	0.762 00	0.198 86
1.19	0.882 98	0.117 02	0.765 95	0.196 52
1.20	0.884 93	0.115 07	0.769 86	0.194 19
1.21	0.886 86	0.113 14	0.773 72	0.191 86
1.22	0.888 77	0.111 23	0.777 54	0.189 54
1.23	0.890 65	0.109 35	0.781 30	0.187 24
1.24	0.892 51	0.107 49	0.785 02	0.184 94
1.25	0.894 35	0.105 65	0.788 70	0.182 65
1.26	0.896 17	0.103 83	0.792 33	0.180 37
1.27	0.897 96	0.102 04	0.795 92	0.178 10
1.28	0.899 73	0.100 27	0.799 45	0.175 85
1.29	0.901 47	0.098 53	0.802 95	0.173 60
1.30	0.903 20	0.096 80	0.806 40	0.171 37
1.31	0.904 90	0.095 10	0.809 80	0.169 15
1.32	0.906 58	0.093 42	0.813 17	0.166 94
1.33	0.908 24	0.091 76	0.816 48	0.164 74
1.34	0.909 88	0.090 12	0.819 75	0.162 56
1.35	0.911 49	0.088 51	0.822 98	0.160 38
1.36	0.913 09	0.086 92	0.826 17	0.158 22
1.37	0.914 66	0.085 34	0.829 31	0.156 08
1.38	0.916 21	0.083 79	0.832 41	0.153 95
1.39	0.917 74	0.082 26	0.835 47	0.151 83
1.40	0.919 24	0.080 76	0.838 49	0.149 73
1.41	0.920 73	0.079 27	0.841 46	0.147 64
1.42	0.922 20	0.077 80	0.844 39	0.145 56
1.43	0.923 64	0.076 36	0.847 28	0.143 50
1.44	0.925 07	0.074 93	0.850 13	0.141 46
1.45	0.926 47	0.073 53	0.852 94	0.139 43
1.46	0.927 86	0.072 15	0.855 71	0.137 42
1.47	0.929 22	0.070 78	0.858 44	0.135 42
1.48	0.930 56	0.069 44	0.861 13	0.133 44
1.49	0.931 89	0.068 11	0.863 78	0.131 47
1.50	0.933 19	0.066 81	0.866 39	0.129 52

Table A1. (Continued)

u	$G(u)$	$Q(u)$	$G(u)-Q(u)$	$g(u)$
1.51	0.934 48	0.065 52	0.868 96	0.127 58
1.52	0.935 74	0.064 26	0.871 49	0.125 66
1.53	0.936 99	0.063 01	0.873 98	0.123 76
1.54	0.938 22	0.061 78	0.876 44	0.121 88
1.55	0.939 43	0.060 57	0.878 86	0.120 01
1.56	0.940 62	0.059 38	0.881 24	0.118 16
1.57	0.941 79	0.058 21	0.883 58	0.116 32
1.58	0.942 95	0.057 05	0.885 89	0.114 50
1.59	0.944 08	0.055 92	0.888 17	0.112 70
1.60	0.945 20	0.054 80	0.890 40	0.110 92
1.61	0.946 30	0.053 70	0.892 60	0.109 15
1.62	0.947 38	0.052 62	0.894 77	0.107 41
1.63	0.948 45	0.051 55	0.896 90	0.105 67
1.64	0.949 50	0.050 50	0.898 99	0.103 96
1.65	0.950 53	0.049 47	0.901 06	0.102 26
1.66	0.951 54	0.048 46	0.903 09	0.100 59
1.67	0.952 54	0.047 46	0.905 08	0.098 92
1.68	0.953 52	0.046 48	0.907 04	0.097 28
1.69	0.954 49	0.045 51	0.908 97	0.095 66
1.70	0.955 43	0.044 57	0.910 87	0.094 05
1.71	0.956 37	0.043 63	0.912 73	0.092 46
1.72	0.957 28	0.042 72	0.914 57	0.090 89
1.73	0.958 18	0.041 82	0.916 37	0.089 33
1.74	0.959 07	0.040 93	0.918 14	0.087 80
1.75	0.959 94	0.040 06	0.919 88	0.086 28
1.76	0.960 80	0.039 20	0.921 59	0.084 78
1.77	0.961 64	0.038 36	0.923 27	0.083 29
1.78	0.962 46	0.037 54	0.924 92	0.081 83
1.79	0.963 27	0.036 73	0.926 55	0.080 38
1.80	0.964 07	0.035 93	0.928 14	0.078 95
1.81	0.964 85	0.035 15	0.929 70	0.077 54
1.82	0.965 62	0.034 38	0.931 24	0.076 14
1.83	0.966 38	0.033 63	0.932 75	0.074 77
1.84	0.967 12	0.032 88	0.934 23	0.073 41
1.85	0.967 84	0.032 16	0.935 69	0.072 06
1.86	0.968 56	0.031 44	0.937 11	0.070 74
1.87	0.969 26	0.030 74	0.938 52	0.069 43
1.88	0.969 95	0.030 05	0.939 89	0.068 14
1.89	0.970 62	0.029 38	0.941 24	0.066 87
1.90	0.971 28	0.028 72	0.942 57	0.065 62
1.91	0.971 93	0.028 07	0.943 87	0.064 38
1.92	0.972 57	0.027 43	0.945 14	0.063 16
1.93	0.973 20	0.026 80	0.946 39	0.061 95
1.94	0.973 81	0.026 19	0.947 62	0.060 77
1.95	0.974 41	0.025 59	0.948 82	0.059 59
1.96	0.975 00	0.025 00	0.950 00	0.058 44
1.97	0.975 58	0.024 42	0.951 16	0.057 30
1.98	0.976 15	0.023 85	0.952 30	0.056 18
1.99	0.976 70	0.023 30	0.953 41	0.055 08
2.00	0.977 25	0.022 75	0.954 50	0.053 99
2.01	0.977 78	0.022 22	0.955 57	0.052 92
2.02	0.978 31	0.021 69	0.956 62	0.051 86
2.03	0.978 82	0.021 18	0.957 64	0.050 82
2.04	0.979 32	0.020 68	0.958 65	0.049 80
2.05	0.979 82	0.020 18	0.959 64	0.048 79
2.06	0.980 30	0.019 70	0.960 60	0.047 80
2.07	0.980 77	0.019 23	0.961 55	0.046 82
2.08	0.981 24	0.018 76	0.962 47	0.044 86
2.09	0.981 69	0.018 31	0.963 38	0.044 91
2.10	0.982 14	0.017 86	0.964 27	0.043 98
2.11	0.982 57	0.017 43	0.965 14	0.043 07
2.12	0.983 00	0.017 00	0.965 99	0.042 17
2.13	0.983 41	0.016 59	0.966 83	0.041 28
2.14	0.983 82	0.016 18	0.967 65	0.040 41
2.15	0.984 22	0.015 78	0.968 44	0.039 55
2.16	0.984 61	0.015 39	0.969 23	0.038 71
2.17	0.985 00	0.015 00	0.969 99	0.037 88
2.18	0.985 37	0.014 63	0.970 74	0.037 06
2.19	0.985 74	0.014 26	0.971 48	0.036 26
2.20	0.986 10	0.013 90	0.972 19	0.035 47
2.21	0.986 45	0.013 55	0.972 89	0.034 70
2.22	0.986 79	0.013 21	0.973 58	0.033 94
2.23	0.987 13	0.012 87	0.974 25	0.033 19
2.24	0.987 45	0.012 55	0.974 91	0.032 46
2.25	0.987 78	0.012 22	0.975 55	0.031 74
2.26	0.988 09	0.011 91	0.976 18	0.031 03
2.27	0.988 40	0.011 60	0.976 79	0.030 34
2.28	0.988 70	0.011 30	0.977 39	0.029 65
2.29	0.988 99	0.011 01	0.977 98	0.028 98
2.30	0.989 28	0.010 72	0.978 55	0.028 33
2.31	0.989 56	0.010 44	0.979 11	0.027 68
2.32	0.989 83	0.010 17	0.979 66	0.027 05
2.33	0.990 10	0.009 90	0.980 19	0.026 43
2.34	0.990 36	0.009 64	0.980 72	0.025 82
2.35	0.990 61	0.009 39	0.981 23	0.025 22
2.36	0.990 86	0.009 14	0.981 73	0.024 63
2.37	0.991 11	0.008 89	0.982 21	0.024 06
2.38	0.991 34	0.008 66	0.982 69	0.023 49
2.39	0.991 58	0.008 42	0.983 15	0.022 94
2.40	0.991 80	0.008 20	0.983 61	0.022 39

Table A1. (Continued)

u	$G(u)$	$Q(u)$	$G(u)-Q(u)$	$g(u)$
2.41	0.992 02	0.007 98	0.984 05	0.021 86
2.42	0.992 24	0.007 76	0.984 48	0.021 34
2.43	0.992 45	0.007 55	0.984 90	0.020 83
2.44	0.992 66	0.007 34	0.985 31	0.020 33
2.45	0.992 86	0.007 14	0.985 71	0.019 84
2.46	0.993 05	0.006 95	0.986 11	0.019 36
2.47	0.993 24	0.006 76	0.986 49	0.018 89
2.48	0.993 43	0.006 57	0.986 86	0.018 42
2.49	0.993 61	0.006 39	0.987 23	0.017 97
2.50	0.993 79	0.006 21	0.987 59	0.017 53
2.51	0.993 96	0.006 04	0.987 93	0.017 09
2.52	0.994 13	0.005 87	0.988 26	0.016 67
2.53	0.994 30	0.005 70	0.988 59	0.016 25
2.54	0.994 46	0.005 54	0.988 91	0.015 85
2.55	0.994 61	0.005 39	0.989 23	0.015 45
2.56	0.994 77	0.005 23	0.989 53	0.015 06
2.57	0.994 92	0.005 08	0.989 83	0.014 68
2.58	0.995 06	0.004 94	0.990 12	0.014 30
2.59	0.995 20	0.004 80	0.990 40	0.013 94
2.60	0.995 34	0.004 66	0.990 68	0.013 58
2.61	0.995 47	0.004 53	0.990 95	0.013 23
2.62	0.995 60	0.004 40	0.991 21	0.012 89
2.63	0.995 73	0.004 27	0.991 46	0.012 56
2.64	0.995 85	0.004 15	0.991 71	0.012 23
2.65	0.995 98	0.004 04	0.991 95	0.011 91
2.66	0.996 09	0.003 91	0.992 19	0.011 60
2.67	0.996 21	0.003 79	0.992 41	0.011 30
2.68	0.996 32	0.003 68	0.992 64	0.011 00
2.69	0.996 43	0.003 57	0.992 85	0.010 71
2.70	0.996 53	0.003 47	0.993 07	0.010 42
2.71	0.996 64	0.003 36	0.993 27	0.010 14
2.72	0.996 74	0.003 26	0.993 47	0.009 87
2.73	0.996 83	0.003 17	0.993 67	0.009 61
2.74	0.996 93	0.003 07	0.993 86	0.009 35
2.75	0.997 02	0.002 98	0.994 04	0.009 09
2.76	0.997 11	0.002 89	0.994 22	0.008 85
2.77	0.997 20	0.002 80	0.994 39	0.008 61
2.78	0.997 28	0.002 72	0.994 56	0.008 37
2.79	0.997 36	0.002 64	0.994 73	0.008 14
2.80	0.997 44	0.002 56	0.994 89	0.007 92
2.81	0.997 52	0.002 48	0.995 05	0.007 70
2.82	0.997 60	0.002 40	0.995 20	0.007 48
2.83	0.997 67	0.002 33	0.995 35	0.007 27
2.84	0.997 74	0.002 26	0.995 49	0.007 07
2.85	0.997 81	0.002 19	0.995 63	0.006 87
2.86	0.997 88	0.002 12	0.995 76	0.006 68
2.87	0.997 95	0.002 05	0.995 90	0.006 49
2.88	0.998 01	0.001 99	0.996 02	0.006 31
2.89	0.998 07	0.001 93	0.996 15	0.006 13
2.90	0.998 13	0.001 87	0.996 27	0.005 95
2.91	0.998 19	0.001 81	0.996 39	0.005 78
2.92	0.998 25	0.001 75	0.996 50	0.005 62
2.93	0.998 31	0.001 69	0.996 61	0.005 45
2.94	0.998 36	0.001 64	0.996 72	0.005 30
2.95	0.998 41	0.001 59	0.996 82	0.005 14
2.96	0.998 46	0.001 54	0.996 92	0.004 99
2.97	0.998 51	0.001 49	0.997 02	0.004 85
2.98	0.998 56	0.001 44	0.997 12	0.004 71
2.99	0.998 61	0.001 39	0.997 21	0.004 57
3.00	0.998 65	0.001 35	0.997 30	0.004 43
3.01	0.998 69	0.001 31	0.997 39	0.004 30
3.02	0.998 74	0.001 26	0.997 47	0.004 17
3.03	0.998 78	0.001 22	0.997 55	0.004 05
3.04	0.998 82	0.001 18	0.997 63	0.003 93
3.05	0.998 86	0.001 14	0.997 71	0.003 81
3.06	0.998 89	0.001 11	0.997 79	0.003 70
3.07	0.998 93	0.001 07	0.997 86	0.003 58
3.08	0.998 97	0.001 04	0.997 93	0.003 48
3.09	0.999 00	0.001 00	0.998 00	0.003 37
3.10	0.999 03	0.000 97	0.998 06	0.003 27
3.11	0.999 06	0.000 94	0.998 13	0.003 17
3.12	0.999 10	0.000 90	0.998 19	0.003 07
3.13	0.999 13	0.000 87	0.998 25	0.002 98
3.14	0.999 16	0.000 84	0.998 31	0.002 88
3.15	0.999 18	0.000 82	0.998 37	0.002 79
3.16	0.999 21	0.000 79	0.998 42	0.002 71
3.17	0.999 24	0.000 76	0.998 48	0.002 62
3.18	0.999 26	0.000 74	0.998 53	0.002 54
3.19	0.999 29	0.000 71	0.998 58	0.002 46
3.20	0.999 31	0.000 69	0.998 63	0.002 38
3.21	0.999 34	0.000 66	0.998 67	0.002 31
3.22	0.999 36	0.000 64	0.998 72	0.002 24
3.23	0.999 38	0.000 62	0.998 76	0.002 16
3.24	0.999 40	0.000 60	0.998 80	0.002 10
3.25	0.999 42	0.000 58	0.998 85	0.002 03
3.26	0.999 44	0.000 56	0.998 89	0.001 96
3.27	0.999 46	0.000 54	0.998 92	0.001 90
3.28	0.999 48	0.000 52	0.998 96	0.001 84
3.29	0.999 50	0.000 50	0.999 00	0.001 78
3.30	0.999 52	0.000 48	0.999 03	0.001 72

Table A1. (Continued)

u	$G(u)$	$Q(u)$	$G(u)-Q(u)$	$g(u)$
3.31	0.999 53	0.000 47	0.999 07	0.001 67
3.32	0.999 55	0.000 45	0.999 10	0.001 61
3.33	0.999 57	0.000 43	0.999 13	0.001 56
3.34	0.999 58	0.000 42	0.999 16	0.001 51
3.35	0.999 60	0.000 40	0.999 19	0.001 46
3.36	0.999 61	0.000 39	0.999 22	0.001 41
3.37	0.999 62	0.000 38	0.999 25	0.001 36
3.38	0.999 64	0.000 36	0.999 28	0.001 32
3.39	0.999 65	0.000 35	0.999 30	0.001 27
3.40	0.999 66	0.000 34	0.999 33	0.001 23
3.41	0.999 68	0.000 32	0.999 35	0.001 19
3.42	0.999 69	0.000 31	0.999 37	0.001 15
3.43	0.999 70	0.000 30	0.999 40	0.001 11
3.44	0.999 71	0.000 29	0.999 42	0.001 07
3.45	0.999 72	0.000 28	0.999 44	0.001 04
3.46	0.999 73	0.000 27	0.999 46	0.001 00
3.47	0.999 74	0.000 26	0.999 48	0.000 97
3.48	0.999 75	0.000 25	0.999 50	0.000 94
3.49	0.999 76	0.000 24	0.999 52	0.000 90
3.50	0.999 77	0.000 23	0.999 53	0.000 87
3.51	0.999 78	0.000 22	0.999 55	0.000 84
3.52	0.999 78	0.000 22	0.999 57	0.000 81
3.53	0.999 79	0.000 21	0.999 58	0.000 79
3.54	0.999 80	0.000 20	0.999 60	0.000 76
3.55	0.999 81	0.000 19	0.999 61	0.000 73
3.56	0.999 81	0.000 19	0.999 63	0.000 71
3.57	0.999 82	0.000 18	0.999 64	0.000 68
3.58	0.999 83	0.000 17	0.999 66	0.000 66
3.59	0.999 83	0.000 17	0.999 67	0.000 63
3.60	0.999 84	0.000 16	0.999 68	0.000 61
3.61	0.999 85	0.000 15	0.999 69	0.000 59
3.62	0.999 85	0.000 15	0.999 71	0.000 57
3.63	0.999 86	0.000 14	0.999 72	0.000 55
3.64	0.999 86	0.000 14	0.999 73	0.000 53
3.65	0.999 87	0.000 13	0.999 74	0.000 51

u	$G(u)$	$Q(u)$	$G(u)-Q(u)$	$g(u)$
3.66	0.999 87	0.000 13	0.999 75	0.000 49
3.67	0.999 88	0.000 12	0.999 76	0.000 47
3.68	0.999 88	0.000 12	0.999 77	0.000 46
3.69	0.999 89	0.000 11	0.999 78	0.000 44
3.70	0.999 89	0.000 11	0.999 78	0.000 42
3.71	0.999 90	0.000 10	0.999 79	0.000 41
3.72	0.999 90	0.000 10	0.999 80	0.000 39
3.73	0.999 90	0.000 10	0.999 81	0.000 38
3.74	0.999 91	0.000 09	0.999 82	0.000 37
3.75	0.999 91	0.000 09	0.999 82	0.000 35
3.76	0.999 92	0.000 09	0.999 83	0.000 34
3.77	0.999 92	0.000 08	0.999 84	0.000 33
3.78	0.999 92	0.000 08	0.999 84	0.000 31
3.79	0.999 92	0.000 08	0.999 85	0.000 30
3.80	0.999 93	0.000 07	0.999 86	0.000 29
3.81	0.999 93	0.000 07	0.999 86	0.000 28
3.82	0.999 93	0.000 07	0.999 87	0.000 27
3.83	0.999 94	0.000 06	0.999 87	0.000 26
3.84	0.999 94	0.000 06	0.999 88	0.000 25
3.85	0.999 94	0.000 06	0.999 88	0.000 24
3.86	0.999 94	0.000 06	0.999 89	0.000 23
3.87	0.999 95	0.000 05	0.999 89	0.000 22
3.88	0.999 95	0.000 05	0.999 90	0.000 21
3.89	0.999 95	0.000 05	0.999 90	0.000 21
3.90	0.999 95	0.000 05	0.999 90	0.000 20
3.91	0.999 95	0.000 05	0.999 91	0.000 19
3.92	0.999 96	0.000 04	0.999 91	0.000 18
3.93	0.999 96	0.000 04	0.999 92	0.000 18
3.94	0.999 96	0.000 04	0.999 92	0.000 17
3.95	0.999 96	0.000 04	0.999 92	0.000 16
3.96	0.999 96	0.000 04	0.999 93	0.000 16
3.97	0.999 96	0.000 04	0.999 93	0.000 15
3.98	0.999 97	0.000 03	0.999 93	0.000 14
3.99	0.999 97	0.000 03	0.999 93	0.000 14
4.00	0.999 97	0.000 03	0.999 94	0.000 13

Approximation for $u > 4$:

$$g(u) = \frac{1}{\sqrt{2\pi}} e^{-u^2/2}$$

$$Q(u) = \frac{g(u)}{u} \left\{ 1 - \frac{1}{u^2} + \frac{3}{u^4} - \frac{15}{u^6} + \frac{105}{u^8} \right\}$$

G	Q	u	G	Q	u
0.50	0.50	0	0.995	0.005	2.575 8
0.60	0.40	0.253 3	0.997 5	0.002 5	2.807 0
0.70	0.30	0.524 4	0.999 0	0.001 0	3.090 2
0.80	0.20	0.841 6	0.999 5	0.000 5	3.290 5
0.90	0.10	1.281 6	0.999 9	0.000 1	3.719 0
0.95	0.05	1.644 9	0.999 95	0.000 05	3.890 6
0.975	0.025	1.960 0	0.999 99	0.000 01	4.264 9
0.990	0.010	2.326 3			

Table A2. Factors for the estimation of standard deviation of total populations $\hat{\sigma} = \bar{s}/a_n$ or $\hat{\sigma} = \bar{R}/d_n$

n	a_n	d_n
2	0.789	1.128
3	0.886	1.693
4	0.921	2.059
5	0.940	2.326
6	0.952	2.534
7	0.959	2.704
8	0.965	2.847
9	0.969	2.970
10	0.973	3.078
11	0.975	3.173
12	0.978	3.258
13	0.979	3.336
14	0.981	3.407
15	0.982	3.472
16	0.984	3.532
17	0.985	3.588
18	0.985	3.640
19	0.986	3.689
20	0.987	3.735

Table A3. Probability sums for insertion of x-th values of an ordered sample test with range n in the probability grid (probability sums of the relevant expected values)

x	$n =$ 2	3	4	5	6	7	8	9	10	11	12	13	x
1	0.286	0.199	0.152	0.122	0.102	0.089	0.078	0.068	0.062	0.056	0.052	0.048	1
2	0.714	0.500	0.383	0.310	0.261	0.244	0.198	0.176	0.159	0.145	0.131	0.123	2
3		0.801	0.617	0.500	0.421	0.363	0.319	0.284	0.255	0.233	0.215	0.198	3
4			0.849	0.690	0.579	0.500	0.440	0.394	0.352	0.323	0.295	0.274	4
5				0.878	0.739	0.637	0.560	0.500	0.452	0.413	0.378	0.348	5
6					0.898	0.776	0.681	0.606	0.548	0.500	0.460	0.425	6
7						0.912	0.802	0.716	0.648	0.587	0.540	0.500	7
8							0.922	0.824	0.745	0.677	0.622	0.575	8
9								0.932	0.841	0.767	0.705	0.652	9
10									0.938	0.855	0.785	0.726	10
11										0.944	0.869	0.802	11
12											0.949	0.877	12
13												0.953	13

Table A3. (Continued)

x	$n =$ 14	15	16	17	18	19	20	21	22	23	24	25	x
1	0.045	0.041	0.039	0.037	0.034	0.033	0.031	0.029	0.028	0.027	0.026	0.024	1
2	0.113	0.106	0.100	0.093	0.089	0.084	0.079	0.076	0.072	0.069	0.067	0.064	2
3	0.184	0.171	0.161	0.152	0.142	0.136	0.129	0.123	0.117	0.113	0.107	0.104	3
4	0.255	0.239	0.224	0.209	0.198	0.187	0.179	0.171	0.164	0.156	0.149	0.142	4
5	0.323	0.302	0.284	0.268	0.251	0.239	0.227	0.218	0.206	0.198	0.189	0.181	5
6	0.394	0.367	0.348	0.326	0.309	0.291	0.278	0.264	0.251	0.242	0.233	0.224	6
7	0.464	0.433	0.409	0.382	0.363	0.345	0.326	0.312	0.298	0.284	0.274	0.261	7
8	0.536	0.500	0.468	0.440	0.417	0.397	0.378	0.359	0.341	0.326	0.316	0.302	8
9	0.606	0.567	0.532	0.500	0.472	0.448	0.425	0.405	0.386	0.371	0.356	0.341	9
10	0.677	0.633	0.591	0.560	0.528	0.500	0.476	0.452	0.433	0.413	0.397	0.382	10
11	0.745	0.698	0.652	0.618	0.583	0.552	0.524	0.500	0.476	0.456	0.436	0.421	11
12	0.816	0.761	0.716	0.674	0.637	0.603	0.575	0.548	0.524	0.500	0.480	0.460	12
13	0.887	0.829	0.776	0.732	0.691	0.655	0.622	0.595	0.567	0.544	0.520	0.500	13
14	0.955	0.894	0.839	0.791	0.749	0.709	0.674	0.641	0.614	0.587	0.564	0.540	14
15		0.959	0.900	0.848	0.802	0.761	0.722	0.688	0.659	0.629	0.603	0.579	15
16			0.961	0.907	0.858	0.813	0.773	0.736	0.702	0.674	0.644	0.618	16
17				0.963	0.912	0.864	0.821	0.782	0.749	0.716	0.684	0.659	17
18					0.966	0.916	0.871	0.829	0.794	0.758	0.726	0.698	18
19						0.967	0.921	0.877	0.836	0.802	0.767	0.739	19
20							0.969	0.924	0.883	0.844	0.811	0.776	20
21								0.971	0.928	0.887	0.851	0.819	21
22									0.972	0.931	0.893	0.858	22
23										0.973	0.933	0.896	23
24											0.974	0.936	24
25												0.976	25

Table A3. (Continued)

x	n = 26	27	28	29	30	31	32	33	34	35	36	37	x
1	0.024	0.023	0.022	0.021	0.021	0.020	0.019	0.019	0.018	0.018	0.017	0.017	1
2	0.062	0.059	0.057	0.055	0.053	0.051	0.050	0.048	0.047	0.045	0.044	0.043	2
3	0.099	0.095	0.092	0.089	0.087	0.083	0.081	0.078	0.076	0.074	0.072	0.070	3
4	0.138	0.134	0.127	0.123	0.119	0.115	0.112	0.108	0.105	0.102	0.099	0.097	4
5	0.176	0.169	0.164	0.159	0.152	0.148	0.143	0.139	0.135	0.131	0.127	0.124	5
6	0.215	0.206	0.198	0.192	0.187	0.180	0.174	0.169	0.164	0.159	0.155	0.151	6
7	0.251	0.242	0.233	0.227	0.218	0.212	0.205	0.199	0.193	0.188	0.182	0.177	7
8	0.291	0.281	0.271	0.261	0.251	0.244	0.236	0.229	0.222	0.216	0.210	0.204	8
9	0.330	0.316	0.305	0.295	0.284	0.276	0.267	0.259	0.251	0.244	0.238	0.231	9
10	0.367	0.352	0.341	0.330	0.319	0.308	0.298	0.289	0.281	0.273	0.265	0.258	10
11	0.405	0.390	0.374	0.363	0.352	0.340	0.329	0.319	0.310	0.301	0.293	0.285	11
12	0.444	0.425	0.413	0.397	0.386	0.371	0.360	0.349	0.339	0.330	0.321	0.312	12
13	0.480	0.464	0.448	0.433	0.417	0.404	0.391	0.379	0.368	0.358	0.348	0.339	13
14	0.520	0.500	0.484	0.464	0.452	0.436	0.422	0.410	0.398	0.386	0.376	0.366	14
15	0.556	0.536	0.516	0.500	0.484	0.468	0.453	0.440	0.427	0.415	0.403	0.392	15
16	0.595	0.575	0.552	0.536	0.516	0.500	0.484	0.470	0.456	0.443	0.431	0.419	16
17	0.633	0.610	0.587	0.567	0.548	0.532	0.516	0.500	0.485	0.472	0.459	0.446	17
18	0.670	0.648	0.626	0.603	0.583	0.564	0.547	0.530	0.515	0.500	0.486	0.473	18
19	0.709	0.684	0.659	0.637	0.614	0.596	0.578	0.560	0.544	0.528	0.514	0.500	19
20	0.749	0.719	0.695	0.670	0.648	0.628	0.609	0.590	0.573	0.557	0.541	0.527	20
21	0.785	0.758	0.729	0.705	0.681	0.660	0.640	0.621	0.602	0.585	0.569	0.554	21
22	0.824	0.794	0.767	0.739	0.716	0.692	0.671	0.651	0.632	0.614	0.597	0.581	22
23	0.862	0.831	0.802	0.773	0.749	0.724	0.702	0.681	0.661	0.642	0.624	0.608	23
24	0.902	0.866	0.836	0.808	0.782	0.756	0.733	0.711	0.690	0.670	0.652	0.634	24
25	0.938	0.905	0.873	0.841	0.813	0.788	0.764	0.741	0.719	0.699	0.679	0.661	25
26	0.976	0.941	0.908	0.877	0.848	0.820	0.795	0.771	0.749	0.727	0.707	0.688	26
27		0.977	0.943	0.912	0.881	0.852	0.826	0.801	0.778	0.756	0.735	0.715	27
28			0.978	0.945	0.913	0.885	0.857	0.831	0.807	0.784	0.762	0.742	28
29				0.979	0.947	0.917	0.888	0.861	0.836	0.812	0.790	0.769	29
30					0.979	0.949	0.919	0.892	0.865	0.841	0.818	0.796	30
31						0.980	0.950	0.922	0.895	0.869	0.845	0.823	31
32							0.981	0.952	0.924	0.898	0.873	0.849	32
33								0.981	0.953	0.926	0.901	0.876	33
34									0.982	0.955	0.928	0.903	34
35										0.982	0.956	0.930	35
36											0.983	0.957	36
37												0.983	37

Table A3. (Continued)

x	$n =$ 38	39	40	41	42	43	44	45	46	47	48	49	50	x
1	0.016	0.016	0.015	0.015	0.015	0.014	0.014	0.014	0.013	0.013	0.013	0.013	0.012	1
2	0.042	0.041	0.040	0.039	0.038	0.037	0.036	0.035	0.035	0.034	0.033	0.032	0.032	2
3	0.068	0.066	0.065	0.063	0.062	0.060	0.059	0.057	0.056	0.055	0.054	0.053	0.052	3
4	0.094	0.092	0.090	0.087	0.085	0.083	0.081	0.080	0.078	0.076	0.075	0.073	0.072	4
5	0.120	0.117	0.114	0.112	0.109	0.106	0.104	0.102	0.100	0.097	0.095	0.093	0.092	5
6	0.147	0.143	0.139	0.136	0.133	0.130	0.127	0.124	0.121	0.119	0.116	0.114	0.112	6
7	0.173	0.168	0.164	0.160	0.156	0.153	0.149	0.146	0.143	0.140	0.137	0.134	0.131	7
8	0.199	0.194	0.189	0.185	0.180	0.176	0.172	0.168	0.165	0.161	0.158	0.155	0.151	8
9	0.225	0.219	0.214	0.209	0.204	0.199	0.195	0.190	0.186	0.182	0.178	0.175	0.171	9
10	0.251	0.245	0.239	0.233	0.228	0.222	0.217	0.212	0.208	0.204	0.199	0.195	0.191	10
11	0.278	0.271	0.264	0.257	0.251	0.245	0.240	0.235	0.229	0.225	0.220	0.215	0.211	11
12	0.304	0.296	0.289	0.282	0.275	0.269	0.262	0.257	0.251	0.246	0.241	0.236	0.231	12
13	0.330	0.321	0.314	0.306	0.299	0.292	0.285	0.279	0.273	0.267	0.261	0.256	0.251	13
14	0.356	0.347	0.338	0.330	0.322	0.315	0.308	0.301	0.294	0.288	0.282	0.277	0.271	14
15	0.382	0.373	0.363	0.355	0.346	0.338	0.330	0.323	0.316	0.309	0.303	0.297	0.291	15
16	0.408	0.398	0.388	0.379	0.370	0.361	0.353	0.345	0.338	0.331	0.324	0.317	0.311	16
17	0.435	0.424	0.413	0.403	0.393	0.384	0.376	0.367	0.359	0.352	0.345	0.337	0.331	17
18	0.461	0.449	0.438	0.427	0.417	0.407	0.398	0.389	0.381	0.373	0.365	0.358	0.351	18
19	0.487	0.474	0.463	0.451	0.441	0.431	0.421	0.411	0.402	0.394	0.386	0.378	0.371	19
20	0.513	0.500	0.488	0.476	0.464	0.454	0.443	0.434	0.424	0.415	0.407	0.398	0.390	20
21	0.539	0.526	0.512	0.500	0.488	0.477	0.466	0.456	0.446	0.436	0.427	0.419	0.410	21
22	0.565	0.551	0.537	0.524	0.512	0.500	0.489	0.478	0.467	0.458	0.448	0.439	0.430	22
23	0.592	0.576	0.562	0.549	0.536	0.523	0.511	0.500	0.489	0.479	0.469	0.459	0.450	23
24	0.618	0.602	0.587	0.573	0.559	0.546	0.534	0.522	0.511	0.500	0.490	0.480	0.470	24
25	0.644	0.627	0.612	0.597	0.583	0.569	0.557	0.544	0.533	0.521	0.510	0.500	0.490	25
26	0.670	0.653	0.637	0.621	0.607	0.593	0.579	0.566	0.554	0.542	0.531	0.520	0.510	26
27	0.696	0.679	0.662	0.645	0.630	0.616	0.602	0.589	0.576	0.564	0.552	0.541	0.530	27
28	0.722	0.704	0.686	0.670	0.654	0.639	0.624	0.611	0.597	0.585	0.573	0.561	0.550	28
29	0.749	0.729	0.711	0.694	0.678	0.662	0.647	0.633	0.619	0.606	0.593	0.581	0.570	29
30	0.775	0.755	0.736	0.718	0.701	0.685	0.670	0.655	0.641	0.627	0.614	0.602	0.590	30
31	0.801	0.781	0.761	0.743	0.725	0.708	0.692	0.677	0.662	0.648	0.635	0.622	0.610	31
32	0.827	0.806	0.786	0.767	0.749	0.731	0.715	0.699	0.684	0.669	0.655	0.642	0.629	32
33	0.853	0.832	0.811	0.791	0.772	0.755	0.738	0.721	0.706	0.691	0.676	0.663	0.649	33
34	0.880	0.857	0.836	0.815	0.796	0.778	0.760	0.743	0.727	0.712	0.697	0.683	0.669	34
35	0.906	0.883	0.861	0.840	0.820	0.801	0.783	0.765	0.749	0.733	0.718	0.703	0.689	35
36	0.932	0.908	0.886	0.864	0.844	0.824	0.805	0.788	0.771	0.754	0.739	0.723	0.709	36
37	0.958	0.934	0.910	0.888	0.867	0.847	0.828	0.810	0.792	0.775	0.759	0.744	0.729	37
38	0.984	0.959	0.935	0.913	0.891	0.870	0.851	0.832	0.814	0.796	0.780	0.764	0.749	38
39		0.984	0.960	0.937	0.915	0.894	0.873	0.854	0.835	0.818	0.801	0.785	0.769	39
40			0.985	0.961	0.938	0.917	0.896	0.876	0.857	0.839	0.822	0.805	0.789	40
41				0.985	0.962	0.940	0.919	0.898	0.879	0.860	0.842	0.825	0.809	41
42					0.985	0.963	0.941	0.920	0.900	0.881	0.863	0.845	0.829	42
43						0.986	0.964	0.943	0.922	0.903	0.884	0.866	0.849	43
44							0.986	0.966	0.944	0.924	0.905	0.886	0.869	44
45								0.986	0.965	0.945	0.925	0.907	0.888	45
46									0.987	0.966	0.946	0.927	0.908	46
47										0.987	0.967	0.947	0.928	47
48											0.987	0.968	0.948	48
49												0.987	0.968	49
50													0.988	50

Data for $n \leqq 5$ calculated with the aid of expected values given in the Biometrika tables, data for $6 \leqq n \leqq 30$ from Henning/Wartman and for $31 \leqq n \leqq 50$ from Göldner.

Approximation for $n > 50$: $\dfrac{x - 0.5}{n}$.

Table A4. Format for calculation of tolerances

error proportion p %	10	5	4.55	2.5	2	1	0.27	0.1	0.01
conforming prop. 1-p %	90	95	95.45	97.5	98	99	99.73	99.9	99.99
variable of ND u_{1-p}	1.6449	1.9600	2	2.2414	2.3263	2.5758	3	3.2905	3.8906
conversion factor	0.9497	1.1316	1.1547	1.2941	1.3431	1.4871	1.7321	1.8998	2.2462
order or choice									

process:

date:		positive score direction			negative score direction			aritnm. calculation		new values, statistical calculated			
column	1	2	3	4	5	6	7	8	9	10	11	12	13
line	size No.	nominal size N_i	up. deviation A_o	low. deviation A_u	nominal size N_i	up. deviation A_o	low. deviation A_u	tolerance T_i	T_i^2	T_{si}^2	T_{si}	centre size C_i	deviations $A_{o,u}$
1													
2													
3													
4													
5													
6													
7													
8													
9													
10													
11													
12													
13													
14	$\sum$ =				①	②	③						
15		- ①	- ③	- ②						= riquired total			
16	N_S =				← check of calculation								
17	C_S =												

provisional calculation:

18	T_a =	T_{sv} =
19	ND: T_{sv}^2 =	$= \sum T_{si}^2$
20	RD: $(T_{sv}/r_u)^2$ =	$= \sum T_{si}^2$

Subject Index

The ordered page numbers indicated by the index words often mean that further mentions of the relevant concept are made on the following pages.

MIX
Papier aus verantwortungsvollen Quellen
Paper from responsible sources
FSC® C105338

If you have any concerns about our products,
you can contact us on
ProductSafety@springernature.com

In case Publisher is established outside the EU,
the EU authorized representative is:
Springer Nature Customer Service Center GmbH
Europaplatz 3, 69115 Heidelberg, Germany

Printed by Libri Plureos GmbH
in Hamburg, Germany